17 Springer Series in Solid-State Sciences

Edited by Peter Fulde

Springer Series in Solid-State Sciences

Editors: M. Cardona P. Fulde H.-J. Queisser

1 **Principles of Magnetic Resonance**
2nd Edition 2nd Printing
By C. P. Slichter

2 **Introduction to Solid-State Theory**
2nd Printing
By O. Madelung

3 **Dynamical Scattering of X-Rays in Crystals** By Z. G. Pinsker

4 **Inelastic Electron Tunneling Spectroscopy**
Editor: T. Wolfram

5 **Fundamentals of Crystal Growth I**
Macroscopic Equilibrium and Transport Concepts. 2nd Printing
By F. Rosenberger

6 **Magnetic Flux Structures in Superconductors** By R. P. Huebener

7 **Green's Functions in Quantum Physics**
By E. N. Economou

8 **Solitons and Condensed Matter Physics**
2nd Printing
Editors: A. R. Bishop and T. Schneider

9 **Photoferroelectrics**
By. V. M. Fridkin

10 **Phonon Dispersion Relations in Insulators** By H. Bilz and W. Kress

11 **Electron Transport in Compound Semiconductors** By B. R. Nag

12 **The Physics of Elementary Excitations**
By S. Nakajima, Y. Toyozawa, and R. Abe

13 **The Physics of Selenium and Tellurium**
Editors: E. Gerlach and P. Grosse

14 **Magnetic Bubble Technology**
By A. H. Eschenfelder 2nd Edition

15 **Modern Crystallography I**
Symmetry of Crystals
Methods of Structural Crystallography
By B. K. Vainshtein

16 **Organic Molecular Crystals**
Their Electronic States
By E. Silinsh

17 **The Theory of Magnetism I**
Statics and Dynamics By D. C. Mattis

18 **Relaxation of Elementary Excitations**
Editors: R. Kubo and E. Hanamura

19 **Solitons,** Mathematical Methods for Physicists By G. Eilenberger

20 **Theory of Nonlinear Lattices**
By M. Toda

21 **Modern Crystallography II**
Structure of Crystals
By B. K. Vainshtein, V. M. Fridkin, and V. L. Indenbom

22 **Point Defects in Semiconductors I**
Theoretical Aspects
By M. Lannoo and J. Bourgoin

23 **Physics in One Dimension**
Editors: J. Bernasconi, T. Schneider

24 **Physics in High Magnetic Fields**
Editors: S. Chikazumi and N. Miura

25 **Fundamental Physics of Amorphous Semiconductors**
Editor: F. Yonezawa

26 **Elastic Media with Microstructure I**
One-Dimensional Models
By I. A. Kunin

27 **Superconductivity in Transition Metals and Compounds** By S. Vonsovsky, Yu. A. Isumov, and E. Z. Kurmaev

28 **The Structure and Properties of Matter**
Editor: T. Matsubara

29 **Electron Correlation and Magnetism in Narrow-Band Systems**
Editor: T. Moriya

30 **Statistical Physics I**
By M. Toda and R. Kubo

31 **Statistical Physics II**
By R. Kubo and M. Toda

32 **Quantum Theory of Magnetism**
By R. M. White

33 **Mixed Crystals**
By A. I. Kitaigorodsky

34 **Phonons: Theory and Experiments I**
Lattice Dynamics and Models of Interatomic Forces
By P. Brüesch

35 **Point Defects in Semiconductors II**
Experimental Aspects
By J. Bourgoin and M. Lannoo

36 **Modern Crystallography III**
Formation of Crystals
By A. A. Chernov et al.

37 **Modern Crystallography IV**
Physical Properties of Crystals
By L. A. Shuvalov et al.

Daniel C. Mattis

The Theory of Magnetism I

Statics and Dynamics

With 58 Figures

Springer-Verlag Berlin Heidelberg New York 1981

Professor *Daniel C. Mattis,* Ph. D.
Department of Physics, The University of Utah, 201 North Physics Bldg.,
Salt Lake City, UT 84112, USA

Series Editors:
Professor Dr. Manuel Cardona
Professor Dr. Peter Fulde
Professor Dr. Hans-Joachim Queisser
Max-Planck-Institut für Festkörperforschung, Heisenbergstrasse 1
D-7000 Stuttgart 80, Fed. Rep. of Germany

ISBN 3-540-10611-1 Springer-Verlag Berlin Heidelberg New York
ISBN 0-387-10611-1 Springer-Verlag New York Heidelberg Berlin

Library of Congress Cataloging in Publication Data. Mattis, Daniel Charles, 1932-. The theory of magnetism. (Springer series in solid-state sciences ; 17). Bibliography: v. 1, p. Includes index. Contents: I, Statics and dynamics --. 1. Magnetism. I. Title. II. Series. QC753.2.M37 538 81-5060 AACR2

Printed in Germany

Offset printing: Beltz Offsetdruck, 6944 Hemsbach/Bergstr. Bookbinding: J. Schäffer oHG, 6718 Grünstadt.
2153/3130-5 4 3 2 1 0

To Noémi, for a multitude of reasons

Foreword

The challenge of writing a clear, concise and exciting account of a field as complicated as the theory of magnetism is one which few would choose to undertake. The author, in his first edition, wisely chose to discuss models of sufficient simplicity, that the reader could follow in detail the theoretical development of model systems which lay a fundamental foundation for understanding magnetism in physical systems. In this considerably revised and expanded second edition of "Theory of Magnetism" the author has again achieved a remarkable degree of clarity while maintaining physical relevance throughout the text. The treatment of many subjects of active research interest, such as the Kondo effect, itinerant ferromagnetism and magnetic solitons bring the reader in contact with some of the most recent work in this active field. As in the first edition an economy of formalism and an elegance of style make reading the text a great pleasure. It is with equal pleasure that I find that my lifelong friend and colleague has again found the time and the will to share his insights into this fascinating subject. We eagerly look forward to the effects of finite temperature promised for the second volume in this treatment.

Santa Barbara, California, March 1981 *J.R. Schrieffer*

Preface

This book is based only in small part on the author's earlier work "The Theory of Magnetism: an introduction to the study of cooperative phenomena"[1], which has recently gone out of print. At the time of writing the first version, there were no contemporary books on research in magnetism theory, a topic barely touched upon in survey courses in solid-state physics. Since that time we have witnessed an explosion in research and understanding in this field, which has become central to what is now known as "condensed-matter physics". As an example, the topics of thermodynamics and statistical mechanics and critical phenomena which occupied only two chapters of the earlier book, now require so much space that the entire companion volume to be entitled "The Theory of Magnetism, Part II: Thermodynamics and Statistical Mechanics" will be devoted to this study. The present "Part I: Statics and Dynamics" deals with all the concepts and methodology that are required for a coherent discipline – the nature of the many-body ground states, the construction of suitable operators, the nature of the elementary excitations (the "dynamics" of the title) – that is a microcosm of the present world of theoretical physics.

In the writing of this book, I have kept in mind a single purpose: to convey the intellectual content of a field that has, not without good reason, fascinated so many creative minds. Magnetism is a marvellous combination of the useful and the fanciful. The theory of interacting spins has been a clarifying concept in modern physics, and has served as a laboratory and tool for theoretical work in all disciplines. The Ising model has branched out into biology and economics as well. But in our own areas, the young particle physicist or field theorist of today will be likely to know, or wish to learn, as much about the "anisotropic Heisenberg model" of magnetism, for example, as his professor of solid-state theory knew, though his studies date back to an earlier, simpler time. The condensed-matter physicist must know all this too, as well as scattering theory, composite particles, solitons and vortices.

---

1 Harper and Row, New York 1965

There is now but one physics, and our book will be devoted to such ideas as can cut across interdisciplinary lines.

In the first chapter, which is to a large extent reproduced from the earlier book, where it evoked a gratifying response from readers and critics, we review the history of magnetism and see how this mirrored the general evolution of physics and the intellectual development of western thought. This review has been extended to bring it more nearly up to date. Obviously it is impossible to list the important research let alone divine which of it will survive the test of time. Still, we endeavor to convey the flavor of the contemporary work which makes the study of magnetism exciting, although we leave to the next volume the description of the most exciting developments of all: the new field of "critical phenomena".

Chapters 2–4 lay the foundations. We review one-, two- and many-electron wavefunctions, their symmetry properties and quantum numbers. We develop the theorems due to Lieb and Mattis concerning the nonexistence of a magnetic ground state in one dimension. Angular momentum is studied in detail in Chap 3. Operator- and field-theoretic representations are stressed, after a careful development starting from first principles. Chapter 3 can provide some of the material for a short course on angular momentum.

Chapter 5 deals with the elementary excitations of interacting spin systems: magnons, solitons, vortices, bound magnons, etc. Various linearized approximations and nonlinear improvements are developed and criticized. Ferromagnetism and antiferromagnetism are contrasted, ferrimagnetism is seen as a blend of the two. The effects of surfaces on spin waves are calculated. The one technique used consistently is scattering theory, and the identification of bound states. It was thought that this is didactically sounder than the use of quantum-mechanical Green's functions, which have waxed and then waned since the earlier book. We believe the present approach is closer to first principles and easier for the beginner to understand, and reserve the use of Green's functions to the second volume, where their advantages will be more evident.

In Chap. 6 the methods developed in the earlier chapters are extended and applied to the study of delocalized electrons in a metal. The conditions for them to exhibit magnetic properties are discussed, then the low-density electron (or hole) gas is solved by the scattering formalism previously elaborated. The study of magnons in metals again involves identification of bound states, similar to what we have already seen in connection with the interacting spin system in Chap 5. Finally, the appearance or disappearance of local moments in metals – the

Friedel-Anderson models of a barely magnetic impurity – are solved in great detail and generality by these methods, although the Kondo problem which results cannot be studied by the methods of this volume and is relegated to the next volume. Such modern concerns as the existence of a spin-glass ground state are touched upon.

Chapter 1 is written for all who enjoy history and are thoughtful about the lessons it may teach us. For the other chapters, only elementary quantum mechanics is needed, and a belief that the space lattice, always underlying magnetic systems, will supply a simplifying magic. Solid-state people have always known that the properties of matter on a regular lattice are simpler than in the continuum; field theorists are only just rediscovering this. I hope physicists of all persuasions will find topics of interest herein. To help self-study, a small number of problems is included as well as copious illustrations, tables and graphs, and a bibliography.

Salt Lake City, March 1981 *Daniel Mattis*

Acknowledgments

A number of people have helped with this book, or provided materials, photographs, or ideas. We are particularly grateful to A.H. Bobeck, M. Pomerantz, H. Fogedby, M.-B. Stearns, J.A. Tjon, J.A. Wright, and D. Betts.

The earlier book benefited from similar help notably from P.W. Anderson and a kind preface by J. Bardeen. Subsequent to publication, suggestions from M.E. Fisher, H.S. Leff and M. Flicker, R.L. Peterson, C.F. Squire and K. Yamazaki were gratefully received. Comments on this new version are similarly welcomed. To those people named above, to J.R. Schrieffer who has provided the foreword, and to Noemi Mattis who co-authored the first chapter, go my deepest appreciation. The reader will agree that the editorial work, for which thanks go to Dr. H. Lotsch, Mr. R. Michels and their staff, has been superb.

Daniel Mattis

Contents

1. History of Magnetism

It was probably the Greeks who first reflected upon the wondrous properties of magnetite, the magnetic iron ore $FeO—Fe_2O_3$ and famed lodestone (leading stone, or compass). This mineral, which even in the natural state often has a powerful attraction for iron and steel, was mined in the province of Magnesia.

> The magnet's name the observing Grecians drew
> From the magnetick region where it grew. [1.1]

This origin is not incontrovertible. According to Pliny's account the magnet stone was named after its discoverer, the shepherd Magnes, "the nails of whose shoes and the tip of whose staff stuck fast in a magnetick field while he pastured his flocks." [1.2]

1.1 Physics and Metaphysics

The lodestone appeared in Greek writings by the year 800 B.C., and Greek thought and philosophy dominated all thinking on the subject for some 23 centuries following this. A characteristic of Greek philosophy was that it did not seek so much to explain and predict the wonders of nature as to force them to fit within a preconceived scheme of things. It might be argued that this seems to be precisely the objective of modern physics as well, but the analogy does not bear close scrutiny. To understand the distinction between modern and classical thought on this subject, suffice it to note the separate meanings of the modern word *science* and of its closest Greek equivalent, *ἐπιστήμη*. We conceive science as a specific activity pursued for its own sake, one which we endeavor to keep free from "alien" metaphysical beliefs. Whereas, *ἐπιστήμη* meant *knowledge* for the Greeks, with aims and methods undifferentiated from those of philosophy.

The exponents of one important school of philosophy, the *animists*, took cognizance of the extraordinary properties of the lodestone by ascribing to it a divine origin. Thales, then later Anaxagoras and others, believed the lodestone to possess a soul. We shall find this idea echoed into the seventeenth century A.D.

The school of the *mechanistic*, or atomistic, philosophers should not be misconstrued as being more scientific than were the animists, for their theories were similarly deductions from general metaphysical conceptions, with little relation to what we would now consider "the facts." Diogenes of Apollonia (about 460 B.C.), a contemporary of Anaxagoras, says there is humidity in iron which the dryness of the magnet feeds upon. The idea that magnets feed upon iron was also a long lived superstition. Still trying to cheek on it, John Baptista Porta, in the sixteenth century, reported as follows:

> I took a Loadstone of a certain weight, and I buried it in a heap of Iron-filings, that I knew what they weighed; and when I had left it there many months, I found my stone to be heavier, and the Iron-filings lighter: but the difference was so small, that in one pound I could finde no sensible declination; the stone being great, and the filings many: so that I am doubtful of the truth. [1.3]

But the more sophisticated theories in this category involved effluvia, which were invisible emanations or a sort of dynamical field. The earliest of these is due to Empedocles, later versions to Epicurus and Democritus. We quote a charming accounting by the Roman poet Lucretius Carus showing that in the four centuries since Empedocles, in an era of high civilization, the theory had not progressed:

> Now sing my muse, for 'tis a weighty cause.
> Explain the Magnet, why it strongly draws,
> And brings rough Iron to its fond embrace.
> This Men admire; for they have often seen
> Small Rings of Iron, six, or eight, or ten,
> Compose a subtile chain, no Tye between;
> But, held by this, they seem to hang in air,
> One to another sticks and wantons there;
> So great the Loadstone's force, so strong to bear! ...
> First, from the Magnet num'rous Parts arise.
> And swiftly move; the Stone gives vast supplies;
> Which, springing still in Constant Stream, displace
> The neighb'ring air and make an empty Space;
> So when the Steel comes there, some Parts begin
> To leap on through the Void and enter in ...
> The Steel will move to seek the Stone's embrace,
> Or up or down, or t'any other place
> Which way soever lies the Empty Space. [1.1]

The first stanza is a vivid enough description of magnetic induction, the power of magnetized iron to attract other pieces of iron. Although this fact was already known to Plato, Lucretius was perhaps among the first to notice, by accident, that magnetic materials could also repel. The phenomenon awaited the discovery of the existence of two types of magnetic poles for an explanation.

There followed many centuries without further progress at a time when only monks were literate and rescarch was limited to theological considerations.

The date of the first magnetic technological invention, the compass, and the place of its birth are still subjects of dispute among historians. Considerable weight of opinion places this in China at some time between 2637 B.C. and 1100 A.D., reflecting a historical *uncertainty principle*, no doubt. Many other sources have it that the compass was introduced into China only in the thirteenth century A.D. and owed its prior invention to Italian or Arab origin. In any event, the compass was certainly known in western Europe by the twelfth century A.D. It was an instrument of marvelous utility and fascinating properties. Einstein has written in his autobiography of its instinctive appeal:

A wonder ... I experienced as a child of 4 or 5 years, when my father showed me a compass. That this needle behaved in such a determined way did not at all fit into the nature of events, which could find a place in the unconscious world of concepts (effects connected with direct "touch"). I can still remember—or at least believe I can remember—that this experience made a deep and lasting impression upon me. [1.4]

Many authors in the middle ages advanced metaphysical explanations of the phenomenon. However, the Renaissance scientist William Gilbert [1.5] said of these writers:

... they have lost both their oil and their pains; for, not being practised in the subjects of Nature, and being misled by certain false physical systems, they adopted as theirs, from books only, without magnetical experiments, certain inferences based on vain opinions, and many things that are not, dreaming old wives' tales.

Doubtless his condemnation was too severe. Before Gilbert and the sixteenth century, there had been some attempts at experimental science, although not numerous. The first and most important was due to Pierre Pélerin de Maricourt, better known under the Latin nom de plume Petrus Peregrinus. His "Epistola Petri Peregrini de Maricourt ad Sygerum de Foucaucourt Militem de Magnete," dated 1269 A.D., is the earliest known treatise of experimental physics. Peregrinus experimented with a spherical lodestone which he called *terrella*. Placing on it an oblong piece of iron at various spots, he traced lines in the direction it assumed and thus found these lines to circle the lodestone the way meridians gird the earth, crossing at two points. These he called the *poles* of the magnet, by analogy with the poles of the earth.

1.2 Gilbert and Descartes

Of the early natural philosophers who studied magnetism the most famous is William Gilbert of Colchester, the "father of magnetism."

Gilbert shall live till loadstones cease to draw
Or British fleets the boundless ocean awe. [1.6]

The times were ripe for him. Gilbert was born in 1544, after Copernicus and before Galileo, and lived in the bloom of the Elizabethan Renaissance. Physics was his hobby, and medicine his profession. Eminent in both, he became Queen Elizabeth's private physician and president of the Royal College of Physicians. It is said that when the Queen died, her only personal legacy was a research grant to Gilbert. But this he had no time to enjoy, for he died a few months after her, carried off by the plague in 1603.

Some 20 years before Sir Francis Bacon, he was a firm believer in what we now call the experimental method. Realizing that "it is very easy for men of acute intellect, apart from experiment and practice, to slip and err," he resolved to trust no fact which he could not prove by his own experience. *De Magnete* was Gilbert's masterpiece, 17 years in the writing and containing almost all his results prior to the date of publication in 1600. There he assembled all the trustworthy knowledge of his time on magnetism, together with his own major contributions. Among other experiments, he reproduced those performed three centuries earlier by Peregrinus with the terrella; but Gilbert realized that his terrella was an actual model of the earth and thus was the first to state specifically that the earth is itself a magnet, "which opinion of his was no sooner broached that it was embraced and wel-commed by many prime wits as well English as Forraine" [1.7]. Gilbert's theory of magnetic fields went as follows: "Rays of magnetick virtue spread out in every direction in an orbe; the center of this orbe is not at the pole (as Porta reckons) but in the center of the stone and of the terrella" [Ref. 1.5, p.95].

Gilbert dispelled superstitions surrounding the lodestone, of which some dated from antiquity, such as, "if a loadstone be anointed with garlic, or if a diamond be near, it does not attract iron." Some of these had already been disproved by Peregrinus in 1269, and even nearer to Gilbert's time, by the Italian scientist Porta, founder of one of the earliest scientific academies. Let Porta recount this:

> It is a common Opinion amongst Sea-men, that Onyons and Garlick are at odds with the Loadstone: and Steersmen, and such as tend the Mariners Card are forbidden to eat Onyons or Garlick, lest they make the Index of the Poles drunk. But when I tried all these things, I found them to be false: for not onely breathing and beleching upon the Loadstone after eating of Garlick, did not stop its Virtues: but when it was all anoynted over with the juice of Garlick, it did perform its office as well as if it had never been touched with it: and I could observe almost not the least difference, lest I should seem to make void the endeavours of the Ancients. And again, When I enquired of the Mariners, Whether it were so, that they were forbid to eat Onyons and Garlick for that reason; they said, They were old Wives fables, and things ridiculous; and that Sea-men would sooner lose their lives, than abstain from eating Onyons and Garlick [Ref. 1.3, p.211].

But the superstitions survived the disproofs of Peregrinus, Porta, and Gilbert, and have left their vestiges in our own time and in common language. Between superstition and fraud there is but a thin line, and Galileo recounts how his natural skepticism protected him in one instance from a premature Marconi:

... a man offered to sell me a secret for permitting one to speak, through the attraction of a certain magnet needle, to someone distant two or three thousand miles, and I said to him that I would be willing to purchase it, but that I would like to witness a trial of it, and that it would please me to test it, I being in one room and he being in another. He told me that, at such short distance, the action could not be witnessed to advantage; so I sent him away, and said that I could not just then go to Egypt or Muscovy to see his experiment, but that if he would go there himself I would stay and attend to the rest in Venice [1.8].

Medical healers of all times have been prompt to invoke magnetism. For example, mesmerism, or animal magnetism, was just another instance of the magnetic fluids, which we shall discuss shortly, invading the human body. Still seeking to disprove such hypotheses, it was in the interest of science that Thomas Alva Edison, as late as 1892, subjected himself "together with some of his collaborators and one dog" to very strong magnetic fields without, however, sensing any effects.

But how could experiments disprove metaphysics? In spite of their own extensive investigations, Gilbert and Porta were themselves believers in an animistic philosophy, and such were their theory and explanations of the phenomena which they had studied. Note in what sensuous terms Porta describes magnetic attraction:

... iron is drawn by the Loadstone, as a bride after the bridegroom, to be embraced; and the iron is so desirous to joyn with it as her husband, and is so sollicitious to meet the Loadstone: when it is hindred by its weight, yet it will stand an end, as if it held up its hands to beg of the stone, and flattering of it, ... and shews that it is not content with its condition: but if it once kist the Loadstone, as if the desire were satisfied, it is then at rest; and they are so mutually in love, that if one cannot come to the other it will hang pendulous in the air ... [Ref. 1.3, p. 201].

His explanation, or theory, for this phenomenon is no less anthropomorphic:

I think the Loadstone is a mixture of stone and iron ... whilst one labors to get the victory of the other, the attraction is made by the combat between them. In that body, there is more of the stone than of the iron; and therefore the iron, that it may not be subdued by the stone, desires the force and company of iron; that being not able to resist alone, it may be able by more help to defend itself. For all creatures defend their being. [Ref. 1.3, p. 191].

To which Gilbert, picking up the dialogue 40 years later, retorts:

As if in the Loadstone the iron were a distinct body and not mixed up as the other metals in their ores! And that these, being so mixed up, should fight with one another, and should extend their quarrel, and that in consequence of the battle auxilliary forces should be called in, is indeed absurd. But iron itself, when excited by the Loadstone, seizes iron no less strongly than the Loadstone. Therefore those fights, seditions, and conspiracies in the stone ... are the ravings of a babbling old woman, not the inventions of a distinguished mage. [Ref. 1.5, p. 63].

Gilbert's own ideas are themselves a curious blend of science and myth. On the one hand he dismisses the effluvia theory of magnetism with cogent reasoning, although he admits this concept might apply to electricity. His arguments are pithy: magnetic force can penetrate objects and the lodestone attracts iron through solid materials other than air, which should act as deterrents to any sort of effluvium. Electricity, on the other hand, is strongly affected by all sorts of materials. But when he comes to give his own explanation for magnetic attraction, he states it arises because "the Loadstone hath a soul." He believed the earth to have one, and therefore the loadstone also, it "being a part and choice offspring of its animate mother the earth. [p. 210 of Ref 1.5].

Notwithstanding the shortcomings of his theory, Gilbert had indeed inaugurated the experimental method. At the other pole stands René Descartes (1596–1650). Here was a philosopher who ignored the facts, but whose merit it was to exorcise the soul out of the lodestone, laying the foundations of a rational theory. Descartes is the author of the first extensive theory of magnetism, stated in his *Principia*, Part IV, sections 133–183.

Since Descartes was among the "prime forraine wits" to embrace Gilbert's hypothesis linking the lodestone to the earth, his theory of ferromagnetism is accessory to his theory of geomagnetism. Both can be summarized as follows: The prime imponderables were not specifically denoted "effiuvia" but rather "threaded parts" (*parties cannelées*). These were channeled in one-way ducts through the earth, entering through pores in one pole while leaving through pores in the other. Two kinds of parts were distinguished: those which could only enter the North Pole and leave by the South, and those which made the inverse voyage. The return trip was, in either case, by air. The parts find this a disagreeable mode of travel, and seize upon the opportunity to cross any lodestone in the way. So much so, indeed, that if they chance to meet a lodestone they will even abandon their ultimate destination and stay with it, crossing it over and over again. This is shown in Fig. 1.1. Vortices are thus created in and around the material. Lodestone, iron, and steel are the only materials having the proper channels to accommodate the parts because of their origin in the inner earth. Of these, lodestone ducts are best, whereas iron is malleable, and therefore the furlike cillia which cause the ducts to be one-way are disturbed in the process of mining. The threaded parts—throwing themselves upon the iron with great speed—can restore the position of the cillia, and thus magnetize the metal. Steel, being harder, retains magnetization better.

With this theory, Descartes claimed to be able to interpret all magnetic phenomena known to his time. From today's perspective it is hard to see how he even met Gilbert's objection to the theory of effluvia, stated a generation earlier, nor how this theory could answer the practical questions which arise in the mind of anyone working directly with magnetism. However, such was Descartes' reputation that this theory came to be accepted as fact, and influenced all thinking on the subject throughout his century, and much of the eighteenth. Two of his more prominent disciples in the eighteenth century were the famed Swiss mathematician Léonard Euler and the Swedish mystic and physicist Emmanuel Swedenborg.

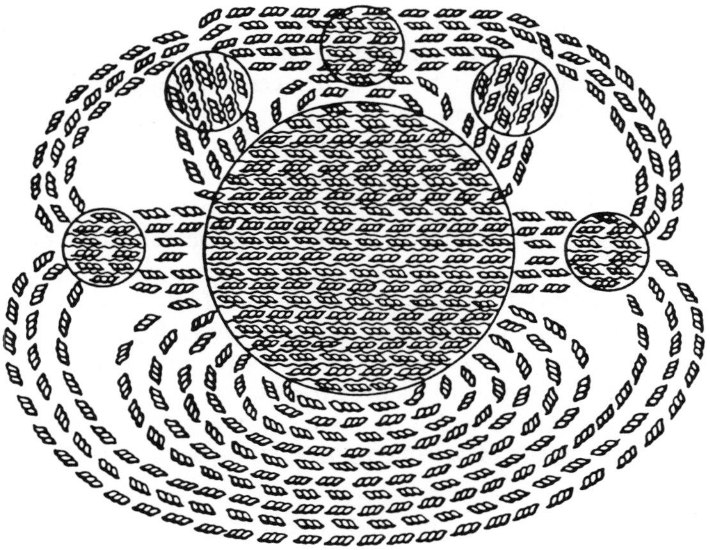

Fig. 1.1. The first theory: Descartes' threaded parts are shown going through the earth (center sphere) and in and around other magnetized bodies

Descartes, in his physics and in his philosophy, marks the transition between metaphysical and scientific thought. First, he re-established confidence in the power of reason, which was an absolute necessity for the birth of theoretical science. Second, he postulated a dichotomy of the soul and of the body, which opened the door to the study of nature on her own terms. In that, he was not alone. The beginning of the seventeenth century had witnessed a widespread mechanistic revolution in the sciences, led by such men as Gassendi, Mersenne, Hobbes, Pascal, Huygens and others. These people often went much further than Descartes in divorcing physics from metaphysics. Descartes still believed that physics could be deduced from unprovable first principles, and his mechanism was thus close to the Greeks'. He invoked metaphysics to ascertain his scientific assumptions: his argument was that, since God created both nature and our reason, we can trust that the certitudes He has instilled in us correspond to truth. The other mechanists, however, were content with a more humble approach to nature: let us describe the phenomena, they would say, and not mind the deeper essence of things.

Probably the most important contribution of mechanism to modern science is the adoption of a separate language to describe nature, that of mathematics. At first, there is an intuition. Galileo had already said in 1590: "Philosophy is written in a great book which is always open in front of our eyes (I mean: the universe), but one cannot understand it without first applying himself to understanding its language and knowing the characters in which it is written. It is written in the *mathematical* language" [1.9]. Descartes, almost 30 years later, receives the same idea as an illumination, though he fails to apply it successfully. The new language having been adopted, physics will receive an impetus from the invention of calculus (by Newton and Leibnitz, in the 1680's) and from a "mathematics that continually shifts its foundations and gets more abstract ... It seems ... that advance in physics is to be associated with continual modification and generalisation of the axioms at the base of mathematics rather than ... any one ... scheme" [3.4]. In this way, mathematics has come to replace metaphysics.

In magnetism, it is the French monk Mersenne, a friend of Descartes', who is the first, in 1644, to quantify many of Gilbert's observations [1.10]. Progress in theory remains slow. The mechanists are reluctant to speculate about deeper causes, and the field is left to the neocartesians. Even so, by the year 1700 there is one (as later there will be many) dissident voice singing a new tune. It belongs to John Keill, Savilian professor of astronomy at Oxford, who in his eighth lecture of that year observes:

It is certain that the magnetic attractions and directions arise from the structure of parts; for if a loadstone be struck hard enough, so that the position of its internal parts be changed, the loadstone will also be changed. And if a loadstone be put into the fire, insomuch as the internal structure of the parts be changed or wholly destroyed, then it will lose also its former virtue and will scarce differ from other stones ... And what some generally boast of, concerning effluvia, a subtile matter, particles adapted to the pores of the loadstone, etc, ... does not in the least lead us to a clear and distinct explication of those operations; but notwithstanding all these things, the magnetick virtues must still be reckoned amongst the occult qualities [1.11].

1.3 Rise of Modern Science

It is until the second half of the eighteenth century that we see the beginnings of a modern scientific attack on the problems of magnetism, characterized by a flexible interplay between theory and experiment and founded in rational hypotheses. For a while, theory becomes variations on the theme of *fluids*. Even Maxwell was to swim in this hypothesis, and it was not until the discovery of the electron that magnetic theory could be placed on more solid ground.

The fluid theory was originally proposed as an explanation for electricity after the discovery by Stephen Gray in 1729 that electricity could be conveyed from one body to another. This was through the medium of metals or other "nonelectric" substances, i.e., substances which conduct electricity and do not lend themselves readily to the accumulation of static charge. An early, and eminent proponent of the "one-fluid" hypothesis was Benjamin Franklin. He interpreted static charge as the lack, or excess, of the electric fluid. One of his scientific disciples, a German *émigré* to St. Petersburg by the name of Franz Maria Aepinus (1724–1802), applied the one-fluid theory to magnetism. His theories of electricity and magnetism appear in *Tentament Theoriae Electricitatis et Magnetismi*, published in St. Petersburg, 1759 [1.12]. This is an important work for it brought many ideas of Franklin into sharp focus and gave the theories of effluvia the *coup de grâve* by dint of mathematical and experimental reasoning.

In 1733, Charles François duFay, superintendent of the French Royal Botanical Gardens, discovered that there were two types of electricity. These he denoted *vitreous* and *resinous*, each of which attracted its opposite and repelled its own kind. As this idea came into competition with the Franklin one-fluid theory, much thought and many experiments were expended on proving one at the expense of the other.

Some years after electricity was granted a second fluid, so was magnetism. The two fluids were denoted *austral* and *boreal* in correspondence with the two poles. It was said that in the natural nonmagnetic state the two fluids saturate iron equally, but that magnetization parts them and leads them slightly towards their respective poles, where they accumulate. The Swede Johan C. Wilcke, a former student and collaborator of Aepinus, and the Dutchman Anton Brugmans presented this two-fluid hypothesis independently in 1778.

The best-known proponent of the two-fluid theory was Charles Augustin Coulomb (1736–1806), who proposed an important modification in the theory, and whose experiments have immortalized his name as the unit of charge and in connection with the law of force [1.13]. It was by means of a torsion balance of his invention that he established with some precision the law which bears his name: that infinitesimals of either fluid, in electricity as in magnetism, attract or repel in the ratio of the inverse of the square of their distance. After establishing this is 1785, he performed many experiments on the thermal properties of magnets. On the theoretical side, his principal contribution was the realization that the magnetic fluids could not be free to flow like their electrical counterpart, but perforce were bound to the individual molecules. Thus, he supposed each molecule to become polarized somewhat in the process of magnetization. In this manner it could be explained why the analogue of Gray's effect had never been found in magnetism, and why two new poles always appeared when a magnet was cut in twain. Coulomb was also aware that the laws of force he had discovered were *not* applicable on an atomic scale, that a solid body was not in stable equilibrium under the effect of the inverse square law of force alone.

However, concerning the unknown laws of molecular repulsion, attraction, and cohesion, he was to write, "It is almost always more curious than useful to seek to know their causes" [1.14], a correct (if defeatist) attitude considering the nearly complete ignorance of atomic structure in his day.

Siméon Denis Poisson (1781–1840) is the man who eventually became the best interpreter of the physical constructs which Coulomb discovered. Here was a brilliant mathematician whose scientific career appears to have been predestined from his early studies, whose school-teacher had predicted, punning on the verse of La Fontaine:

Petit Poisson deviendra grand
Pourvu que Dieu lui prête vie.[1]

To magnetism, Poisson brought the concept of the static potential, with which he had been so successful in solving the problems of static electricity. And having invented the mathematical theory of magnetostatics, he proceeded by solving a considerable number of problems in that field. This work dates from 1824 onwards, a very exciting time in physics as we shall see in the following section. But Poisson ignored all developments that came after the experiments of Coulomb, and while he gave the full theory of all the discoveries by the master of the torsion balance, he did not participate in the exciting movement which followed and which was to lead science into a totally new direction.

We understand the magnetical work of Poisson today mostly in the manner in which it was extended and interpreted by George Green (1793–1841). Poisson's "equivalent volume and surface distributions of magnetization" are but a special case of one of the later Green theorems. Poisson found it convenient to consider instead of $\boldsymbol{H}(\boldsymbol{r})$ the scalar potential $V(\boldsymbol{r})$ with the property that

$$\nabla V(\boldsymbol{r}) = \boldsymbol{H}(\boldsymbol{r}) \tag{1.1}$$

The contribution to $V(\boldsymbol{r})$ from the point $\boldsymbol{r}'$ in a magnetic material is

$$\left(m_x \frac{\partial}{\partial x'} \frac{1}{|\boldsymbol{r} - \boldsymbol{r}'|} + \cdots\right) d_3 r' = \boldsymbol{m} \cdot \nabla' \frac{1}{|\boldsymbol{r} - \boldsymbol{r}'|} d_3 \boldsymbol{r}' \tag{1.2}$$

and the potential from the entire sample is therefore,

$$V(\boldsymbol{r}) = \int_{\text{vol}} \boldsymbol{m}(\boldsymbol{r}') \cdot \nabla' \frac{1}{|\boldsymbol{r} - \boldsymbol{r}'|} d^3 \boldsymbol{r}'. \tag{1.3}$$

By partial integration, this is transformed into the sum of a surface integral and of a volume integral, [1.15]

$$V(\boldsymbol{r}) = \int_{\text{sur}} \frac{1}{|\boldsymbol{r} - \boldsymbol{r}'|} \boldsymbol{m} \cdot d\boldsymbol{S}' - \int_{\text{vol}} \frac{1}{|\boldsymbol{r} - \boldsymbol{r}'|} \nabla' \cdot \boldsymbol{m}(\boldsymbol{r}')\, d^3 \boldsymbol{r}' \tag{1.4}$$

[1] Little Fish will become great if God allows him but to live.

the latter vanishing for the important special case of constant magnetization. In that special case, the sources behaved *as if* they were all on the surface of the north and south poles of the magnet, and from this one might suspect that experiments concerning the nature of magnetic fields surrounding various substances would *never* reveal the slightest information about the mechanism within the material. Therefore, the fluid hypothesis seemed as good a working model as any. Poisson also extended the theory in several directions. For example by means of certain assumptions regarding the susceptibility of magnetic fluids to applied fields he obtained a law of induced magnetization, thereby explaining the phenomenon of which Lucretius had sung.

Poisson, like Coulomb, refused to become excited about any speculation concrening the *nature* of the sources of the field, which is to say, the fluids. This reluctance to discuss the profound nature of things was, of course, an extreme swing of the pendulum away from metaphysics. That attitude was itself developed into an all-embracing philosophy by Auguste Comte. *Positivism*, as it was called, holds that in every field of knowledge general laws can only be induced from the accumulated facts and that it is possible to arrive, in this way, at fundamental truths. Comte believed this to be the only scientific attitude. But, after Poisson, the main theoretical advances will be made by those who ask *why* as well as *how*. That is, by physicists who will make *hypotheses vaster, simpler, and more speculative than the mere facts allow*. The specialization to the facts at hand answers the *how*; the vaster theory, the *why*.

1.4 Electrodynamics

As early as the seventeenth century there was reason to connect the effects of electricity and magnetism. For example, "in 1681, a ship bound for Boston was struck by lightning. Observation of the stars showed that "the compasses were changed'; 'the north point was turn'd clear south.' The ship was steered to Boston with the compass reversed" [1.16]. The fluids had proliferated, and this now made it desirable to seek a relationship among them. The invention of the voltaic pile about 1800 was to stimulate a series of discoveries which would bring relative order out of this chaos.

During the time that Poisson, undisturbed, was bringing mathematical refinements to the theory of fluids, an exciting discovery opened the view to a new science of electrodynamics. In April, 1820, a Danish physicist, Hans Christian Oersted (1777–1851) came upon the long-sought connection between electricity and magnetism. Oersted himself had been seeking to find such a connection since the year 1807, but always unfruitfully, until the fateful day when he directed his assistant Hansteen to try the effect of a current on a delicately suspended magnetic needle nearby. *The needle moved.* The theory which had guided his earlier research had indicated to him that the relation should manifest itself most favorably under open-circuit conditions and not when the electric fluid was allowed to leak away, and one can easily imagine his stupefaction when the unexpected occurred. On July 21, 1820, he published a memoir in Latin (then a more universal tongue than Danish), which was sent to scientists and scientific societies around the world.[2] Translations of his paper were published in the languages and journals of every civilized country.

The reaction was feverish; immediately work started, checking and extending the basic facts of electromagnetism. French and British scientists led the initial competition for discoveries. At first the French came in ahead. The French Academy of Science of that time was a star-studded assembly, and besides Poisson it included such personalities as Laplace, Fresnel, Fourier, and more particularly active in this new field, Biot, Savart, Arago, and Ampère.

Dominique F. J. Arago (1786–1853) was the first to report upon the news of Oersted's discovery, on September 11, 1820. Here was a most remarkable scientist, who had been elected to the Academy 12 years earlier at the age of 23 as a reward for "adventurous conduct in the cause of science." The story of his dedication is worth retelling. In 1806, Arago and Biot had been commissioned to conduct a geodetic survey of some coastal islands of Spain. This was the period of Napoleon's invasion of that country, and the populace took them for a pair of spies. After escaping from a prison, Arago escaped to Algiers, whence he took a boat back to Marseilles. This was captured by a Spanish man of war almost within sight of port! After several years of imprisonment and wanderings about North Africa, he finally made his way back to Paris in the Summer of 1809, and forthwith deposited the precious records of his survey in the *Bureau des Longitudes*, having preserved them intact throughout his vicissitudes. In this, he followed in a great tradition, and today's armchair scientists may well find their predecessors an adventurous lot. Earlier, in 1753, G. W. Richmann of St. Petersburg, had been struck dead by lightning while verifying the experiments of Franklin. The effects of the electricity on his various organs were published in leading scientific journals, and Priestley wrote: "It is not given to every electrician to die in so glorious manner as the justly envied Richmann" [1.17].

Shortly after his initial report to the Academy, Arago performed experiments of his own and established that a current acts like an ordinary magnet, both in attracting iron filings and in its ability to induce permanent magnetism in iron needles.

Seven days after Arago's report, André Marie Ampère (1775–1836) read a paper before the Academy, in which he suggested that internal electrical currents were responsible for the existence of ferromagnetism, and that these currents flowed perpendicular to the axis of the magnet. By analogy, might not steel needles magnetized in a solenoid show a stronger degree of magnetization than

[2] This memoir refers to the experiment performed "last year," by which Oersted meant "last academic year," viz., "last April." This has often been misinterpreted, and the date 1819 incorrectly given for the discovery.

those exposed to a single current-carrying wire? Ampère proposed this idea to Arago, and they jointly performed the successful experiment on which Arago reported to the Academy on November 6, 1820.

The English were not for behind. It took Sir Humphrey Davy (1778–1829) until November 16 of that year to report on his similar experiments. Everyone took particular pains to witness and record important experiments and the dates thereof, and thus establish priority. No scientist on either side of the Channel underestimated the importance of Oersted's discovery nor of the results which flowed therefrom.

After Ampère's death, correspondence to him from Fresnel (one letter undated, the other dated 5 June, 1821) was found among his papers, containing the suggestion that the "Amperian currents" causing ferromagnetism should be molecular rather than macroscopic in dimension. Ampère had hesitated on this point. Fresnel wrote that the lack of (Joule) heating, and arguments similar to those advanced above in connection with Coulomb's theory, suggested the existence of elementary, atomic or molecular currents [1.18]. This must have accorded well with Ampère's own ideas, and he made some (unpublished) calculations on the basis of such a model. This work was carried on subsequently by W. E. Weber (1804–1891), who assumed the molecules of iron or steel to be capable of movement around their fixed centers. These molecules, in unmagnetized iron or steel, lie in various directions such as to neutralize each other's field; but under the application of an external force, they turn around so that their axes lie favorably oriented with respect to this external field. Precisely the same concept had been stated by Ampère, in a letter addressed to Faraday dated 10 July, 1822, and it proved superior to Poisson's theory in explaining the saturation of magnetization, for example. The final evolution of this idea may be traced to J. A. Ewing (1855–1935), who pivoted tiny magnets arranged in geometric arrays so that they might be free to turn and assume various magnetic configurations. If these magnets could be assumed each to represent a magnetic molecule, and the experimental distance scaled to molecular size, the experiments of Ewing might have been expected to yield quantitative as well as qualitative information about ferromagnets. As Digby had written long before, in connection with Gilbert's terrella, "any man that hath an ayme to advance much in naturall science, must endeavour to draw the matter he inquireth of, into some such modell, or some kind of manageable methode; which he may turne and winde as he pleaseth ... " But for the unfortunate outcome, see ahead, pp. 27–28.

The fact that current loops had been found by Ampère to behave in every manner like elementary magnets did not logically prove that ferromagnetism is caused by internal electric currents. Nevertheless, this was the hypothesis most economical in concepts which could be put forward, and as it turned out, the most fruitful in stimulating new discoveries and in creating "insight" into the "physics" of magnetism. It was Ampère and his followers, rather than Poisson and his school, who acted in the modern style, the harmonious union of theory and experiment, economical in hypotheses.

The modern scientific method, which found in electrodynamics one of its early applications, seeks to imbed every phenomenon into a vaster mathematical and conceptual framework and, within this context, to answer the question of why it occurs, and not merely of how it happens. The long list of discoveries by the electrodynamicists is sufficient tribute to the efficacy of this method, even though their explanations of the cause of magnetism were to be proved wrong. If we dwell on the subject, it is because of the popular misconception that modern science is positivistic, an idea which is not justified by the methods with which science is carried forward today. Einstein wrote: "There is no inductive method which could lead to the fundamental concepts of physics ... in error are those theorists who believe that theory comes inductively from experience." [1.19].

The nineteenth century was so rich in interrelated theories and discoveries in the fields of atomic structure, thermodynamics, electricity, and magnetism, that it is quite difficult to disentangle them and neatly pursue our history of the theory of magnetism. Fortunately many excellent and general accounts exist of the scientific progress made in that epoch, in which magnetism is discussed in the proper perspective, as one of the many areas of investigation. Here we concentrate only on the conceptual progress which was made and distinguish between progress in the physical theory of magnetism and progress in the understanding of the nature of the magnetic forces.

Much is owed to the insight of Michael Faraday (1791–1867), the humble scientist who is often called the greatest experimental genius of his century. Carrying forward the experiments of the Dutchman Brugmans, who had discovered that (paramagnetic) cobalt is attracted, whereas (diamagnetic) bismuth and antimony are repelled from the single pole of a magnet, Faraday studied the magnetic properties of a host of ordinary materials and found that all matter has one magnetic property or the other, although usually only to a very small degree. It is in describing an experiment with an electromagnet, in which a diamagnetic substance set its longer axis at right angles to the magnetic flux, that Faraday first used the term, "magnetic field" (December, 1845). He was not theoretically minded and never wrote an equation in his life. Nevertheless, his experiments led him unambiguously to the belief that magnetic substances acted upon one another by means of intermediary fields and not by "action at a distance." This was best explained by Maxwell, who

> ... resolved to read no mathematics on the subject till I had first read through Faraday's *Experimental Researches in Electricity*. I was aware that there was supposed to be a difference between Faraday's way of conceiving phenomena and that of the mathematicians. ... As I proceeded with the study of Faraday, I perecived that his method ... [was] capable of being expressed in ordinary mathematical forms. ... For instance, Faraday, in his mind's eye, saw lines of force traversing all space where the mathematicians saw centres of force attracting at a distance: Faraday saw a medium where they saw nothing

but distance. ... I also found that several of the most fertile methods of research discovered by the mathematicians could be expressed much better in terms of ideas derived from Faraday than in their original form. [1.20].

Faraday's concept of fields led him to expect that they would influence light, and after many unfruitful experiments he finally discovered in 1845 the effect which bears his name. This was a rotation of plane-polarized light upon passing through a medium in a direction parallel to the magnetization. Beside the Faraday effect, several other magneto-optic phenomena have been found: Kerr's magneto-optic effect is the analogue of the above, in the case of light reflected off a magnetic or magnetized material. Magnetic double refraction, of which an extreme example is the Cotton-Mouton effect, is double refraction of light passing perpendicular to the magnetization. But the effect which was to have the greatest theoretical implications was that discovered by Zeeman, which we shall discuss subsequently.

By virtue of the physicomathematical predictions to which it led, the hypothesis of fields acquired a reality which it has never since lost. Henry Adams wrote *ca.* 1900:

For a historian, the story of Faraday's experiments and the invention of the dynamo passed belief; it revealed a condition of human ignorance and helplessness before the commonest forces, such as his mind refused to credit. He could not conceive but that someone, somewhere, could tell him all about the magnet, if one could but find the book. ... [1.21]

But of course there was such a book: James Clerk Maxwell (1831–1879) had summarized Faraday's researches, his own equations, and all that was known about the properties of electromagnetic fields and their interactions with ponderable matter [1.20].

With Faraday, Maxwell believed the electric field to represent a real, physical stress in the ether (vacuum). Because electrodynamics had shown the flow of electricity to be responsible for magnetic fields, the latter must therefore have a physical representation as rates of change in the stress fields. Thus the energy density $\boldsymbol{E}^2/8\pi$ stored in the electric field was of necessity potential energy; and the energy density $\boldsymbol{H}^2/8\pi$ stored in the magnetic field was necessarily kinetic energy of the field. This interpretation given by Maxwell showed how early he anticipated the Hamiltonian mechanics which was later to provide the foundations of the quantum theory, and how modern his outlook was.

The harmonic solutions of Maxwell's equations were calculated to travel with a velocity close to that of light. When the values of the magnetic and electrical constants were precisely determined, it was found by H. R. Hertz in 1888 that these waves were *precisely* those of light, radio, and those other disturbances that we now commonly call *electromagnetic waves*. Later, the special theory of relativity was invented by Einstein with the principal purpose of giving to the material sources of the field the same beautiful properties of invariance which Maxwell had bestowed on the fields alone. One formulation of Maxwell's equations which has turned out most useful introduces the vector potential $\boldsymbol{A}(\boldsymbol{r},t)$ in terms of the magnetic field $\boldsymbol{H}(\boldsymbol{r}, t)$ and the magnetization $\boldsymbol{M}(\boldsymbol{r}, t)$ as

$$\nabla \times \boldsymbol{A} = \boldsymbol{H} + 4\pi\boldsymbol{M} = \boldsymbol{B}. \tag{1.5}$$

The vector $\boldsymbol{B}$ is defined as the curl of $\boldsymbol{A}$, and is therefore solenoidal by definition; that is, $\nabla\cdot\boldsymbol{B} = 0$, as is instantly verified. The next equation introduces the electric field in terms of the potentials $\boldsymbol{A}(\boldsymbol{r},t)$ and the scalar potential $U(\boldsymbol{r})$ without the necessity of describing the sources. More precisely,

$$\boldsymbol{E} = -\nabla U - \frac{1}{c}\frac{\partial \boldsymbol{A}}{\partial t}. \tag{1.6}$$

This is but the equation of the dynamo: the motion or rate of change of the magnetic field is responsible for an electric field, and if this is made to occur in a wire, a current will flow. The familiar differential form of this equation,

$$\nabla \times \boldsymbol{E} = -\frac{1}{c}\frac{\partial \boldsymbol{B}}{\partial t} \tag{1.7}$$

is obtained by taking the curl of both sides. Next, the results of Oersted, Ampère, Arago, and their colleagues almost all were summarized in the compact equation

$$\nabla \times \boldsymbol{H} = \frac{4\pi}{c}\boldsymbol{j} + \frac{1}{c}\frac{\partial \boldsymbol{D}}{\partial t}. \tag{1.8}$$

where $\boldsymbol{j}$ is the real current density, and

$$\boldsymbol{D} = \boldsymbol{E} + 4\pi\boldsymbol{P}. \tag{1.9}$$

relates the electric displacement vector $\boldsymbol{D}$ to the electric field $\boldsymbol{E}$ and the polarization of material substances, $\boldsymbol{P}$. Equation (1.8) can be transformed into more meaningful form by using the definitions of $\boldsymbol{H}$ and $\boldsymbol{D}$ in terms of $\boldsymbol{A}$, $\boldsymbol{P}$, and $\boldsymbol{M}$, and assuming a gauge $\nabla\cdot\boldsymbol{A} = 0$:

$$\left(-\nabla^2 + \frac{1}{c^2}\frac{\partial^2}{\partial t^2}\right)\boldsymbol{A}(\boldsymbol{r}, t) = 4\pi\left(\frac{1}{c}\boldsymbol{j} + \frac{1}{c}\frac{\partial \boldsymbol{P}}{\partial t} + \nabla \times \boldsymbol{M}\right). \tag{1.10}$$

This equation shows that in regions characterized by the absence of all material sources $\boldsymbol{j}$, $\boldsymbol{P}$, and $\boldsymbol{M}$, the vector potential obeys a wave equation, the solutions of which propagate with c = speed of light. The right-hand side of this equation provides a ready explanation for Ampère's famous theorem, that every current element behaved, insofar as its magnetic properties were concerned, precisely like a fictitious magnetic shell which would contain it. For if we replace all constant currents $\boldsymbol{j}$ by a fictitious magnetization $\boldsymbol{M} = \boldsymbol{r} \times \boldsymbol{j}/2c$ the fields would be unaffected. Nevertheless, for the sake of definiteness it will be useful to assume that $\boldsymbol{j}$ always refers to real free currents, $\boldsymbol{P}$ to bound or quasibound charges, and $\boldsymbol{M}$ to real, permanent magnetic moments.

1.5 The Electron

Once these equations were formulated, permitting a conceptual separation of cause and effect, of the fields and of their sources, progress had to be made in understanding the *sources.* This was tied in with the nature of matter itself, a study which at long last became "more useful than curious," to turn about the words of Coulomb. A giant step was taken in this direction by the discovery of the electron, one of the greatest scientific legacies of the nineteenth century to our own.

While Faraday, Maxwell, and many others had noted the likelihood that charge existed in discrete units only, this idea did not immediately make headway into chemistry, and Mendeléev's 1869 atomic table was based on atomic weights rather than atomic numbers. The first concrete suggestion was made in 1874 by G. Johnstone Stoney, the man who was to give the particle its name in 1891 [1.22]. We know the electron as the fundamental particle, carrier of $e = 1.602 \times 10^{-19}$ Coulomb unit of charge, and $m = 9.11 \times 10^{-28}$ gram of rest mass, building block of atoms, molecules, solid and liquid matter. But it is amusing to us to recall that it was first isolated far from its native habitat, streaming from the cathode of the gas discharge tubes of the 1870's. Certainly it was of mixed parentage. To mention but two of the greatest contributors to its "discovery," Jean Perrin found in his 1895 thesis work that the cathode rays consisted of negatively charged particles, and Thomson had obtained the ratio e/m to good precision by 1897 (curvature in a magnetic field). Its existence was consecrated at the Paris International Congress of Physics held in 1900 inaugurating the twentieth century with a survey of the problems which the discovery of the existence of the electron had finally solved, and of those which it now raised. [1.23].

By this time there already was great interest in the spectral lines emitted by incandescent gases, for their discrete nature suggested that the fluids that constituted the atoms were only capable of sustaining certain well-defined vibrational frequencies. In 1896 Zeeman had shown that the spectral lines could be decomposed into sets of lines, *multiplets,* if the radiating atoms were subjected to intense magnetic fields. This experiment had disastrous consequences on various hypotheses of atomic structure which had hitherto appeared in accord with experimental observations, for example, that of Kelvin. Concerning his "gyrostatic" model of the atom as an electrified ring, Kelvin himself, in 1899, was to write its epitaph:

No simplifying suppositions as to the character of the molecule, such as the symmetry of forces and moments of inertia round the axis of the ring, can possibly give Zeeman's normal results of the splitting of a bright line into two sharp lines circularly polarized in opposite directions, when the light is viewed (in a spectrograph) from a direction parallel to the lines of magnetic force; and the dividing of each bright line into three, each plane polarized, when the light is viewed from a direction perpendicular to the lines of force. Hence, although from 1856 till quite lately I felt quite satisfied in knowing that it sufficed to explain Faraday's magnetooptic discovery, I now, in the light of Zeeman's recent discovery, discard my old tempting gyrostatic hypothesis for an irrefragable reason. ... [1.24].

It was Zeeman's teacher, the Dutch theoretician Hendrik Antoon Lorentz (1853–1928), who provided the first reasonable theory of the phenomenon, and he based it on the electron theory. Later he extended his electron theory to give a physical basis for all of electrodynamics. "The judgment exhibited by him here is remarkable," wrote *Sommerfeld* [1.25], "he introduced only concepts which retained their substance in the later theory of relativity." Consider Lorentz' formulation of the force exerted by electric and magnetic fields on a (nonrelativistic) particle of charge e and velocity $\boldsymbol{v}$

$$\boldsymbol{F} = e\left(\boldsymbol{E} + \frac{\boldsymbol{v}}{c} \times \boldsymbol{B}\right). \tag{1.11}$$

In the Lorentz theory of the Zeeman effect, one assumes the electron to be held to the atom by a spring (of strength K) and a weak magnetic field $\boldsymbol{H}(=\boldsymbol{B})$ applied along the z direction. Thus, by the laws of Newton and Lorentz,

$$m\ddot{\boldsymbol{r}} = \boldsymbol{F} = -K\boldsymbol{r} + \frac{e}{c}\dot{\boldsymbol{r}} \times \boldsymbol{H}. \tag{1.12}$$

The parallel motion $z(t)$ proceeds at the unperturbed frequency

$$z(t) = z(0)\cos\omega_0 t \qquad \text{with} \qquad \omega_0 = \sqrt{\frac{K}{m}} \tag{1.13}$$

but in the perpendicular direction there is a frequency shift in the amount

$$\Delta\omega \cong \pm\frac{1}{2}\frac{eH}{mc} + O(H^2) \text{ (assuming } \Delta\omega \ll \omega_0) \tag{1.14}$$

according to whether the angular motion is clockwise or counterclockwise in the plane perpendicular to the magnetic field. If this is equated to the width of the splittings observed, it yields a value for e/m within a factor of 2 from that which had been established for the cathode rays. This factor was not to be explained for another quarter century.

The thermal properties of magnetic substances were first investigated in a systematic manner by Pierre Curie (1859–1906), who found M to be proportional to the applied field H. He studied χ, the constant of proportionality, known as the magnetic susceptibility, finding for paramagnetic substances:

$$\chi = \lim_{H \to 0} \frac{\mathcal{M}}{H} = \frac{C}{T}. \tag{1.15}$$

Curie's constant C assumed different values depending on the material, with T the temperature measured from the absolute zero. In diamagnetic substances χ is negative and varies little with T. In all ferromagnetic materials, he found a relatively rapid decrease of the magnetization as the temperature was raised to a critical value, now known as the Curie temperature; above this temperature, the ferromagnets behaved much like ordinary paramagnetic substances [1.26]. While many of these results had been known for one or another material, the scope of his investigations and the enunciation of these general laws gave particular importance to Curie's research.

The diamagnetism was explained a decade after Curie's experiments, in a famous paper [1.27] by *Langevin* (1872–1946), as a natural development of Lorentz's electronic theory of the Zeeman effect. As far as Langevin could see, diamagnetism was but another aspect of the Zeeman effect. Without entering into the details of his calculation, we may merely observe that this phenomenon is apparently already contained in one of the Maxwell equations: Eqs. (1.6,7) indicate that when the magnetic field is turned on, an electric field will result; this accelerates the electron, producing an incremental current loop, which in turn is equivalent to a magnetization opposed to the applied field. Thus Lenz' law, as this is called in the case of circuits, was supposed valid on an atomic scale, and held responsible for the universal diamagnetism of materials. Paramagnetism was explained by Langevin as existing only in those atoms which possessed a permanent magnetic moment. The applied magnetic field succeeded in aligning them against thermal fluctuations. Standard thermodynamic reasoning led Langevin to the relationship

$$\mathcal{M} = f\left(\frac{H}{T}\right) \tag{1.16}$$

where f = odd function of its argument. Then, in weak fields the leading term in a Taylor series expansion of the right-hand side yields Curie's law, without further ado.

1.6 The Demise of Classical Physics

As we know, the great development of physics in our century is indebted mainly to the invention of quantum mechanics. Progress in the modern understanding of magnetism has depended, to a great extent, on progress in quantum theory. Conversely, the greatest contributions of the theory of magnetism to general physics have been in the field of quantum statistical mechanics and thermodynamics. Whereas understanding in this important branch was limited in the nineteenth century to the theory of gases, the study of *magnetism as a cooperative phenomenon* has been responsible for the most significant advances in the theory of thermodynamic phase transitions. This has transformed statistical mechanics into one of the sharpest and most significant tools for the study of solid matter.

The initial, and perhaps the greatest, step in this direction was taken in 1907 when Pierre Weiss (1865–1940) gave the first modern theory of magnetism [1.28]. Long before, Coulomb already knew that the ordinary laws of electrostatics and magnetostatics could not be valid on the atomic scale; and neither did Weiss presume to guess what the microscopic laws might be. He merely assumed that the *interactions* between magnetic molecules could be described empirically by what he called a "molecular field." This molecular field H_m would act on each molecule just as an external field did, and its magnitude would be proportional to the magnetization and to a parameter N which would be a constant physical property of the material. Thus, Weiss' modification of the Langevin formula was

$$\mathscr{M} = f\left(\frac{H + \mathbb{N}\mathscr{M}}{T}\right). \tag{1.17}$$

If the molecular field were due to the demagnetizing field caused by the free north and south poles on the surface of a spherical ferromagnet, the Weiss constant would be $\mathbb{N} \approx -4\pi/3$ in some appropriate units. In fact, experiments gave to $\mathbb{N}$ the value $\approx +10^4$ for iron, cobalt, or nickel, as could be determined by solving the equation above for $\mathscr{M}(\mathbb{N},T)$ and fitting $\mathbb{N}$ to secure best agreement with experiment. This is easy to do at high temperature, where f can be replaced by the leading term in the Taylor series development. That is,

$$\mathscr{M} \doteq \frac{C}{T}(H + \mathbb{N}\mathscr{M}) \tag{1.18}$$

predicting a Curie temperature (where $\mathscr{M}/H = \infty$) at

$$T_c = C\mathbb{N}. \tag{1.19}$$

Above the Curie temperature, the combination of the two equations above yields

$$\chi = \frac{\mathscr{M}}{H} = \frac{C}{T - T_c} \tag{1.20}$$

the famous Curie-Weiss law which is nearly, if not perfectly, obeyed by all ferromagnets. This agreement with experiment was perhaps unfortunate, for it meant that the gross features of magnetism could be explained without appeal to any particular mechanism and with the simplest of quasithermodynamic arguments. Therefore *any* model which answered to those few requirements would explain the gross facts, and a correct theory could be tested only on the basis of its predictions for the small deviations away from the laws of Weiss. It was to be some time before the deviations were systematically measured and interpreted, but the large value of Weiss' constant $\mathbb{N}$ was intriguing enough in his day and pointed to the existence of new phenomena on the molecular scale.

To the mystery of the anomalously large molecular fields was added that of the *anomalous Zeeman effect*, such as that observed when the sodium D lines were resolved in a strong magnetic field. Unlike the spectrum previously described, these lines split into quartets and even larger multiplets, which could no more be explained by Lorentz' calculation than the normal Zeeman effect could be explained by Kelvin's structures. Combined with many other perplexing facts of atomic, molecular, and solid-state structure, these were truly mysteries.

A development now took place which should have effected a complete and final overthrow of the Langevin-Lorentz theory of magnetism. It was the discovery of an important theorem in statistical mechanics by Niels Bohr (1885–1962), contained in his doctoral thesis of 1911. In view of the traditional obscurity of such documents, it is not surprising that this theorem should have been rediscovered by others, in particular by Miss J. H. van Leeuwen in her thesis work in Leyden, 8 years later. In any event, with so many developments in quantum theory occurring in that period, the Bohr-van Leeuwen theorem was not universally recognized as the significanf landmark it was, until so pointed out by Van Vleck in 1932 [1.29].[3] Consider the following paraphrase of it, for classical nonrelativistic electrons: *At any finite temperature, and in all finite applied electrical or magnetical fields, the net magnetization of a collection of electrons in thermal equilibrium vanishes identically.* Thus, this theorem demonstrates the lack of relevance of classical theory and the need for a quantum theory. We now proceed to a proof to of it.

The Maxwell-Boltzmann thermal distribution function gives the probability that the nth particle have momentum $\boldsymbol{p}_n$ and coordinate $\boldsymbol{r}_n$ as the following function:

$$dP(\boldsymbol{p}_1, \ldots, \boldsymbol{p}_N; \boldsymbol{r}_1, \ldots, \boldsymbol{r}_N) = \exp\left[-\frac{1}{kT}\mathscr{H}(\boldsymbol{p}_1, \ldots ; \ldots, \boldsymbol{r}_N)\right] d\boldsymbol{p}_1 \ldots d\boldsymbol{r}_N \tag{1.21}$$

where k = Boltzmann's constant
T = temperature
$\mathscr{H}$ = Hamilton's function, total energy of the system.

The thermal average (TA) of any function $F(\boldsymbol{p}_1, \ldots ; \ldots \boldsymbol{r}_N)$ of these generalized coordinates is then simply

[3] Summarizing the consequences of this theorem, Van Vleck quips: "... when one attempts to apply classical statistics to electronic motions within the atom, the less said the better ..."

$$\langle F \rangle_{\text{TA}} = \frac{\int F\, dP}{\int dP} \tag{1.22}$$

with the integration carried out over all the generalized coordinates, or "phase space."

Next we consider the solution of Maxwell's equation, (1.8), giving the magnetic field in terms of the currents flowing. As the current density created by the motion of a single charge e_n is $\boldsymbol{j}_n = e_n\boldsymbol{v}_n$, the integral solution of (1.8) becomes

$$\boldsymbol{H}(\boldsymbol{r}) = \sum_{n=1}^{N} e_n \frac{\boldsymbol{v}_n \times \boldsymbol{R}_n}{cR_n^3} \tag{1.23}$$

where $\boldsymbol{R}_n = \boldsymbol{r} - \boldsymbol{r}_n$. This is the law of Biot-Savart in electrodynamics. The quantity $\boldsymbol{v}_n \times \boldsymbol{R}_n$ which appears above is closely related to the *angular momentum*, to be discussed in Chap. 3. To calculate the expected magnetic field caused by the motion of charges in a given body, it suffices to take the thermal average of $\boldsymbol{H}(\boldsymbol{r})$ over the thermal ensemble dP characteristic of that body. For this purpose it is necessary to know something about Hamilton's function $\mathscr{H}$, also denoted the *Hamiltonian*, for particles in electric and magnetic fields. In the absence of magnetic fields, this function is

$$\mathscr{H} = \sum_n \tfrac{1}{2} m_n v_n^2 + U(\boldsymbol{r}_1, \ldots, \boldsymbol{r}_N) \tag{1.24}$$

where $\boldsymbol{v}_n = \boldsymbol{p}_n/m_n = d\boldsymbol{r}_n/dt$ as a consequence of Lagrange's equations; the first term being the kinetic energy of motion and the second the potential energy due to the interactions of the particles amongst themselves and with any fixed potentials not necessarily restricted to Coulomb law forces. We observe from Maxwell's equation, (1.6), that Newton's second law $e_n\boldsymbol{E} = \dot{\boldsymbol{p}}_n$ assumes once more the form in which forces are derivable from potentials if we incorporate the magnetic field by the substitution $\boldsymbol{p}_n \to \boldsymbol{p}_n + e_n\boldsymbol{A}(\boldsymbol{r}_n, t)/c$. This is in fact correct, and the appropriate velocity for a particle under the effects of a vector potentials is

$$m_n\boldsymbol{v}_n = \boldsymbol{p}_n + \frac{e_n}{c}\boldsymbol{A}(\boldsymbol{r}_n, t) \tag{1.25}$$

For a more rigorous derivation, see the text by *Goldstein* [1.30]. Substitution of (1.25) into (1.24) yields the correct Hamiltonian in a field.

It shall be assumed that a magnetic field as described by a vector potential $\boldsymbol{A}(\boldsymbol{r}, t)$, is applied to the particles constituting a given substance, and the resultant motions produce in turn a magnetic field $\boldsymbol{H}(\boldsymbol{r})$ as given by (1.23). The thermal average of this field can be computed with three possible outcomes:

1) $\langle \boldsymbol{H} \rangle_{\mathrm{TA}}$ has a finite value which is independent of A in the limit $A \to 0$. In this case, the substance is evidently a *ferromagnet.*
2) $\langle \boldsymbol{H} \rangle_{\mathrm{TA}}$ is parallel to the applied magnetic field $\nabla \times \boldsymbol{A}$, but its magnitude is proportional to the applied field and vanishes when the latter is turned off. This is the description of a *paramagnetic* substance.
3) $\langle \boldsymbol{H} \rangle_{\mathrm{TA}}$ is proportional but *antiparallel* to the applied field. Such a substance is repelled by a magnetic pole and is a *diamagnet.*

The actual calculation is simplicity itself. In calculating the thermal average of $\boldsymbol{H}(\boldsymbol{r})$, we note that the convenient variables of integration are the $\boldsymbol{v}_n$ and not the $\boldsymbol{p}_n$. A transformation of variables in the integrals (1.22) is then made from the $\boldsymbol{p}_n$ to the $\boldsymbol{v}_n$, and according to the standard rules for the transformation of integrals, the integrands must be divided by the Jacobian of the transformation

$$J = \det \left\| \frac{m_n \partial v_n}{\partial p_m} \right\| = 1. \tag{1.26}$$

As is seen, the Jacobian equals unity, and therefore $\boldsymbol{A}(\boldsymbol{r}, t)$ simply disappears from the integrals, both from the Maxwell-Boltzmann factor and from the quantity $\boldsymbol{H}(\boldsymbol{r})$. Finally, the thermal expectation value of the latter vanishes because it is odd under the inversion $\boldsymbol{v}_n \to -\boldsymbol{v}_n$. This completes the proof that the currents and associated magnetic moments induced by an external field all vanish identically.

Therefore, by actual calculation the classical statistical mechanics of charged particles resulted in none of the three categories of substances described by Faraday, but solely in a *fourth* category, viz., substances having *no* magnetic properties whatever. This stood in stark conflict with experiment.

1.7 Quantum Theory

In the period between 1913 and 1925 the "old quantum theory" was in its prime and held sway. Bohr had quantized Rutherford's atom, and the structure of matter was becoming well understood. Spatial quantization was interpretable on the basis of the old quantum theory, and a famous experiment by *Stern* and *Gerlach* [1.31] allowed the determination of the angular momentum quantum number and magnetic moment of atoms and molecules. The procedure was to pass an atomic beam through an inhomogeneous magnetic field, which split it into a discrete number of divergent beams. In 1911 the suggestion was made that all elementary magnetic moments should be an integer multiple, of what later came to be called the *Weiss magneton*, in honor of its inventor. Although the initial ideas were incorrect, this suggestion was taken up again by Pauli in 1920, the unit magnetic moment given a physical interpretation in terms of the Bohr atom, a new magnitude some five times larger

$$\mu_B = \frac{e\hbar}{2mc} = 0.927 \times 10^{-20}\ \text{erg/G} \tag{1.27}$$

and renamed the *Bohr magneton*. In 1921, *Compton* [1.32] proposed that the electron possesses an intrinsic spin and magnetic moment, in addition to any orbital angular momentum and magnetization. This was later proven by Goudsmit and Uhlenbeck, then students of Ehrenfest. In a famous paper in 1925 they demonstrated that the available evidence established the spin of the electron as $\hbar/2$ beyond any doubt [1.33]. The magnetic moment was assigned the value in (1.27), i.e., twice larger than expected for a charge rotating with the given value of angular momentum. The reason for this was yet purely empirical: the study of the anomalous Zeeman effect had led Landé some two years earlier to his well-known formula for the g factor, the interpretation of which in terms of the spin quantum number obliged Goudsmit and Uhlenbeck to assign the anomalous factor $g = 2$ to the spin of the electron. But a satisfactory explanation of this was to appear within a few years, when Dirac wedded the theory of relativity to quantum mechanics.

Developments in a new quantum mechanics were then proceeding at an explosive rate, and it is impossible in this narrative to give any detailed accounting of them. In 1923 De Broglie made the first suggestion of wave mechanics [1.34], and by 1926 this was translated into the wave equation through his own work and particularly that of Schrödinger. Meantime, Heisenberg's work with Kramers on the quantum theory of dispersion, in which the radiation field formed the "virtual orchestra" of harmonic oscillators, suggested the power of noncommutative matrix mechanics. This was partly worked out, and transformation scheme for the solution of general quantum-mechanical problem given, in a paper by *Heisenberg* et al. [1.35]. *Born* and *Wiener* [1.36] then collaborated in establishing the general principle that to every physical quantity there corresponds an operator. From this it followed that Schrödinger's equation and that of matrix mechanics were identical, as was actually shown by *Schrödinger* [1.37] also in 1926. It is the time-independent formulation of his equation which forms the basis for much of the succeeding work in solid-state physics and in quantum statistical mechanics, when only equilibrium situations are contemplated, a generalization of $\mathscr{H} = E$ of classical dynamics:

$$\mathscr{H}\left(\frac{\hbar}{\mathrm{i}}\nabla_n, \boldsymbol{r}_n\right)\psi = E\psi \tag{1.28}$$

with the operator $\hbar\nabla_n/\mathrm{i}$ replacing the momentum $\boldsymbol{p}_n$ in Hamilton's function. By radiation or otherwise, a system will always end up with the lowest possible, or ground state, energy eigenvalue E_0. When thermal fluctuations are important, however, the proper balance between maximizing the entropy (i.e., the logarithm of the probability) and minimizing the energy is achieved by minimizing an appropriate combination of these, the *free energy*. This important connection between quantum theory and statistical mechanics was forged from the very beginnings, in the introduction of quantization to light and heat by Planck in 1900, through Einstein's theory of radiation (1907), its adaptation by Debye to the specific heat of solids (1912), and most firmly in the new quantum mechanics. For Heisenberg's uncertainty principle,

$$\Delta p_n \Delta r_n \gtrsim \hbar \tag{1.29}$$

which expressed the lack of commutativity of $\partial/\partial x_n$ with x_n, gave a natural size to the cells in phase space in terms of Planck's constant $2\pi\hbar$. More pertinently perhaps, the interpretation of processes in quantum mechanics as statistical, aleatory events seemed an important necessary step in applying this science to nature. This required unravelling the properties of ψ

The importance of the energy eigenfunction ψ in the subsequent development of quantum theory was overriding, and is discussed in some detail in Chaps. 2 and 4 of this volume. Here a bare outline will suffice. In 1927 Pauli invented his spin matrices; the next year Heisenberg and (almost simultaneously) Dirac explained ferromagnetism by means of *exchange*, the mysterious combination of the Pauli exclusion principle and physical overlap of the electronic wavefunctions. The next development was to consider the wavefunction itself as a field operator, i.e., *second quantization*. The importance of this was perhaps not immediately recognized. But the quantization of the electromagnetic field by Planck, leading to the statistics of Bose and Einstein, had here its precise analogue: the quantization of the electronic field by *Jordan* and *Wigner* [1.38] in 1928. This incorporated Pauli's exclusion principle an led to the statistics of Fermi and Dirac and to the modern developments in field theory and statistical mechanics. In 1928 also, *Dirac* [1.39] incorporated relativity into Hamiltonian wave mechanics, and the first and immediate success of his theory was the prediction of electron spin in precisely the form postulated by Pauli by means of his 2×2 matrix formulation, and an explanation of the anomalous factor $g = 2$ for the spin. The anomalous Zeeman effect finally stood stripped of all mystery. The Jordan-Wigner field theory permitted one loophole to be eliminated from the theory, viz., the existence of negative energy states. These were simply postulated to be filled [1.40]. The striking confirmation came with the 1932 discovery by C.D. Anderson of the positron—a *hole* in the negative energy Dirac sea.

Side by side with these fundamental developments, Hartree, Fock, Heitler, London, and many others were performing atomic and molecular calculations that were the direct applications of the new theory of the quantum electron. In 1929 *Slater* [1.41] showed that a single determinant, with entries which are individual electronic wavefunctions of space *and* spin, provides a variational many-electron wavefunction for use in problems of atomic and molecular structure. Heisenberg, Dirac, Van Vleck, Frenkel, Slater, and many others contributed to the notions of exchange, and to setting up the operator formalism to deal with it. And so, without further trying to list individual contributions, we can note that by 1930, after four years of the most exciting and brilliant set of discoveries in the history of theoretical physics, the foundations of the modern electronic theory of matter were definitively laid, and an epoch of consolidation and calculation based thereupon was started, in which we yet find ourselves at the date of this writing. We note that 1930 was the year of a Solvay conference devoted to magnetism; it was time to pause and review the progress which had been made, time to see if "someone ... could tell all about the magnet. ... "

It is not out of place to note the youth of the creators of this scientific revolution, most of them under 30 years of age. None of them had ever known a world without electrons or without the periodic table of the atoms. On the one hand their elders saw determinism and causality crumbling: "God does not throw dice!" Einstein was to complain [1.42]. But determinism was just the scientific substitute for God's prescience [1.43] and the elders forgot that long before, in the very foundations of statistical mechanics and thermodynamics, such as had been given by J. Willard Gibbs and Einstein himself, determinism had already been washed out to sea. The alarm about causality was equally unwarranted, for it has retained the same status in quantum theory as in classical theory in spite of all the assults upon it. On the other hand, the young workers only saw in the new theory a chance finally to understand *why* Bohr's theory, *why* the classical theory worked when they did—and also why they failed when they did. They saw in quantum theory a greater framework with which to build the universe, no less than in the theory of relativity; and being devoid of severe metaphysical bias, they did not interpret this as philosophical retrogression. There is a psychological truth in this which had not escaped the perspicacious Henry Adams: "Truly the animal that is to be trained to unity must be caught young. Unity is vision; it must have been part of the process of *learning to see* [1.21] (italics ours).

1.8 Modern Foundations

At the same time as the theoretical foundations were becoming firmer, the experimental puzzles concerning magnetism were becoming numerous. Let us recount these with utmost brevity. Foremost was the question, why is not iron spontaneously ferromagnetic? Weiss proposed that his molecular field had various directions in various elementary crystals forming a solid; thus the magnetic circuits are all closed, minimizing the magnetostatic energy. Spectacular evidence was provided by Barkhausen in 1919, who by means of newly developed electronic amplifiers, heard distinct *clicks* as an applied field aligned the various Weiss domains. This irreversible behavior also explained hysteresis phenomena. Measurements of the gyromagnetic ratio gave $g \cong 2$ for most ferromagnetic substances, showing that unlike in atomic or molecular magnetism only the spins participated in the magnetic properties of the solids. In atoms, the mounting spectroscopic data permitted Stoner to assign the correct number of equivalent electrons to each atomic shell and Hund to enunciate his rules concerning the spontaneous magnetic moment of a free atom or ion.

In the study of metals it was found that alloying magnetic metals with non-magnetic ones resulted in a wide spectrum of technical properties. In metals, unlike insulators, it was also found that the number of magnetic electrons *per* atom was not, in general, an integer. In many respects, knowledge of the magnetic properties of many classes of solids was fairly definitive by 1930. Only the class of solids which are magnetic, but non*ferro*magnetic, ordered structures remained to be discovered, and only two crucial tools of investigation were lacking: neutron diffraction, and magnetic resonance, each of which has permitted modern investigators to study solids from within.

The progress of the theory of magnetism in the first third of the twentieth century can be followed in the proceedings of the Sixth Solvay Conference of 1930, which was entirely devoted to this topic, and in two important books: Van Vleck's *The Theory of Electric and Magnetic Susceptibilities* (Oxford, 1932), and E. C. Stoner's *Magnetism and Matter* (Methuen, London, 1934). In the first of these books, but one chapter out of thirteen is devoted to the study of magnetism as a cooperative phenomenon. In the other, barely more emphasis is given to this field of study; after all, Heisenberg had written "it seems that till now, Weiss' theory is a sufficient basis, even for the deduction of second-order effects," in his contribution to the Solvay conference. Nevertheless, some attempts at understanding magnetism as a collective phenomenon which had already been made by this time were to lay the groundwork for our present understanding of the subject.

First came the Lenz-Ising formulation [1.44] of the problem of ferromagnetism; spins were disposed at regular intervals along the length of a *one-dimensional* chain. Each spin was allowed to take on the values ± 1, in accordance with the laws of Goundsmit and Uhlenbeck. This model could be solved exactly (*vide-infra*), and as long as each spin interacted with only a finite number of neighbors, the Curie temperature could be shown to vanish identically. Did this signify that forces of indefinitely large range were required to explain ferromagnetism?

The introduction by Pauli of his spin matrices established spin as a vector quantity; the requirements that the interaction between two spins be an isotropic scalar and the theory of permutations led Dirac in 1929 to the explicit formulation of the "exchange operator":

$$\boldsymbol{S}_i \cdot \boldsymbol{S}_j \tag{1.30}$$

as the essential ingredient in the magnetic interaction. The coefficient (usually denoted J) of this operator was a function of the electrostatic force between the electrons, a force so large that the magnetic dipole—dipole interactions (the Amperian forces) could even be neglected to a first approximation. *The theory of Weber and Ewing was now rendered obsolete.* Quoting Stoner:

... As soon as the elementary magnets are interpreted as corresponding to electrons in orbits, or as electron spins, it is found that the magnetic forces between neighbouring molecules are far too small to give rise to constraints of the magnitude required in the Ewing treatment of ferromagnetism. Atoms or molecules may in certain cases have some of the characteristics of small bar magnets—in possessing a magnetic moment; but an interpretation of the properties of a bar of iron as consisting of an aggregate of atomic bar magnets ceases to be of value, whatever its superficial success, once it is known that the analogy between atoms and bar magnets breaks down just at those points which are essential to the interpretation [1.45].

With (1.30) or its equivalent as the basic ingredient in the magnetic Hamiltonian, *Bloch* [1.46] and *Slater* [1.47] discovered that *spin waves* were the elementary excitations. Assigning to them Bose-Einstein statistics, Bloch showed that an indefinitely large number of them would be thermally excited at any finite positive temperature, no matter how small, in one or two dimensions; but that a three-dimensional ferromagnet possessed a finite Curie temperature. For the magnetization in three dimensions, Bloch derived his "three-halves' power law"

$$\mathscr{M}(T) = \mathscr{M}(0)\left[1 - \left(\frac{T}{T_c}\right)^{3/2}\right] \qquad (1.31)$$

with T_c calculable from the basic interaction parameters. *Now*, at last, Ising's result, $T_c = 0$, could be understood to have resulted primarily from the one-dimensionality of his array and not at all from the old-quantum-theoretical formulation of the spins.

The properties of metallic conduction electrons were fairly well understood by the 1930's. In 1926 Pauli had calculated the spin paramagnetism of conduction electrons obeying Fermi statistics; and not much later Landau obtained their motional diamagnetism; one of the primary differences between classical and quantum-mechanical charged particles arose in this violation of the Bohr-van Leeuwen theorem. Bloch then considered the Coulomb repulsion among the carriers in a very dilute gas of conduction electrons in a monovalent metal. In the Hartree-Fock approximation, he found that for sufficiently low concentrations (or sufficiently large effective mass, we would add today) this approximation gave lowest energy to the ferromagnetic configuration, in even greater contrast to classical theory. However, Pauli's criticized the calculation:

Under those conditions (of low concentration, etc.), in fact, the approximation used by Bloch is rather bad; one must, however, consider as proved his more general result, which is, that ferromagnetism is possible under circumstances very different from those in which the Heitler-London method is applicable; and that it is not sufficient, in general, to consider merely the signs of the exchange integrals. [1.48].

More modern work has fully borne out Pauli's skepticism; for example, it is now widely believed that the charged electrons of a metal behave, in most respects, as ideal, noninteracting fermions. Nevertheless, many detailed studies are still anchored on this frail Hartree-Fock ferromagnetism due to the natural reluctance of physicists to abandon the only model of ferromagnetism other than the Heitler-London theory, which could be characterized as truly *simple*. A modern many-body theory took another 20 years to derelop (see Chap. 6).

In 1932, *Néel* [1.49] put forward the idea of *anti*ferromagnetism to explain the temperature independent paramagnetic susceptibility of such metals as Cr and Mn, too large to be explained by Pauli's theory. He proposed the idea of two compensating sublattices undergoing negative exchange interactions, resulting in (1.20) with a *negative* Curie temperature—now known as the Néel temperature.

In 1936, Slater and Wannier both found themselves at Princeton, one at the Institute for Advanced Studies, and the other at the University. This overlap must have had positive results, for in an issue of the *Physical Review* of the succeeding year there are two consecutive papers of some importance to the theory of magnetic solids. In the first of these, *Wannier* [1.50] introduced the set of orthogonal functions bearing his name, of which we shall have much more to say. In the second of these, *Slater* [1.51] gave a nontrivial theory of ferromagnetism of metals partly based on the use of Wannier functions. He discussed the case of a half-filled band which, while least favorable to ferromagnetism, is amenable to analysis (one electron + one hole is *exactly* soluble):

Starting with the theory of energy bands we have set up the perturbation problem and solved approximately the case of a band containing half enough electrons to fill it, all having parallel spins but one. This problem is a test for ferromagnetism: if the lowest energy of the problem is lower than the energy when all have parallel spins, the system will tend to reduce its spin, and will not be ferromagnetic; whereas if all energies of the problem are higher ... we shall have ferromagnetism. [1.51].

This followed hard upon his band theory of the ferromagnetism of nickel [1.52].

The study of nickel-copper alloys was strong evidence for the band theory of ferromagnetism in metals. A single copper atom dissolved in nickel does not succeed in binding the extra electron which it brings into the metal by virtue of its higher valency. This electron finds its way into the lowest unfilled band, which belongs to the minority spins; and therefore decreases the magnetization of the entire crystal in the amount of one Bohr magneton, homogeneously distributed. As pure nickel possesses 0.6 Bohr magnetons per atom, the magnetization should decrease linearly with copper concentration, extrapolating to zero at 60 percent concentration. The experiments accorded beautifully with this hypothesis [1.53a].

Continuing study of Weiss domains soon indicated that they did not necessarily coincide with the physical crystals, that their size was determined by such effects as magnetostriction, quantum-mechanical exchange energy, etc., as well as the magnetostatic energy. The first complete theory (for Co) was sketched by Heisenberg in 1931, and subsequent work by Bloch, Landau, Bozorth, Becker, and many others have laid the experimental and theoretical framework of a theory of domains. Recent formulations by Brown and collaborators, Landau and Ginzburg, and others, have produced the interesting differential equations that regulate the slow distortions in space and time associated with domains, spinwaves, and the like. This field [1.53b] has even acquired a name: micromagnetics, and spawned a new industry: magnetic "bubbles."

1.9 Magnetic Bubbles

In the 1960's a number of experimental and theoretical studies of strip- and cyclindrical domains [1.54] were followed in 1967 by *Bobeck*'s suggestion of the device potential of the cylindrical domains [1.55]. This invention by Bobeck, Shockley, Sherwood and Gianola has become known as *magnetic bubbles*; they are ubiquitous today, with applications in telephone switching, computer memory banks—that is wherever a high density, low energy rapidly switchable information storage is required. In view of their great intrinsic interest, and as a simple demonstration of micromagnetics, we now outline some of the basic facts and theory and refer the reader to the recent monograph by *Eschenfelder* [1.56] for supplementary information.

Magnetic bubbles are small, mobile cylindrical domains in thin films magnetized perpendicular to the film surface. Bubble memory chips with capacities of 10^6 bits and data rates of the order of megabits/s are now technological commonplace, with other applications in the offing. Yet the field is so recent that two excellent, modern books on thin magnetic films [1.57] have no mention of this phenomenon at all! Magnetic thin films are themselves a recent development, dating to Blois' fabrication of 1000 Å thick permalloy films (80% Ni, 20% Fe), chosen for the low value of magnetocrystalline anisotropy [1.58]. Magnetostatic considerations favor the magnetization in the plane of the film, and the early literature [1.57] discussed mainly applications of this type. But already by 1960 there were photographs of bubble domains [1.59] and attempts at a theory [1.54]. By 1967 *Bobeck* and colleagues [1.55] had invented a bubble device (capable of generating, storing and counting the units) and shortly thereafter a complete theory was developed by *Thiele* [1.60]. Such a theory must predict the range of parameters (film parameters and thickness, external field parameter) over which bubbles are stable, and take into account the factors affecting their motion when subjected to magnetic field gradients. So first, what are the materials and parameters?

Magnetic rare earth garnets—orthoferrites such as $TmFeO_3$ or $DyFeO_3$—oriented with c axis perpendicular to the film have very small barriers (as small 0.01 Oe) against the motion of a domain wall and have been found eminently suitable. Other materials, including amorphous magnetic metal alloys, have been used.

Figure. 1.2 shows a demagnetized sample, with contrasting shades indicating domains of opposite magnetization (into or out of the film). In such a configuration, strip domains are most stable, the sole bubble being the exception. An isolated bubble is always unstable and will grow indefinitely unless an external magnetic field is present, in such a direction as to collapse the bubble. In Fig. 1.2, the bubble is stabilized by the demagnetizing field of the neighboring domains. An external field applied to the demagnetized sample causes the domains in the direction of the field to grow and the others to shrink until they assume the cir-

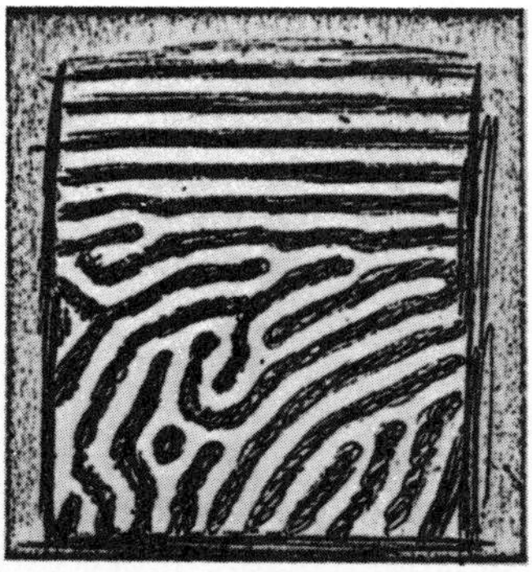

◁ **Fig. 1.2.** Magnetic domains in a thin platelet of orthoferrite $TmFeO_3$ in the absence of any magnetic field. A lone "bubble" domain is seen—the strip domains are in stable equilibrium. Any resemblance to Fig. 1.1 is purely coincidental

▽ **Fig. 1.3.** The left-hand sides of **(a)** and **(b)** are as-grown 6μm $(YGdTm)_3(GaFe)_5O_{12}$ garnet film, the right-hand sides have been treated by superficial ($\frac{1}{2}$μm) hydrogen ion implantation to improve the quality of the domains. **(a)** is in zero bias—note the total absence of bubble domains; **(b)** is in 170 Oe applied magnetic field, sufficient to produce *only* bubble domains. [1.55]

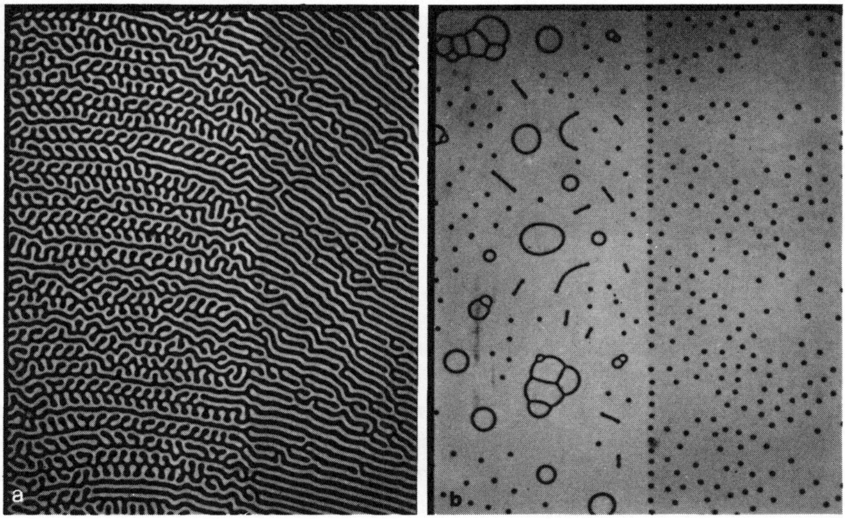

cular shape denoted as bubbles. Increasing the field causes the bubbles to shrink until at a second critical field they disappear totally. This is illustrated in Fig. 1.3 [(a) is no bias, (b) is for 170 Oe external field)], which also shows the importance of the proper choice of materials. The material is 6-μm $(YGdTm)_3$ $(GaFe)_5O_{12}$ garnet film, untreated in the left-hand side of the pictures and after hydrogen ion implantation on the right-hand sides.

Thiele's theory was subsequently nicely simplified by *Callen* and *Josephs* [1.61] and it is this version we now present starting with the conditions for static equilibrium.

We consider a film of magnetic material of thickness h 0.1 mm or less. The bubble radius r will turn out to be of approximately the same magnitude as h and both are related to a characteristic length $l \equiv \sigma_w/4\pi M^2$. The magnetization has the magnitude $|M|$ everywhere, pointing in the $+z$ direction (normal to the film) within the domain, and $-z$ direction elsewhere. An applied field $\boldsymbol{H}$ stabilizes the domain.

The total energy $E_T = E_w + E_H - E_D$ comprises the following: the wall energy,

$$E_w = 2\pi rh\sigma_w \tag{1.32}$$

in which the important parameter is σ_w, the wall energy per unit surface; the interaction with the external field

$$E_H = 2MH\,\pi r^2 h, \tag{1.33}$$

and the demagnetizing energy of a single cylindrical domain in an infinite plate E_D, given by *Thiele* [1.60] in terms of elliptic functions, for which Callen and Josephs found an excellent—and simple—approximation. Before proceeding with the calculation, it is convenient to express all the quantities in dimensionless form: a dimensionless energy $\mathscr{E} \equiv E/(16\pi^2 M^2 h^3)$, a dimensionless field $\mathscr{H} = H/4\pi M$ a dimensionless radius $x = r/h$, and a dimensionless characteristic length $\lambda = l/h = \sigma_w/4\pi M^2 h$.

Then, $d\mathscr{E}_T/dx = 0$, the condition for bubble stability, yields

$$\frac{1}{2}\lambda + \mathscr{H}x - x/\left(1 + \frac{3}{2}x\right) = 0 \tag{1.34}$$

where the Callen-Josephs approximation to Thiele's E_D yields the simple term, $-x/(1 + 3x/2)$. The equation is solved graphically as shown in Fig. 1.4. with the regions of stability being indicated, as derived from the sign of $d^2\mathscr{E}_T/dx^2$. Above a critical field $\mathscr{H}_0 = 1 + 3\lambda/4 - (3\lambda)^{1/2}$ *no* bubble domain is stable (the straight line is tangent to the curve). It is similarly possible to calculate a criterion for the transformation of bubbles into strip domains [1.60].

In bubble dynamics, one assumes that the rate of change of the energy in the bubble is balanced by drag (the friction force is proportional to the velocity

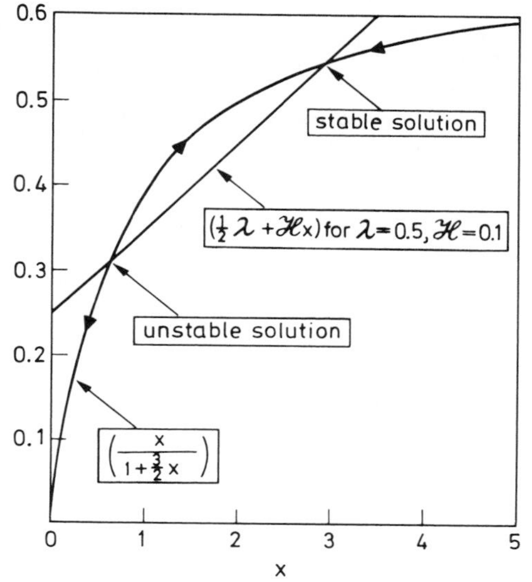

Fig. 1.4. Graphical solution of (1.34). Tilt of straight-line segment depends on external field. Stable and unstable solutions are shown, with the direction of bubble growth or shrinkage shown by arrows. [1.60]

so the frictional loss is proportional to the *square* of $\dot{x}$) and by a second mechanism dependent only on the distance traversed per unit time (i.e., proportional to $|\dot{x}|$). Writing the energy balance per unit wall:

$$\left[\frac{\lambda}{2x} + \mathscr{H} - \left(1 + \frac{3}{2}x\right)^{-1}\right] x\dot{x} = \eta^{-1} x\dot{x}^2 - \mathscr{H}_c x |\dot{x}| \tag{1.35}$$

$\mathscr{H}_c$ is the coercive field for wall motion, and η is the wall mobility parameter (in our units of $h/4\pi M$). The equation of bubble dynamics is then

$$\dot{x} = -\eta\left[\left(\frac{\lambda}{2x}\right) - \left(1 + \frac{3}{2}x\right)^{-1} + \mathscr{H} + \mathscr{H}_c \operatorname{sgn}(\dot{x})\right] \tag{1.36}$$

determining the rates at which the bubbles can be grown or shrunk.

A study of the interaction energy between two or more bubble domains does not seem to have been carried out at this time, perhaps because here cooperative phenomena—the correlated behavior of several domains—are not important.

1.10 Ultimate Thin Films

The search for the ultimate thin film led Pomerantz and collaborators to the fabrication of literally two-dimensional ferromagnets [1.62]. Two types are shown in Fig. 1.5; in either case, the magnetic stratum consists of Mn^{2+} ions.

Despite this *tour de force*, two-dimensional systems stand in peculiar relation to the rest of physics. Difficult of resolution, they seem capable of exhibiting

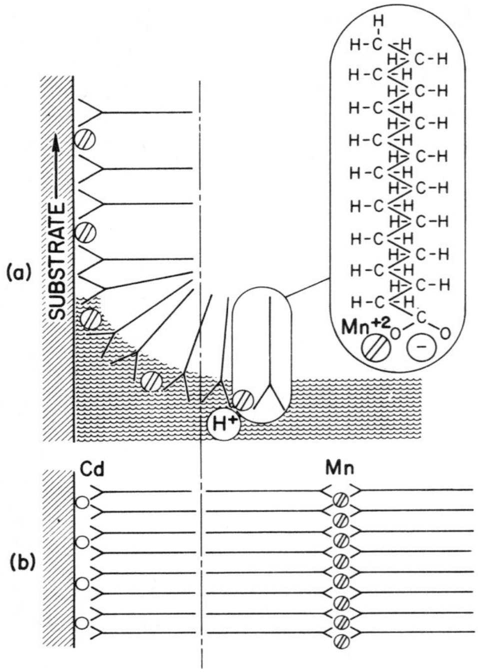

Fig. 1.5. (**a**) Schematic diagram of the deposition of a literally 2-D magnet by the Langmuir-Blodgett technique. The hydrophilic substrate (e.g., quartz) is pulled through the water surface. A monolayer of manganese stearate that was floating on the water adheres to the quartz to produce a "Structure I" film. The inset shows the structure of a stearate anion, and a Mn^{2+} ion. (**b**) Schematic cross section of a 2-D magnetic of structure II. The Mn ions are shown in a plane, but the actual structure may be nonplanar. [1.62]

some of the properties of the "real" three-dimensional world, but not all. In any event, they have been of some interest for many years now. In 1942, *Onsager* developed an exact solution to the problem of Ising spins in a plane, the "two-dimensional Ising model" [1.63]. This work stands, to this day, as a pinnacle of the achievements of theoretical physics of our time. Onsager's solution yielded the thermodynamic properties of the interacting system, and demonstrated the phase transition at T_c but in a form quite unlike that of Curie-Weiss. In particular, the infinite specific-heat anomaly at T_c is a challenge for approximate, simpler theories to reproduce. Onsager's discovery was not without an amusing sequel. The original solution was given by Onsager as a discussion remark, following a paper presented to the New York Academy of Science in 1942 by *Gregory Wannier*, but the paper, based on an application of Lie algebras, only appeared two years later. [1.64]. However, his formula for the spontaneous magnetization below T_c, which requires substantial additional analysis,

$$\mathscr{M} = (1 - x^{-2})^{1/8}, \; x = \sinh\frac{2J_1}{kT}\sinh\frac{2J_2}{kT} \tag{1.37}$$

was *never* published by him, but merely "disclosed". It

... required four years for its decipherment. It was first exposed to the public on 23 August 1948 on a blackboard at Cornell University on the occasion of a conference on phase transitions. Laslo Tisza had just presented a paper on The General Theory of Phase Transitions.

Gregory Wannier opened the discussion with a question concerning the compatibility of the theory with some properties of the Ising model. Onsager continued this discussion and then remarked that—incidentally, the formula for the spontaneous magnetization of the two-dimensional model is just that (given above.) To tease a wider audience, the formula was again exhibited during the discussion which followed a paper by Rushbrooke at the first postwar IUPAP (International Union of Pure and Applied Physics) statistical mechanics meeting in Florence in 1948; it finally appeared in print as a discussion remark... However, Onsager never published his derivation. The puzzle was finally solved by C.N. Yang and its solution published in 1952. Yang's analysis is very complicated ... [1.65].

Of course the Ising model is peculiar in that it does not accommodate spin-waves. Bloch had already established that in two dimensions, as well as in one, the number of spinwaves excited at finite temperature is sufficient to destroy the long range order. In 1966, *Mermin* and *Wagner* [1.66] succeeded in proving rigorously the absence of long range order in both the Heisenberg ferromagnet and antiferromagnet in one and two dimensions, i.e., $\mathcal{M} = 0$ at all temperatures. This by itself is, however, not sufficient to rule out a phase transition at finite temperature T_c! This interesting possibility came out as a result of numerical work by *Stanley* and *Kaplan* [1.67], later confirmed by a remarkable theory of the *xy* model due to *Kosterlitz* and *Thouless* [1.68], in which vortices—a topological disorder—were seen to be an important ingredient, an adjunct to the spin-waves which are the more common excitations of a magnetic substance. The two-dimensional phase transition which they obtained was quite different from the usual. Indeed, an entire field of physics has now developed around the properties of low-dimensional systems, principally two-and one-dimensional (fractional dimensionality being also admitted).

1.11 Dilute Magnetic Alloys

At first sight, no topic could appear less intellectually challenging than that of dilute magnetic alloys. Yet some of the great recent conceptual advances of of theoretical physics have had their inception in this seemingly modest subject!

The original studies by *Friedel* [1.69] centered about the question of formation of a local magnetic moment-how do iron or manganese maintain their atomic magnetic moments when imbedded in a nonmagnetic host matrix such as copper metal? *Anderson* then suggested a model [1.70] with the intention of resolving or closing this question in a semiquantitative way. Far from it! Anderson's model, which was not rigorously soluble either, opened an entire new field of investigation, seeking to explain or predict the occurrence and non-occurrence of local moments, as in the variety of examples in Fig. 1.6 and Table 1.1.

If the impurity atom does possess a magnetic moment this polarizes the conduction electrons in its vicinity by means of the exchange interaction and thereby influences the spin orientation of a second magnetic atom at some distance.

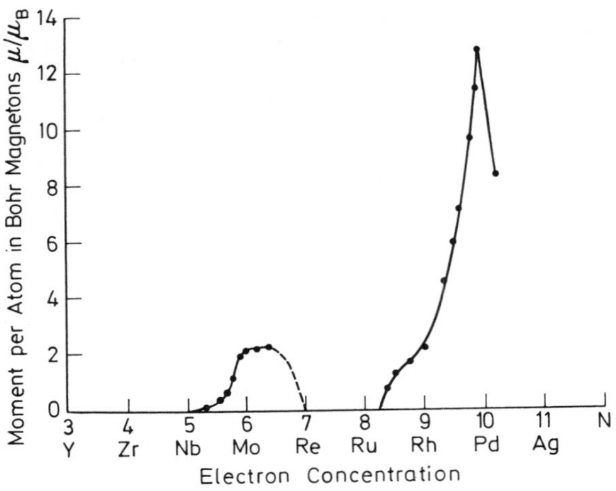

Fig. 1.6. Magnetic moment on Fe impurity ion as function of electron number of the host. (From [1.79])

Table 1.1. Which "magnetic" atoms retain their moments in dilute alloys (adapted from [1.78])

Magnetic Species	Host Metal			
	Au	Cu	Ag	Al
Ti	no	?	?	no
V	?	?	?	no
Cr	yes	yes	yes	no
Mn	yes	yes	yes	no
Fe	yes	yes	?	no
Co	?	?	?	no
Ni	no	no	?	no

Owing to quantum oscillations in the conduction electrons' spin polarization the resulting effective interaction between two magnetic impurities at some distance apart can be ferromagnetic (tending to align their spins) or antiferromagnetic (tending to align them in opposite directions). Thus a given magnetic impurity is subject to a variety of ferromagnetic and antiferromagnetic interactions with the various neighboring impurities. What is the state of lowest energy of such a system? This is the topic of a new very active field of studies entitled "spin glasses" the magnetic analog to an amorphous solid, which we touch upon in a later chapter, to expand on it in the next Volume.

When the impurity atoms are very few, in the *very* dilute magnetic alloys, each impurity atom correlates with just the conduction electrons in its own vicinity. This has a surprising consequence: for many decades now, an anomalous resistivity of *high-purity* gold and cooper samples has been remarked, in which the resistance *increased* with decreasing temperature. This contradicts the usual behavior, for generally thermal disorder decreases as the temperature is lowered. These anomalies were finally found to be caused by the trace impurities of iron deposited by contact with the steel used in the fabrication.

The resistance mystery was ultimately pierced by *J. Kondo* in the early 1960's, with his explanation of what is now called, the "Kondo effect" [1.71]. It occurs only when the magnetic species is *anti* ferromagnetically coupled with the conduction electrons, causing a scattering resonance at the Fermi level, which becomes the sharper the lower the temperature. A completely satisfactory theory of this many-body effect is not yet known, but recent developments on the ground state and thermodynamic properties have led some to believe that an explanation of the transport anomalies is close to hand. First, the thermodynamics has been obtained by analytical, numerical and renormalization-group type methods [1.72]. The symmetry of the ground state—in many cases, a nondescript *singlet*—is known [1.73]. The thermodynamic and transport properties have been proved, rigorously, to be analytic functions of the temperature by *Hepp* [1.74], despite indications, from the initial calculations of Kondo and his followers, that logarithmic singularities of the form log (T/T_K) were the *signature* of this phenomenon down to the lowest temperatures.

Finally, a breakthrough has appeared. Within months or weeks of each other, in 1980 *Andrei* [1.75] in the United States and *Wiegmann* [1.76] of the Soviet Union have independently found almost identical solutions to the Kondo Hamiltonian, which they believe to be exact. These solution are broadly based on a method first proposed by *H. Bethe* in 1932 to solve the one-dimensional Heisenberg antiferromagnet, now known as *Bethe's ansatz*. This method is thoroughly detailed in Chap. 5 of this book, but aside from a brief introduction to the Kondo phenomena in Chap. 6, we defer the detailed study of this problem until Vol. 2, devoted to finite temperature effects.

1.12 New Directions

To quote Dryden, on the Canterbury Pilgrims: "... But enough of this; there is such a variety of game springing up before me, that I am distracted in my choice and know not which to follow ... "

The world of magnetism has been expanding. In biological systems, the minute presence of magnetite-the loadstone-has been found to explain such remarkable properties as the orientational ability of homing pigeons. Indeed, the production and detection of minute magnetic fields has become an industry, thanks to a remarkable development in electrical science, viz., superconductivity. The theory of superconductivity, jointly invented by *J. Bardeen, J. R. Schrieffer,* and *L. N. Cooper* in 1957 explained how, and why it might be possible to produce infinitesimal magnetic fields in amperian current loops, and devices acronymed SQUIDS are now available for this purpose. This many-body theory also introduced the use of field-theoretic methods into condensed-matter physics.

Nowadays, we find the traffic of information is sometimes reversed; quantum statistical methods specifically designed for the solution of magnetic problems have become the prototypes even in field theory and particle physics [1.77]. Recent Nobel prizes have included the above-named theorists of superconductivity, as well as *L. Onsager* just prior fo his recent death, and *P. W. Anderson* and *J. H. Van Vleck* for research motivated by problems in magnetism and electrical conductivity. We hope to convey some of this excitement in the pages that follow.

2. Exchange

With the simultaneity which has characterized so many of the greatinventions of modern science, *Dirac* [2.1] and *Heisenberg* [2.2] independently dicovered exchange. This effect appeared at first to be mysterious, for its origins in the Pauli exclusion principle of quantum mechanics had no classical analog. But it must be remembered that classical mechanics had failed to provide any explanation of magnetism, and now one could see why. Ferromagnetism cannot exist in the classical correspondence limit $\hbar \to 0$; it is one of the results of quantum mechanics, a manifestation of "exchange."

Loosely speaking, exchange was explained as follows. The Pauli exclusion principle keeps electrons with parallel spins apart, and so reduces their Coulomb repulsion. The difference in energy between the parallel spin configuration and the antiparallel one is the exchange energy. This, however, is favorable to ferromagnetism only in exceptional circumstances because the increase in kinetic energy associated with parallel spins outweighs the favorable decrease in potential energy—as is the case in the familiar examples of the helium atom or the hydrogen molecule. In rare cases, such as with metallic iron, a large number of parallel spins produces ferromagnetism because the cost in kinetic energy is not as great as in some other metals, whereas the gain in potential energy is significant. Thus, the forces which are involved are *electrostatic* Coulomb forces and *not* the far weaker Ampère current, or *magnetic dipole*, forces. These electrostatic forces regulate the spin configurations in atoms, molecules, and solids via the Pauli principle, so that in constructing the foundations of the theory of magnetism, *one could to a very good approximation completely ignore the magnetic fields proper*. It is fortunate that cause and effect could be so neatly disentangled.

It is necessary to anticipate formally some of the material in the next chapter by introducing three common exchange operators, named, respectively, for Majorana, Bartlett, and Heisenberg.[1]

$$\left.\begin{array}{l}\mathcal{P}^{\mathrm{M}}\\ \mathcal{P}^{\mathrm{B}}\\ \mathcal{P}^{\mathrm{H}}\end{array}\right\}\psi(r_1\xi_1;\, r_2\xi_1;\, \dots) = \left\{\begin{array}{l}\psi(r_2\xi_1;\, r_1\xi_2;\, \dots)\\ \psi(r_1\xi_2;\, r_2\xi_1;\, \dots)\\ \psi(r_2\xi_2;\, r_1\xi_1;\, \dots)\end{array}\right.$$

[1] See, e.g. [2.3]. Note that the operator $\mathcal{P}^{\mathrm{B}}$ was invented by *Dirac* [2.4].

where r_1, r_2 are the spatial and ξ_1, ξ_2 the spin coordinates of two electrons. Clearly,

$$\mathscr{P}^{\mathrm{H}} = \mathscr{P}^{\mathrm{M}}\mathscr{P}^{\mathrm{B}}$$

and later we shall see that on wavefunctions allowable to electrons, $\mathscr{P}^{\mathrm{H}}$ always has eigenvalue $\equiv -1$, so that $\mathscr{P}^{\mathrm{M}}$ and $\mathscr{P}^{\mathrm{B}}$ are effectively inverses of one another. Let us construct the simpler one, $\mathscr{P}^{\mathrm{B}}$, out of the spin one-half operators at our disposal. It must be invariant under the inconsequential rotations in spin space and have eigenvalue $+1$ in triplet states and -1 in singlet states. These requirements are satisfied uniquely by the following function of spinoperators $\boldsymbol{S}_1$ and $\boldsymbol{S}_2$:

$$\mathscr{P}^{\mathrm{B}} = \frac{1}{2}(1 + 4\boldsymbol{S}_1 \cdot \boldsymbol{S}_2) \tag{2.1}$$

Assuming there is an energy $-J_{12}$ associated with the exchange, we obtain an effective Hamiltonian $\propto -J_{12}\boldsymbol{S}_1 \cdot \boldsymbol{S}_2$ (omitting the uninteresting constant part). It is also interesting to remember that just such an effective Hamiltonian had been postulated in the "vector model" of the atom, before the invention of wave mechanics, and that it is in effect a sort of semiempirical rule. We shall later see how it correlated with Hund's rules of atomic structure. In the present chapter we explore molecular structure and the role of exchange, its limitations and its successes in laying the groundwork of a theoretical understanding of magnetism.

The work of Dirac, Heisenberg, and others, and particularly the book of *Van Vleck* first focused attention on an elementary interaction of the type in (2.1) as the fundamental object of study of the theory of magnetism. It may be supposed that this is the reason the Heisenberg exchange Hamiltonian [also called the vector model, or the Heisenberg-Dirac-Van Vleck (HDVV) Hamiltonian]

$$\mathscr{H}_{\mathrm{Heis}} = -\sum_{i,j} J_{ij}\boldsymbol{S}_i \cdot \boldsymbol{S}_j \tag{2.2}$$

occupied so many theorists over the span of 50 years, calculating energy levels and eigenfunctions and the statistical mechanics which flowed therefrom. No less effort went into calculating the magnitudes of the exchange constants J_{ij} from atomistic considerations.

In spite of all this effort, and perhaps because of what it revealed, it was *not* possible to elevate exchange to the rank of a universal principle, to make of it a force of nature such as Coulomb's law or Newton's laws. For example, even though the explanation of Hund's rules by the vector model was one of the earliest successes of the theory, it turned out that the idealized Hamiltonian of (2.2) was only semiquantitatively accurate in complicated atoms and that a more exact Hamiltonian had to be solved in a higher approximation than the one in which exchange is well defined. And in general, notions of electronic correlations and other effects not directly related to the Pauli principle came to cloud the naïve picture.

Even today, many difficulties, both in principle and in practice, still plague those who would derive (2.2) from first principles. This will be made evident in the present chapter as well as in succeeding ones. As Henry Wadsworth Longfellow once wrote, "So nature deals with us and takes away our playthings one by one."

2.1 Exchange Equals Overlap

In order for exchange to occur, the paths of the interacting electrons must inevitably and inextricably overlap. Let us prove this, with the aid of a model in which electrons are confined to nonoverlapping regions [2.6]. Consider N electrons, with arbitrary interactions, and divide coordinate space into N separate boxes such that electron number 1 is confined to box number 1, electron number 2 to the second box, etc. The wavefunction is thus subject to the boundary condition that it vanish whenever a particle reaches the surface of the volume assigned to it. We illustrate a reasonable choice of boundary conditions in the case of N electrons + N fixed protons in Fig. 2.1. Subject to the boundary condition we have specified, the spatial eigenfunctions obey the Schrödinger equation

$$\mathscr{H}\varphi = \sum_i \mathscr{H}(\boldsymbol{r}_i)\varphi + \sum_{i\,j} V(\boldsymbol{r}_i, \boldsymbol{r}_j)\varphi = E\varphi \tag{2.3}$$

in which the Hamiltonian is generally invariant under any permutation of the coordinates of the indistinguishable particles, but is otherwise quite arbitrary.

If the electrons were truly noninteracting, an arbitrary solution of (2.3) could be written as a product function,

$$\varphi = f_1(\boldsymbol{r}_1) f_2(\boldsymbol{r}_2) \ldots \tag{2.4}$$

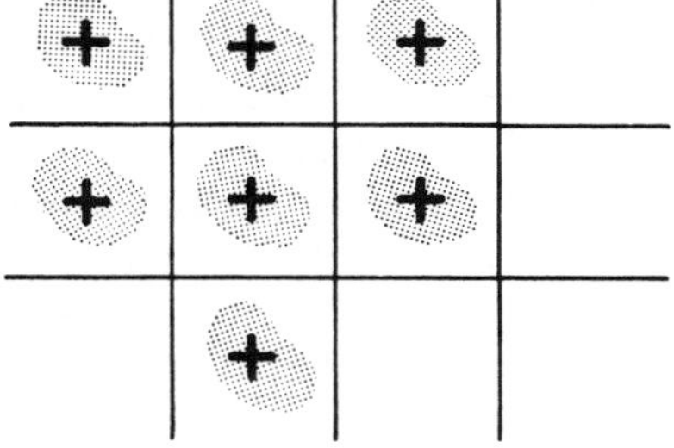

Fig. 2.1. Crosses indicate protons; shading, electrons. The wavefunction is constrained to vanish on the potential barriers indicated by solid lines, i.e., the nodal surfaces are *imposed*

the Hartree product wavefunction. In effect, $f_i(r_j)$ would be the one-particle wavefunction of the jth electron, when constrained to the jth box. But before discussing the product wavefunction, we must stress that it is *not* an eigenfunction of $\mathscr{H}$. For even though the particles are constrained to remain in separate compartments, they do interact, and their correlated motion cannot in general be factored. Therefore if we do not wish to sacrifice generality, we must write the true eigenfunction φ in the most general way as,

$$\varphi = \varphi(1/2/ \dots / N) \tag{2.5}$$

indicating by this notation that particle number 1 is in box number 1, etc. Because of the symmetry of the Hamiltonian under permutations, we can transpose particles 1 and 2, and the wavefunction

$$\mathscr{P}^{M}_{1,2}\varphi = \varphi(2/1 \dots / N)$$

will also be an eigenfunction of the Hamiltonian, (2.3), as indeed are any of the $N!$ permutations

$$\mathscr{P}_{p}\varphi \equiv \varphi_p \qquad p = 1, 2, \dots, N! \tag{2.6}$$

of the N electrons amongst themselves, in the original wavefunction of (2.5). It is trivial to check that all these φ_p functions are mutually orthogonal, as particles are constrained to be in separate boxes. Every energy level of the many-body Hamiltonian is therefore intrinsically $N!$-fold degenerate for the boundary conditions specified. But all $N!$ are not admissible: for example, the totally symmetric linear combination of functions φ_p,

$$\Psi_{\text{sym}} \equiv \sum_{p=1}^{N!} \varphi_p$$

while it might serve in the theory of a system of similar Bosons, is never suitable for $N \geqq 3$ Fermi-Dirac particles. This is but one of the consequences of the Pauli exclusion principle.

Rules for constructing totally antisymmetric wavefunctions of space and spin are given in Chap. 4. What is most important in the present analysis, is that there are two spin degrees of freedom for each electron, therefore 2^N orthonormal spin functions which may be combined with the $N!$ space functions, to give *precisely 2^N wavefunctions of space and spin obeying the Pauli principle, for every eigenvalue E.* If N is even (odd) the resultant total spin ranges in magnitude from $S_{\text{tot}} = 0(\frac{1}{2})$ to a maximum of $\frac{1}{2}N$. There is no interference befween spatial motion and spin degrees of freedom.

Therefore, even in an applied external magnetic field H the free energies of the electronic and spin degrees of freedom are additive: $F = F_{\text{el}} + F_{\text{sp}}$. In particular the spin contribution is

$$F_{sp} = -kTN \ln\left[2\cosh\left(\frac{g\mu_B H}{2kT}\right)\right] \tag{2.7}$$

identically that of the noninteracting spins studied in a later chapter. F_{el} contains a *dia*magnetic contribution, F_{sp} is *para*magnetic, but *neither* is *ferro*magnetic. Thus, spontaneous alignment of magnetic moments in the absence of an applied field is impossible in the present context.

When the fictitious potential barrier separating the electrons is removed, the 2^N-fold degeneracy of each eigenvalue E is lifted, as depicted in Fig. 2.2 where E_0 is the ground state, E_1 the first excited state, etc. The manner in which E_0 fans out is perhaps the most crucial: if the exchange Hamiltonian is capable of reproducing the true spectrum of E_0, we shall have achieved a significant simplification of the problem. (Even if there is considerable overlap between the spectrum of E_0 and that of the higher E_j, the wavefunctions may be distinguished, and classified according to whether they represent magnetic degrees of freedom or electronic degrees of freedom so a whole set of 2^N low-lying states can conceivably be described by the Heisenberg Hamiltonian, i.e., by an "effective" interaction among the spins.)

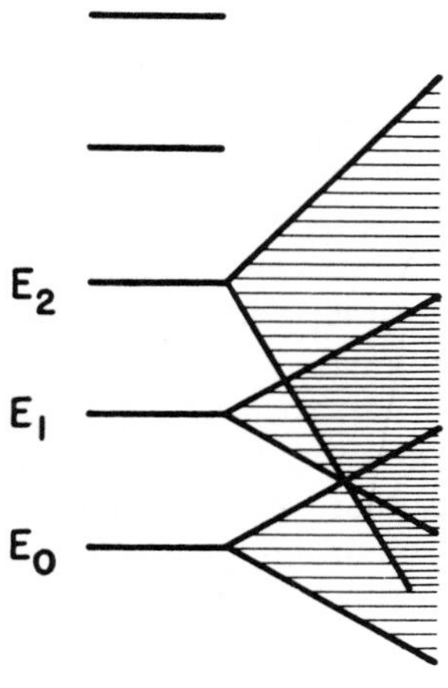

Fig. 2.2. Fanning out of 2^N–fold degenerate levels, when potential barriers are removed. Note that levels may cross

In the chapter on the theory of magnetism in metals, we shall examine at least two instances where this simplification occurs. When electrons are allowed to co-mingle, and their interactions are treated in low order of perturbation theory, it is found that the manner in which the degeneracy of E_0 is lifted *is* precisely described by the Heisenberg Hamiltonian. This is the case both in certain insulators, and in those metals regulated by the indirect exchange mechanism.

However, we are anticipating too far ahead, and it is important to develop the plot gradually.

Suffice it to say that there is in general no guarantee that the levels will fan out in the desired manner, and in many substances they do not do so and therefore defy simple-minded descriptions. Hopefully, a detailed analysis based on physically admissible approximations to the Schrödinger equation will give us some sharper criteria. It is to this task that we turn our attention in this and subsequent chapters.

2.2 Hydrogen Molecule

We study molecules first to appreciate some of the difficulties which are absent in our subsequent study of the atom. But these are by now almost classical subjects on which many books have been written. The actual treatment will therefore be addressed to the students who are yet unacquainted with, or have forgotten atomic and molecular physics ("quantum chemistry", as it is now known). The more experienced reader can omit all but a few remarks towards the end of the chapter concerning subjects with which he might be less familiar. The method with which we break ground was invented by *Heitler* and *London*[2] shortly after the discoveries of Schrödinger, *Heisenberg* [2.2] and *Dirac* [2.1] gave impetus to the quantitative study of the many-body problem in quantum theory. It is still the simplest approach to the problems of molecular binding and interatomic exchange.

Assume nuclei fixed at $\boldsymbol{R}_a$ and $\boldsymbol{R}_b$, with interatomic spacing $\boldsymbol{R}_{ab}$ = several atomic radii. (We set the mass of the proton $= \infty$ so as to be able to localize it. Taking the finite mass into account does not change the results much, but leads to an interesting exchange effect: the existence of two species of H_2, ortho- and parahydrogen. The energy splitting is related to the overlap between protons rotating about a common axis, the overlap due to vibrational motion being negligible. So when the molecule is hindered in its rotation, e.g., by a crystal field, the splitting between ortho- and parahydrogen must disappear.) To a first a approximation each neutral atom exerts no force on the other, and each electron "sees" only the central force field of its own proton. According to this hypothesis, we simplify the two-particle Schrödinger equation,

$$\mathscr{H}\Psi_1 = E\Psi_1$$

with

$$\mathscr{H} = \left(\frac{p_1^2}{2m} - \frac{e^2}{r_{1a}} + \frac{p_2^2}{2m} - \frac{e^2}{r_{2b}}\right) + \left(\frac{e^2}{R_{ab}} + \frac{e^2}{r_{12}} - \frac{e^2}{r_{1b}} - \frac{e^2}{r_{2a}}\right) \tag{2.8}$$

[2] See [2.7] or any good text on molecular physics or chemistry.

by choosing $\Psi_1(\boldsymbol{r}_1, \boldsymbol{r}_2)$ to be a product function $\varphi_a(\boldsymbol{r}_1)\varphi_b(\boldsymbol{r}_2)$, where each factor obeys the one-particle Schrödinger equation, (there should be no confusion between the ground state eigenvalue e_0 and the charge of the electron)

$$\left(\frac{p_1^2}{2m} - \frac{e^2}{r_{1a}}\right)\varphi_a(\boldsymbol{r}_1) = e_0\varphi_a(\boldsymbol{r}_1)$$

and

$$\left(\frac{p_2^2}{2m} - \frac{e^2}{r_{2b}}\right)\varphi_b(\boldsymbol{r}_2) = e_0\varphi_b(\boldsymbol{r}_2). \tag{2.9}$$

At the total Hamiltonian is invariant under the interchange of the two coordinates $\boldsymbol{r}_1$ and $\boldsymbol{r}_2$, an equally good choice must be $\Psi_{\text{II}} = \varphi_a(\boldsymbol{r}_2)\varphi_b(\boldsymbol{r}_1)$. Therefore, let us diagonalize the Hamiltonian of (2.8) within the subspace of these two simple functions. Assume the atomic orbitals $\varphi(\boldsymbol{r})$ to be normalized, and define various overlap integrals l, V, U as follows:

$$1 = \int d^3r\,|\varphi_a(\boldsymbol{r})|^2 = \int d^3r\,|\varphi_b(\boldsymbol{r})|^2 \qquad l \equiv \int d^3r\varphi_a^*(\boldsymbol{r})\varphi_b(\boldsymbol{r})$$

$$V \equiv \int d^3r_1\, d^3r_2\,|\Psi_{\text{I}}|^2\left(\frac{e^2}{R_{ab}} + \frac{e^2}{r_{12}} - \frac{e^2}{r_{1b}} - \frac{e^2}{r_{2a}}\right)$$

$$= \int d^3r_1\, d^3r_2\,|\Psi_{\text{II}}|^2\left(\frac{e^2}{R_{ab}} + \frac{e^2}{r_{12}} - \frac{e^2}{r_{1a}} - \frac{e^2}{r_{2b}}\right)$$

and an "exchange integral":

$$U \equiv \int d^3r_1\, d^3r_2\Psi_{\text{I}}^*\Psi_{\text{II}}\left(\frac{e^2}{R_{ab}} + \frac{e^2}{r_{12}} - \frac{e^2}{r_{1b}} - \frac{e^2}{r_{2a}}\right). \tag{2.10}$$

Let us take a variational wavefunction

$$\Psi = c_{\text{I}}\Psi_{\text{I}} + c_{\text{II}}\Psi_{\text{II}} \tag{2.11}$$

and determine the coefficients c_{I} and c_{II} so as to make the variational energy stationary

$$E_{\text{var}} = \frac{\int d^3r_1\, d^3r_2\Psi^*\mathscr{H}\Psi}{\int d^3r_1\, d^3r_2\Psi^*\Psi}, \quad \frac{\partial E_{\text{var}}}{\partial c_{\text{I,II}}} = 0. \tag{2.12}$$

The solutions to this are best expressed in matrix notation. Let Ψ_{I} correspond to the vector (1, 0) and Ψ_{II} to the vector (0, 1). In this notation, the variational wavefunction Ψ is merely $(c_{\text{I}}, c_{\text{II}})$, and the variational equations can be expressed in compact matrix form as an eigenvalue problem:

$$\begin{bmatrix} V & U \\ U^* & V \end{bmatrix}\begin{bmatrix} c_{\text{I}} \\ c_{\text{II}} \end{bmatrix} = (E - 2e_0)\begin{bmatrix} 1 & l^2 \\ l^{2*} & 1 \end{bmatrix}\begin{bmatrix} c_{\text{I}} \\ c_{\text{II}} \end{bmatrix}. \tag{2.13}$$

As in fact all functions under consideration are (or can be made) real, we shall henceforth omit the asterisk. It is not difficult to guess that the solutions to this euqation are the symmetric and antisymmetric functions corresponding to

$$c_{\mathrm{I}} = \pm c_{\mathrm{II}} \tag{2.14}$$

and that the respective eigenvalues are

$$E_{\pm} = 2e_0 + \frac{V \pm U}{1 \pm l^2}\,. \tag{2.15}$$

The space symmetric (+) solution calls for the spin antisymmetric "singlet" function, and the space antisymmetric function (−) for any of the three symmetric "spin-triplet" functions. The triple-singlet separation is

$$\Delta E = E_{-} - E_{+} = 2\,\frac{Vl^2 - U}{1 - l^4} \tag{2.16}$$

and can be used to define an effective exchange force in the Heisenberg Hamiltonian. For the energy levels of

$$\mathscr{H}_{\mathrm{Heis}} = -J_{12}\boldsymbol{S}_1 \cdot \boldsymbol{S}_2 = -J_{12}\left[\frac{(\boldsymbol{S}_1 + \boldsymbol{S}_2)^2}{2} - \frac{3}{4}\right] \tag{2.17}$$

are $-\frac{1}{4}J_{12}$ in the triplet states, and $+\frac{3}{4}J_{12}$ in the singlet state. (See Chap. 3). By comparison with (2.16), the exchange constant is deduced to be

$$J_{12} = -2\,\frac{Vl^2 - U}{1 - l^4}\,. \tag{2.18}$$

Ferromagnetism, of an embryonic molecular sort, would occur if the exchange constant J_{12} turned out positive. Antiferromagnetism would be the consequence of an antiferromagnetic bond $J_{12} < 0$. What the actual sign turns out to be depends on the relative magnitudes of the "Coulomb integral" V, the "overlap integral" l, and the "exchange integral" U. In the primitive calculation of Heitler and London unperturbed hydrogen 1*s* orbitals were used in these various integrals—yielding results in satisfactory agreement both with experiment and more accurate modern calculations (see Problem 1, below). The exchange constant turned out *negative*, corresponding to a singlet (i.e. embryonic antiferromagnetic) ground state, and varied with internuclear distance R_{ab} as shown in Fig. 2.3. The triplet state was found, correctly, to be *unbound*, to have energy greater than the energy of two separate atoms, $2e_0$.

A second method for treating the H_2 molecule owes its origin to the work of

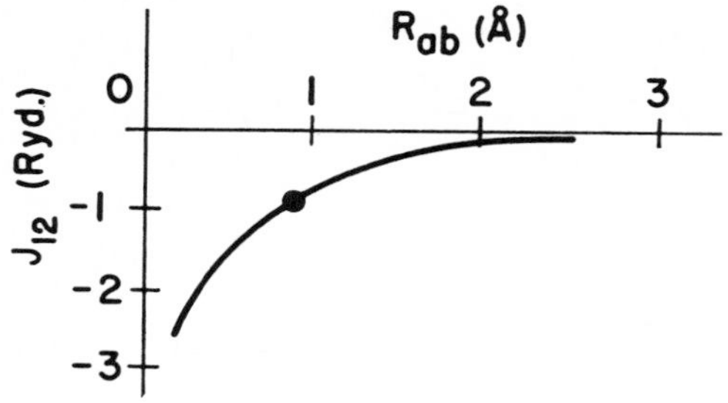

Fig. 2.3. Exchange parameter J_{12} as function of internuclear distance R_{ab} in hydrogen molecule (H_2). Dot shows equilibrium value

F. Hund and *R. S. Mulliken*,[3] and is known as the method of molecular orbitals (M-O). The one-electron functions are chosen, in this method, to be eigenfunctions of some of the symmetry operations which leave an appropriately defined zeroth order Hamiltonian invariant. In this case, the eigenstates of the ion H_2^+ could be chosen as the molecular orbitals. These fall into two classes, the even ones and odd ones. A further refinement would be to take for the one-electron functions the solutions of a Hamiltonian in which the ionic potentials are screened in a self-consistent way. But even the following crude choice gives reasonable agreement with the exact solution, at the equilibrium value of R_{ab}. Let

$$\varphi_g = \frac{\varphi_a(\boldsymbol{r}) + \varphi_b(\boldsymbol{r})}{\sqrt{2(1+l)}} \tag{2.19}$$

and

$$\varphi_u = \frac{\varphi_a(\boldsymbol{r}) - \varphi_b(\boldsymbol{r})}{\sqrt{2(1-l)}} \tag{2.19a}$$

approximating the even and odd (*gerade*, *ungerade*, in the time-honored notation g, u) eigenfunctions of H_2^+ by linear combinations of the atomic orbitals. This generates four functions with which to diagonalize the Hamiltonian,

$$\varphi_g(\boldsymbol{r}_1)\cdot\varphi_g(\boldsymbol{r}_2), \quad \varphi_g(\boldsymbol{r}_1)\cdot\varphi_u(\boldsymbol{r}_2), \quad \varphi_u(\boldsymbol{r}_1)\cdot\varphi_g(\boldsymbol{r}_2), \quad \varphi_u(\boldsymbol{r}_1\cdot\varphi_u(\boldsymbol{r}_2). \tag{2.19b}$$

The M-O method generates four states from two atomic orbitals, twice as many as either the H-L scheme or the Heisenberg Hamiltonian. Molecular orbitals are related to the Bloch functions of solid state theory, discussed elsewhere in the text. There also exists another set of states, more perspicuous than the previous, being mutually orthogonal functions that properly reduce to atomic orbitals in the limit of large separation, $l \to 0$. They are:

---

[3] See, for example, [2.8,9].

$$\psi_a = \frac{\varphi_a - g\varphi_b}{\sqrt{1 + g^2 - 2gl}} \tag{2.20}$$

and

$$\psi_b = \frac{\varphi_b - g\varphi_a}{\sqrt{1 + g^2 - 2gl}} \tag{2.20a}$$

with

$$g = \frac{1}{l}(1 - \sqrt{1 - l^2})$$

to ensure orthogonality. (These functions are prototypes of Wannier functions, treated elsewhere in this book). The eigenvalues of the two-body Hamiltonian can be found with the aid of the four orthonormal functions,

$$\begin{aligned} F_1 &= \psi_a(\boldsymbol{r}_1)\cdot\psi_a(\boldsymbol{r}_2) \qquad & F_2 &= \psi_a(\boldsymbol{r}_1)\cdot\psi_b(\boldsymbol{r}_2) \\ F_3 &= \psi_b(\boldsymbol{r}_1)\cdot\psi_b(\boldsymbol{r}_2) \qquad & F_4 &= \psi_b(\boldsymbol{r}_1)\cdot\psi_a(\boldsymbol{r}_2) \end{aligned} \tag{2.20b}$$

As these functions span the same "function space" as the four M-O functions (2.19b), the four eigenvalues will be the same as if we worked with molecular orbitals. F_2 and F_4 correspond approximately to Ψ_{I} and Ψ_{II} for the nonorthogonal orbitals, F_1 and F_3 to "ionized configurations." The two lowest eigenvalues will correspond to $E_\pm$ of (2.15); see Problem 2.1.

Problem 2.1. Find the eigenvectors and eigenvalues of the 4×4 matrix

$$\mathscr{H}_{ij} = \int F_i \mathscr{H} F_j \, d^3r_1 \, d^3r_2.$$

In addition to the definitions in (2.10), this requires the definition of additional integrals W and X, etc. Define them.

(a) Show that there is one triplet (space antisymmetric) eigenfunction, with energy $E_- = \mathscr{H}_{22} - \mathscr{H}_{24} =$ identically as given in (2.15). Find the lowest of the three space symmetric, singlet, solutions. Prove that it has energy *lower* than E_+, the H-L ground state.

(b) Second-order perturbation theory can also be used to calculate the singlet ground state energy,

$$E_+ \sim \mathscr{H}_{22} + \mathscr{H}_{24} - \frac{(\mathscr{H}_{12} + \mathscr{H}_{14})^2}{\mathscr{H}_{11} - \mathscr{H}_{22} - \mathscr{H}_{24}}.$$

Express this formula in terms of U, V, W, X, and l, and by comparison with the exact result found in part (a) determine the physical region of approximate validity of the perturbation theory.

The importance of this calculation is that it shows that the Heisenberg Hamiltonian can be derived qualitatively on the basis of first- and second-order perturbation theory, using orthogonalized orbitals; and therefore many of the results of the theory (exchange constant, spin waves, existence of Curie temperature, etc.) are valid even when H-L theory is not.

2.3 Three Hydrogen Atoms

The total Hamiltonian is separated into terms appropriate to individual hydrogen atoms $\mathscr{H}_0$ + the interaction terms $\mathscr{H}'$:

$$\begin{aligned}\mathscr{H} = &\left[\left(\frac{p_1^2}{2m} - \frac{e^2}{r_{1a}}\right) + \left(\frac{p_2^2}{2m} - \frac{e^2}{r_{2b}}\right) + \left(\frac{p_3^2}{2m} - \frac{e^2}{r_{3c}}\right)\right] \\ &+ e^2\left[\left(\frac{1}{R_{ab}} + \frac{1}{R_{bc}} + \frac{1}{R_{ac}}\right) + \left(\frac{1}{r_{12}} + \frac{1}{r_{23}} + \frac{1}{r_{13}}\right) - \left(\frac{1}{r_{1b}} + \frac{1}{r_{1c}}\right)\right. \\ &\left. - \left(\frac{1}{r_{2a}} + \frac{1}{r_{2c}}\right) - \left(\frac{1}{r_{3a}} + \frac{1}{r_{3b}}\right)\right] \\ = &\{\mathscr{H}_0\} + \{\mathscr{H}'\} \end{aligned} \tag{2.21}$$

and as before, we use products of the nonorthogonal atomic functions of (2.9) of which there are $3! = 6$ in total. Let us label them according to the following table:

$$\begin{aligned}\psi_1 &= \varphi_a(1)\varphi_b(2)\varphi_c(3)\\ \psi_2 &= \varphi_a(2)\varphi_b(1)\varphi_c(3)\\ \psi_3 &= \varphi_a(3)\varphi_b(2)\varphi_c(1)\\ \psi_4 &= \varphi_a(1)\varphi_b(3)\varphi_c(2)\\ \psi_5 &= \varphi_a(2)\varphi_b(3)\varphi_c(1)\\ \psi_6 &= \varphi_a(3)\varphi_b(1)\varphi_c(2).\end{aligned} \tag{2.22}$$

How are these related to each other under the permutations of various particles? For typographical simplicity, let us omit the Greek symbols, and write ψ_1 as 1, ψ_2 as 2, etc. Then we can draw up simple tables (showing into which functions any of the six transform) under transpositions (permutations of two particles):

$$\begin{aligned}1 &\to 2,3, \quad \text{or} \quad 4\\ 2 &\to 1,5, \quad \text{or} \quad 6\\ 3 &\to 1,5, \quad \text{or} \quad 6\\ 4 &\to 1,5, \quad \text{or} \quad 6\\ 5 &\to 2,3, \quad \text{or} \quad 4\end{aligned} \tag{2.23}$$

$$6 \to 2,3, \quad \text{or} \quad 4$$

and under nontrivial permutations of all *three* particles:

$$\begin{aligned} 1 &\to 5,6 \\ 2 &\to 3,4 \\ 3 &\to 2,4 \\ 4 &\to 2,3 \\ 5 &\to 1,6 \\ 6 &\to 1,5 \end{aligned} \tag{2.24}$$

Thus, for equidistant atoms at the vertices of an equilateral triangle (Fig. 2.4)

$$\int \psi_1^* \psi_2 \, d\tau = \int \psi_1^* \psi_3 \, d\tau = \int \psi_1^* \psi_4 \, d\tau = l^2 = \int \psi_2^* \psi_5 \, d\tau, \quad \text{etc.} \tag{2.25}$$

and

$$\int \psi_1^* \psi_5 \, d\tau = \int \psi_1^* \psi_6 \, d\tau = l^3 = \int \psi_2^* \psi_3 \, d\tau, \tag{2.26}$$

etc. Or in general if we use these integrals to define an overlap matrix $\boldsymbol{\Omega}$

$$\boldsymbol{\Omega}_{ij} = \int \psi_i^* \psi_j \, d\tau \tag{2.27}$$

then we have simply,

$$\boldsymbol{\Omega} = \begin{pmatrix} 1 & l^2 & l^2 & l^2 & l^3 & l^3 \\ l^2 & 1 & l^3 & l^3 & l^2 & l^2 \\ l^2 & l^3 & 1 & l^3 & l^2 & l^2 \\ l^2 & l^3 & l^3 & 1 & l^2 & l^2 \\ l^3 & l^2 & l^2 & l^2 & 1 & l^3 \\ l^3 & l^2 & l^2 & l^2 & l^3 & 1 \end{pmatrix} \tag{2.28}$$

a real symmetric, matrix. And as for the matrix elements of the Hamiltonian, we do not have to examine all 36 possibilities, but in fact just 6:

$$\int \psi_1^* \mathscr{H} \psi_i \, d\tau = H_{1,i} \quad , \quad i = 1, 2, \ldots, 6 \tag{2.29}$$

for we can obtain all the others by appropriate permutations. And of these only three are independent. They are,

$$H_{1,1} = 3e_0 + H'_{1,1} \tag{2.30a}$$

$$H_{1,2} = H_{1,3} = H_{1,4} = 3e_0 l^2 + H'_{1,2} \tag{2.30b}$$

$$H_{1,5} = H_{1,6} = 3e_0 l^3 + H'_{1,5}. \tag{2.30c}$$

From (2.30a) we also get $H_{2,2} = H_{3,3} = \cdots = H_{1,1}$. From (2.30b) we get the matrix elements of $\mathscr{H}$ between any function in (2.23) and its transform, for example, $H_{2,5} = H_{1,2}$, etc.; and (2.30c) gives the archetype matrix element between functions of (2.24) and their transform. A slight change of notation will greatly simplify the appearance of the interaction matrix. We define A, b, and c by

$$A = H'_{1,1}, \qquad bl^2 \cdot A = H'_{1,2} \qquad \text{and} \qquad cl^3 \cdot A = H'_{1,5}. \tag{2.31}$$

This replaces the matrix elements as parameters by A, b, c, and the previously defined overlap integral l [cf. (2.10)]. The Hamiltonian matrix is then particularly convenient to derive: $3e_0$ times the overlap matrix, plus A times an interaction matrix (which can be derived from the overlap matrix merely by replacing l^2 by bl^2 and l^3 by cl^3 in the latter), so that the eigenvalue equation now reads:

$$A\begin{bmatrix} 1 & bl^2 & bl^2 & bl^2 & cl^3 & cl^3 \\ bl^2 & 1 & cl^3 & cl^3 & bl^2 & bl^2 \\ bl^2 & cl^3 & 1 & cl^3 & bl^2 & bl^2 \\ bl^2 & cl^3 & cl^3 & 1 & bl^2 & bl^2 \\ cl^3 & bl^2 & bl^2 & bl^2 & 1 & cl^3 \\ cl^3 & bl^2 & bl^2 & bl^2 & cl^3 & 1 \end{bmatrix} \cdot v = (E - 3e_0) \begin{bmatrix} 1 & l^2 & l^2 & l^2 & l^3 & l^3 \\ l^2 & 1 & l^3 & l^3 & l^2 & l^2 \\ l^2 & l^3 & 1 & l^3 & l^2 & l^2 \\ l^2 & l^3 & l^3 & 1 & l^2 & l^2 \\ l^3 & l^2 & l^2 & l^2 & 1 & l^3 \\ l^3 & l^2 & l^2 & l^2 & l^3 & 1 \end{bmatrix} \cdot v \tag{2.32}$$

We denote eigenvectors by $\boldsymbol{v}$. We construct the six distinct eigenvectors by use of the permutation tables given just previously. One starts with ψ_1, or

$$\begin{bmatrix} 1 \\ 0 \\ 0 \\ 0 \\ 0 \\ 0 \end{bmatrix} \tag{2.33}$$

in the vector notation. A totally symmetric un-normalized function is constructed by adding to this vector all the vectors obtained by permutations of the particles; for example,

$$\boldsymbol{v}_{\text{sym}} = (1, 1, 1, 1, 1, 1). \tag{2.34}$$

(For typographical reasons, we now write the vectors as row vectors. Technically, therefore, they are *left* eigenvectors.) For a totally antisymmetric but unnormalized function, we again start with ψ_1, subtract all the odd permutations, which are given in (2.23), and add the even permulations, given in (2.24)

$$v_{a-s} = (1, -1, -1, -1, +1, +1). \tag{2.35}$$

We next seek vectors antisymmetric in particles 2 and 3, but not *totally* antisymmetric, that is, orthogonal to v_{a-s}. We find

$$v_{23} = \left(1, \frac{1}{2}, \frac{1}{2}, -1, -\frac{1}{2}, -\frac{1}{2}\right) \text{ and } v'_{23} = (0, 1, -1, 0, +1, -1). \tag{2.36}$$

Next, we look for vectors symmetric in particles 2 and 3, but orthogonal to v_{sym}

$$v_{23sym} = \left(1, -\frac{1}{2}, -\frac{1}{2}, 1, -\frac{1}{2}, -\frac{1}{2}\right) \text{ and}$$
$$v'_{23sym} = (0, 1, -1, 0, -1, +1) \tag{2.37}$$

These choices are not unique. However, because of the invariance of the Hamiltonian under permutations, functions of different symmetries do not mix. The totally symmetric and the totally antisymmetric function both stand alone in their own symmetry class and therefore must be eigenvectors. The remaining four eigenvalues are obtained by diagonalizing the matrices of the eigenvalue equation (2.32), in the 2×2 subspaces of the functions of (2.36, 37), respectively. This use of symmetry saves us from the tedium of diagonalizing 6×6 matrices: the importance of the permutation operators should be already abundantly clear.

In the present problem, hidden symmetries simplify the eigenvalue equation further. For it happens that each of the last four vectors is simultaneously an eigenvector of both the overlap and the interaction matrices, and all are degenerate. The eigenvalues can be found almost by inspection now.

$$E_{sym} = 3e_0 + A\,\frac{1 + 3bl^2 + 2cl^3}{1 + 3l^2 + 2l^3} \tag{2.28}$$

$$E_{a-s} = 3e_0 + A\,\frac{1 - 3bl^2 + 2cl^3}{1 - 3l^2 + 2l^3} \tag{2.39}$$

and the four-fold degenerate eigenvalue,

$$E_{23} = 3e_0 + A\,\frac{1 - cl^3}{1 - l^3}\,. \tag{2.40}$$

Elsewhere we discuss how to combine space with spin to obtain wavefunctions of space and spin, obeying the Pauli principle. At present we need only the following which the reader must accept on faith: $\boldsymbol{v}_{\text{sym}}$ is not an allowable eigenfunction for electrons; $\boldsymbol{v}_{\text{a-s}}$ will be used to construct the function of spin three-halves and $\boldsymbol{v}_{23}$ the function of spin one-half, known as quartet and doublet states, respectively. No other values of total spin angular momentum can be obtained with three electrons.

How do these results compare with the solutions of the Heisenberg Hamiltonian? To the two-spin Hamiltonian of (2.17) which corresponded to the hydrogen molecule, $-J_{12}\boldsymbol{S}_1 \cdot \boldsymbol{S}_2$, we now must add two more equal bonds to connect all sides of the equilateral triangle, as shown in Fig. 2.4. That is,

$$\mathscr{H}_{\text{Heis}} = -J_{12}^{\ast}(\boldsymbol{S}_1 \cdot \boldsymbol{S}_2 + \boldsymbol{S}_2 \cdot \boldsymbol{S}_3 + \boldsymbol{S}_3 \cdot \boldsymbol{S}_1). \tag{2.41}$$

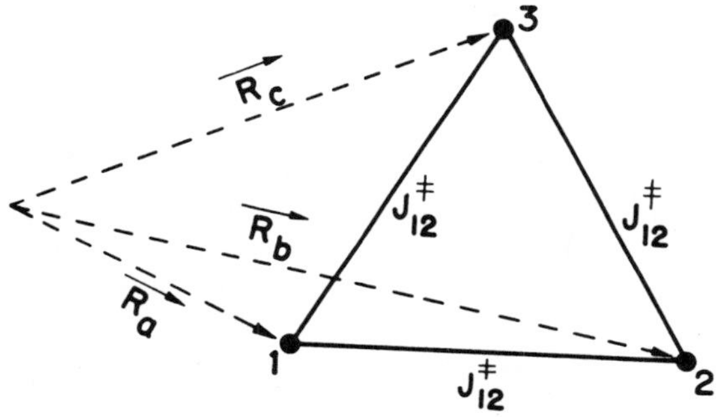

Fig. 2.4. Three-atom molecule, or three equivalent Heisenberg spins, with solid lines indicating bonds

We use a superscript ($\ast$), however, to warn that the exchange "constant" might be rather more variable than its name suggests, and that the value we shall find in the present calculation may not agree with the result previously obtained for two atoms.

The distinct eigenvalues of the above Heisenberg Hamiltonian can be calculated by diagonalizing $\mathscr{H}_{\text{Heis}}$ in the subspace of the three functions belonging to $M = +\frac{1}{2}$. Because of the rotational symmetry, it is found moreover that the two solutions belonging to $M = S = \frac{1}{2}$ are degenerate. The quartet solution is of course unique and, therefore, automatically an eigenfunction of the Hamiltonian. The reader is encouraged to construct these states, starting with a basis

$$\chi_1 \equiv \uparrow\downarrow\downarrow \qquad \chi_2 \equiv \downarrow\uparrow\downarrow \qquad \chi_3 \equiv \downarrow\downarrow\uparrow \tag{2.42}$$

in an obvious notation. One can also obtain the eigenvalues more simply by completing the square in (2.41). Either way, he finds for the solutions of (2.41)

$$E_{\text{quart}} - E_{\text{doubl}} = -\frac{3}{2}J_{12}^{\ast}. \tag{2.43}$$

Now this must be set equal to

$$E_{a-s} - E_{23} = A\left(\frac{1 - 3bl^2 + 2cl^3}{1 - 3l^2 + 2l^3} - \frac{1 - cl^3}{1 - l^3}\right) = -\frac{3}{2}J_{12}^{*} \tag{2.44}$$

the calculated level separation, so as to obtain finally a value for the exchange parameter

$$J_{12}^{*} = -\,2Al^2\frac{(1 - b) + (l + l^2)(c - b)}{(1 - l^3)(1 + l - 2l^2)}\,. \tag{2.45}$$

Unless the overlap is very small, this expression has terms in l^3, etc., and is in no way comparable to the formula of (2.18) which we obtained previously. However, there is no reason to take the H-L scheme seriously if the overlap is that great. [Note that if the atoms are exceedingly close, then the eigenfunctions of atomic lithium are a better zeroth-order set than the three hydrogen atomic functions, whereas those of helium best approximate the solution of the H_2 molecule. In vain can we expect the H-L picture to describe these physical systems when $l \sim 1$. In solids, the H-L picture breaks down completely once l approaches or exceeds $1/z$ in magnitude, where z = number of nearest neighbors of each atom ($z = 6$ for simple cubic structure, etc.) because Ω becomes a singular matrix. That is, the wavefunctions cannot be normalized unless $l < 1/z$; see ahead].

For the reasons above, in a microscopic derivation of the Heisenberg Hamiltonian and of the exchange constant, we must assume that R_{ab} = many atomic units, and calculate only to lowest order in the overlap, l. One is powerless to define interatomic exchange more accurately (although once again in the atom, for *intra*-atomic exchange, the situation is more favorable, as we shall see subsequently).

In accordance with these arguments, we now calculate the exchange parameters *to lowest order in the overlap*: From (2.18),

$$\begin{aligned}\frac{1}{2}J_{12} \doteq -Vl^2 + J = -l^2 \int d^3r_1\, d^3r_2 \varphi_a^2(\boldsymbol{r}_1)\varphi_b^2(\boldsymbol{r}_2)\left(\frac{e^2}{R_{ab}} + \frac{e^2}{r_{12}} - \frac{e^2}{r_{1b}} - \frac{e^2}{r_{2a}}\right)\\ + \int d^3r_1\, d^3r_2 \varphi_a(\boldsymbol{r}_1)\varphi_b(\boldsymbol{r}_1)\varphi_a(\boldsymbol{r}_2)\varphi_b(\boldsymbol{r}_2)\left(\frac{e^2}{R_{ab}} + \frac{e^2}{r_{12}} - \frac{e^2}{r_{1b}} - \frac{e^2}{r_{2a}}\right)\end{aligned} \tag{2.46}$$

and the calculation of (2.45) giving

$$\begin{aligned}\frac{1}{2}J_{12}^{*} &= -Al^2 + Al^2b = -H'_{1,1}l^2 + H'_{1,2}\\ &= -l^2 \int d^3r_1\, d^3r_2\, d^3r_3 \varphi_a^2(\boldsymbol{r}_1)\varphi_b^2(\boldsymbol{r}_2)\varphi_c^2(\boldsymbol{r}_3)\left[\left(\frac{e^2}{R_{ab}} + \frac{e^2}{R_{ac}} + \frac{e^2}{R_{bc}}\right)\right.\\ &\quad \left. + \left(\frac{e^2}{r_{12}} + \frac{e^2}{r_{13}} + \frac{e^2}{r_{23}}\right) - \left(\frac{e^2}{r_{1b}} + \frac{e^2}{r_{1c}} + \frac{e^2}{r_{2a}} + \frac{e^2}{r_{2c}} + \frac{e^2}{r_{3a}} + \frac{e^2}{r_{3b}}\right)\right]\end{aligned}$$

$$+ \int d^3r_1\, d^3r_2\, d^3r_3 \varphi_a(\boldsymbol{r}_1)\varphi_b(\boldsymbol{r}_1)\varphi_a(\boldsymbol{r}_2)\varphi_b(\boldsymbol{r}_2)\varphi_c^2(\boldsymbol{r}_3)\Big[\Big(\frac{e^2}{R_{ab}} + \frac{e^2}{R_{ac}} + \frac{e^2}{R_{bc}}\Big)$$

$$+ \Big(\frac{e^2}{r_{12}} + \frac{e^2}{r_{13}} + \frac{e^2}{r_{23}}\Big) - \Big(\frac{e^2}{r_{1b}} + \frac{e^2}{r_{1c}} + \frac{e^2}{r_{2a}} + \frac{e^2}{r_{2c}} + \frac{e^2}{r_{3a}} + \frac{e^2}{r_{3b}}\Big)\Big] \tag{2.47}$$

recalling the definition of the various parameters in (2.31), and assuming also for simplicity that all the atomic functions are real, and normalized.

Let us try to reduce the second calculation to some of the integrals in the first, so that we may effect a comparison. First, (2.46): The terms in R_{ab} cancel exactly, and the remainder can be put in the form

$$\frac{1}{2}J_{12} = \int d^3r_1\, d^3r_2 \Big(\frac{2e^2}{r_{1b}} - \frac{e^2}{r_{12}}\Big)(\rho_{12}l^2 - \rho_{12\mathrm{ex}}) \tag{2.48}$$

defining ρ_{12} to be the ordinary electron density $\varphi_a^2(\boldsymbol{r}_1)\varphi_b^2(\boldsymbol{r}_2)$ and $\rho_{12\mathrm{ex}}$ to be the "exchange density" $\varphi_a(\boldsymbol{r}_1)\varphi_b(\boldsymbol{r}_1)\varphi_a(\boldsymbol{r}_2)\varphi_b(\boldsymbol{r}_2)$, in an obvious notation.

Some simple manipulations of the terms in the integrals for J_{12}^{*} enable us finally to reach the desired result

$$J_{12}^{*} = J_{12} + 4le^2 \int d^3r_1\, d^3r_2 \varphi_b^2(\boldsymbol{r}_2)\cdot[\varphi_a(\boldsymbol{r}_1)\varphi_c(\boldsymbol{r}_1) - l\varphi_a^2(\boldsymbol{r}_1)]\Big(\frac{1}{r_{12}} - \frac{1}{r_{1b}}\Big) \tag{2.49}$$

The correction term is not necessarily negligible compared to the exchange parameter itself (see Problem 2.2).

Problem 2.2. $\varphi_a(\boldsymbol{r})$ is the unperturbed 1s atomic ground-state wavefunction of an electron belonging to a hydrogen atom at $\boldsymbol{R}_a$, and φ_b is defined correspondingly. Calculate $l(R_{ab})$ and $J_{12}(l)$, $J_{12}^{*}(l)$. Is the exchange parameter increased or decreased by the presence of the third atom? (Assume R_{ab} to be large). See also [Ref. 2.12, Eq. (6)].

The presence of the third atom at $\boldsymbol{R}_c$ modifies the exchange bond, as we have defined it, between spins at $\boldsymbol{R}_a$ and $\boldsymbol{R}_b$. If atoms a and b were imbedded in a solid, the exchange interaction between them would be further modified, and the molecular exchange constant J_{12} or J_{12}^{*} would be of little quantitative use. Those who would calculate exchange parameters must heed the following warning: It is futile to imagine that exchange, even between similar atoms, can be characterized by a universal parameter $J_{12}(R_{12})$ which one might calculate on a molecular model, no matter how refined. It should be regarded as a parameter which depends on all the other environing atoms as well, even to lowest order. When the Heisenberg Hamiltonian (2.1) is not applicable, the exchange parameter loses, of course, even this meaning, and exchange vanishes into the limbo of ill-defined concepts and ideas.

Herring [2.10] points out that the H-L scheme per se is in fact asymptotically wrong for $R_{ab} >$ some 50 atomic distances, where it predicts a ferromagnetic exchange parameter, albeit an exponentially small one. Thus even if the Heisenberg Hamiltonian is applicable, and even if an exchange parameter can be defined and the overlap is small, it still seems a tricky matter to calculate the exchange parameter properly! In the following sections, the difficulties are compounded, not to be resolved until a later chapter.

2.4 Nonorthogonality Catastrophe

In the many-body problem, when one unwittingly expands e^{gN} in a Taylor series,

$$e^{gN} = 1 + gN + \frac{1}{2}(gN)^2 + \cdots \tag{2.50}$$

and tries to take the limit $N \to \infty$, he is overtaken by what is mildly termed a catastrophe. Because of the use of nonorthogonal orbitals in the H-L scheme, just such a nonorthogonality catastrophe occurs when we try to apply the methods which were used in the two-and three-body problems, to the N-body problem of an infinite chain or infinite three-dimensional solid. This difficulty, first pointed out by *Slater* [2.11] and *Inglis* [2.12], has for many years been resolved in principle. Nevertheless it adds considerably to the confusion and difficulties of the problem and to the headaches of those who seek to understand exchange.

The overlap matrix and the interaction matrix will both be $N! \times N!$, although as we have mentioned not all the eigenfunctions will be admissible, and only 2^N will survive the process of antisymmetrization. A row in the overlap matrix may include one 1, N terms in l^2, N^2 terms in l^3, etc., and the question will no longer be, "can we neglect l^2 compared to 1, or l^3 compared to l^2?" but rather, "can we neglect Nl^2 compared to 1? Or N^2l^3 compared to Nl^2?" The answer is an emphatic "No!" no matter how small the overlap. In the present section we give a preliminary qualitative, intuitive analysis resulting in an approximate derivation of the Heisenberg Hamiltonian.

It is not practicable to set down either the overlap or interaction matrices in the explicit form we have previously used, and an operator formalism is required. We expand both as series in permutation operators, decomposed into simple transposition operators and their products, and then take advantage of the Pauli principle to express these in the form of spin exchange operators.

Assume, for simplicity, that the atomic functions are truncated slightly so that there is overlap only with nearest neighbors, in a regular geometric array of N atoms.

$$\int |\varphi_n(\boldsymbol{r})|^2 \, d^3r = 1 \qquad \int \varphi_n^*(\boldsymbol{r})\varphi_m(\boldsymbol{r}) \, d^3r = \begin{cases} l \text{ if } n, m = \text{nearest neighbors} \\ 0 \text{ otherwise} \end{cases} \tag{2.51}$$

Then with the normally ordered functions as the standard, the $N!$ functions can be obtained by various permutations of the particle coordinates

$$\begin{aligned} \psi_1 &= \varphi_1(\boldsymbol{r}_1)\varphi_2(\boldsymbol{r}_2) \cdots \varphi_N(\boldsymbol{r}_N) \\ \mathscr{P}_{12}\psi_1 &= \varphi_1(\boldsymbol{r}_2)\varphi_2(\boldsymbol{r}_1) \cdots \varphi_N(\boldsymbol{r}_N) \\ \mathscr{P}_{12}\mathscr{P}_{2N}\psi_1 &= \varphi_1(\boldsymbol{r}_2)\varphi_2(\boldsymbol{r}_N) \cdots \varphi_N(\boldsymbol{r}_1), \end{aligned} \tag{2.52}$$

etc. The great advantage of the operator formalism is that the eigenvalues of the matrices we obtain are independent of the choice of the initial function ψ_1. We can almost guess that the overlap matrix has the following operator representation:

$$\boldsymbol{\Omega} = 1 + \boldsymbol{h}l^2 + \frac{1}{2}\boldsymbol{h}^2 l^4 + \cdots = e^{\boldsymbol{h}l^2} + \cdots \tag{2.53}$$

with

$$\boldsymbol{h} \equiv \sum_{\substack{(n,m)\\ =\text{nearest neighb.}}} \mathscr{P}_{nm} \tag{2.54}$$

and dots $(\cdots)$ to hide the real difficulties. There are two types of correction: "Kinematical" are those due to counting errors, because a product of x factors of $\mathscr{P}_{nm}$ does not always describe a permutation of x pairs of coordinates. For example, terms of the type

$$\mathscr{P}_{12}\mathscr{P}_{12} = 1 \tag{2.55}$$

should be subtracted from $\frac{1}{2}\boldsymbol{h}^2 l^4$, so that the first kinematical correction is

$$-\frac{1}{4}Nzl^4 \tag{2.56}$$

where z is the number of nearest neighbors of each atom (for example, $z = 2$ in the linear chain, and $z = 6$ in the simple cubic lattice). The next constant correction is

$$-\frac{1}{16}N^2z^2l^8 \tag{2.57}$$

neglecting terms of order Nl^8 which are smaller than (2.56), not to speak of (2.57). The fact is, that the principal corrections enter with ever-increasing powers of N, to form an apparently divergent series. If we add the largest contributions in $\boldsymbol{h}$, $\boldsymbol{h}^2$, etc., the overlap matrix can be resummed as

$$\boldsymbol{\Omega} = Re^{\boldsymbol{h}l^2} + \cdots \tag{2.58}$$

with R representing a nonconvergent (that is, N-dependent) series

$$R = \sum_{n=0}^{N} \left(\frac{Nzl^4}{4}\right)^n C_n \tag{2.59}$$

and the ellipsis $(\cdots)$ higher-order neglected terms.

Among further corrections are the "dynamical corrections," which we shall now define. These are concerned with the nature of the particular law assumed for the overlap, (2.51) in the present case. Unlike the terms in R, the dynamical corrections would be subject to change if we assumed a next-nearest-neighbor overlap, or some other law. They also depend on geometrical niceties, such as: Can an atom have nearest neighbors which are nearest neighbors of each other? If the answer is negative (as in the simple cubic lattice) the first dynamical correction to the overlap occurs in the third term of (2.53), from which we must subtract

$$\frac{1}{2} l^4 \sum_{n \neq n'} \mathscr{P}_{nm} \mathscr{P}_{mn'} \tag{2.60}$$

in order to eliminate what amounts to an effective next-nearest-neighbor exchange, prohibited by our postulates. Further dynamical corrections consist, in part, of eliminating from the series for $\boldsymbol{\Omega}$ all the permutations which, incorrectly, remove an electron to a distance further than its nearest neighbors.

An entirely analogous analysis can be made of the interaction matrix. It is important to recall that there is no exchange interaction between nonoverlapping electrons, as was proved in a preceding section; so that the truncation of the atomic functions, cf. (2.51) *supra*, leaves only a nearest-neighbor interaction. Even the long-ranged Coulomb force cancels to a large extent because of average electrical neutrality. The problem is most simplified if we take for the zero of energy not N times e_0, but $\mathscr{H}_{1,1}$. The interaction term, which is the extension to the present problem of the left-hand side of (2.32), thus can also be guessed to be a series in powers of $\boldsymbol{h}$, with corrections of both the kinematical and dynamical sort, i.e.

$$\text{Int} = u_1 \boldsymbol{h} + \frac{1}{2} u_2 \boldsymbol{h}^2 + \cdots \tag{2.61}$$

with

$$u_1 = \int d^3r_1 \cdots d^3r_N \psi_1^* \mathscr{H}(\mathscr{P}_{12} - l^2)\psi_1 \tag{2.62}$$

$$u_2 = 2u_1 l_1^2, \ldots \qquad u_n = n u_1 l^{2n-2}, \ldots \tag{2.63}$$

Besides corrections of the same nature as in the overlap matrix, there are additional dynamical corrections to the u_n's, particularly when there are successive interchanges of neighbouring pairs. For example, instead of u_2 as defined above, the coefficient of the particular permutation $\mathscr{P}_{12}\mathscr{P}_{34}$ should be precisely

$$\int d^3r_1 \cdots d^3r_N \psi_1^* \mathscr{H}(\mathscr{P}_{12}\mathscr{P}_{34} - l^4)\psi_1 \tag{2.64}$$

or else one counts double the nonnegligible interactions among the four particles when they are nearest neighbors. Again leaving this and other corrections aside, we find

$$\mathrm{Int} = \boldsymbol{h} u_1 (R e^{\boldsymbol{h} l^2} + \cdots). \tag{2.65}$$

Thus, the eigenvalue equation takes the form

$$\boldsymbol{h} u_1 (R e^{\boldsymbol{h} l^2} + \cdots)\varphi = E(R e^{\boldsymbol{h} l^2} + \cdots)\varphi \tag{2.66}$$

with E measured relative to $H_{1,1}$ and φ the appropriate spatial eigenfunction. However, since the latter is eventually to be multiplied by the appropriate spin functions and totally antisymmetrized, it is permissible to replace the particle permutation operators $\mathscr{P}_{nm}$ by their spin conjugate operators,

$$\mathscr{P}_{nm} \to \frac{1}{2}(1 + 4\boldsymbol{S}_n \cdot \boldsymbol{S}_m). \tag{2.67}$$

Let us suppose momentarily that the corrections indicated by dots do not diverge as badly as R, or that they are approximately the same on both sides of the above equation. The parentheses can be factored in that case, and making the above substitution into (2.66) we obtain the HDVV eigenvalue equation

$$-\frac{1}{2} u_1 \sum_{(n,m)} (1 + 4\boldsymbol{S}_n \cdot \boldsymbol{S}_m)\chi = E\chi \tag{2.68}$$

with χ a spin eigenfunction. and (n,m) are nearest-neighbor pairs. Except for the trivial additive constant term, the left-hand side of this eigenvalue equation is the Heisenberg nearest-neighbor exchange Hamiltonian. The exchange parameter is the integral $2u_1$, in the present case, which can be compared to the exchange parameters J_{12} $J_{12}^{\mp}$ which were obtained in the preceding section. Although it is of the same general form, the actual numerical value may differ. Like J_{12}, u_1 may in general be positive or negative. It is normally negative, which leads to an antiferromagnetic ground state. This is in agreement with the observed ground state of N atoms of hydrogen (a molecular nonferromagnetic solid at or near zero absolute temperature).

Although the "catastrophe" has been averted, it has not been shown that the neglect of the dots is justified. Permutations leading to terms quartic or higher-order in spin operators may sometime be found to be of importance.[4] In the following section, we shall discuss a method which has aims somewhat more modest than the construction of a Heisenberg Hamiltonian, but achieves these limited goals quantitatively without uncontrollable approximations. In the chapter on magnetism and magnons in metals, the Heisenberg Hamiltonian will

[4] For a systematic analysis of this approach see [2.13]. For different approaches cf. the next section and Chap. 6. Experimental evidence of higher-order spincoupling is found in [2.14], and also in [2.15].

finally be seen to be the incidental consequence of low-order perturbation theory carried out for insulators. The neglect of three-body permutations will be seen as equivalent to the neglect of third- and higher-order perturbation corrections. However before presenting this simplified interpretation, many other topics will occupy our interest.

2.5 Method of Löwdin and Carr

The nonorthogonality catastrophe in the Heitler-London theory was introduced and discussed qualitatively in the previous section. Here we shall present a constructive method for calculating directıy the eigenvalues of the H-L Hamiltonian. For the range of parameters in which this theory gives reasonable results, these solutions are equivalent to finding the energy of the 2^N lowest eigenstates of $\mathscr{H}$, which can then be fitted to a Heisenberg Hamiltonian. If the fit is good, it provides evidence that the procedure of the previous section works. If the parameters of the Heisenberg Hamiltonian cannot be chosen so as to reproduce the correct eigenvalues, then the evidence is that exchange processes involving three or more spins are important, i.e., the higher-order permutations indicated by dots in the previous section do not factor between the interaction and the overlap, and are of physical importance.

One advantage of the method of *P.O. Löwdin* as discussed by *W.J. Carr*[5] is that it deals with $N \times N$ arrays rather than $N! \times N!$ arrays of the permutation space. We shall here obtain the variational energy of two very important states with it: the ferromagnetic state, for which the spatial function is the totally antisymmetric determinantal function, and the Néel antiferromagnetic state of alternating spins "up" and "down," which we shall define subsequently. Starting with the ferromagnetic state for which the many electron wavefunction is a determinantal function

$$\Psi = \frac{1}{\sqrt{N!}} \begin{vmatrix} \varphi_1(r_1) & \varphi_2(r_1) & \cdots & \varphi_N(r_1) \\ \varphi_1(r_2) & \varphi_2(r_2) & \cdots & \cdot \\ \cdot & \cdot & & \cdot \\ \cdot & \cdot & & \cdot \\ \cdot & \cdot & & \cdot \\ \varphi_1(r_N) & \cdot & \cdots & \varphi_N(r_N) \end{vmatrix}, \tag{2.69}$$

We calculate the energy of this state as simply

$$E_{\text{Ferro}} = \frac{\Psi^* \mathscr{H} \Psi \, d^3r_1 \cdots d^3r_N}{\Psi^* \Psi \, d^3r_1 \cdots d^3r_N}. \tag{2.70}$$

[5] Löwdin's methods for nonorthogonal arrays were applied to the problem of ferromagnetism by *Carr* [2.16]. A variant method was used by *Takano* [2.17].

The Hamiltonian $\mathscr{H}$ may quite generally be taken of the form,

$$\mathscr{H} = \mathscr{H}_0 + \sum_i U(r_i) + \sum_{i \neq j}\sum V(r_i, r_j) \tag{2.71}$$

such that the principal diagonal in the determinantal function is an eigenfunction of $\mathscr{H}_0$ with eigenvalue e_0. Without loss of generality, let us pose $e_0 = 0$, which sets the origin of the scale of energy. Thus,

$$E_{\text{Ferro}} = \sum_i \frac{\int |\Psi|^2 U(r_i)\, d^3r_1 \cdots d^3r_N}{\text{denominator}} + \sum_{i \neq j}\sum \frac{\int |\Psi|^2 V(r_i, r_j)\, d^3r_1 \cdots d^3r_N}{\text{denominator}}. \tag{2.72}$$

The denominator may be manipulated as follows:

$$\text{denominator} = \frac{1}{N!}\int \begin{vmatrix} \varphi_1^*(r_1) & \cdots \\ \vdots & \\ \cdot & \cdots \end{vmatrix} \times \begin{vmatrix} \varphi_1(r_1) & \cdots \\ \vdots & \\ \cdot & \cdots \end{vmatrix} d^3r_1 \cdots$$

$$= \int \varphi_1^*(r_1)\varphi_2^*(r_2) \cdots \varphi_N^*(r_N) \begin{vmatrix} \varphi_1(r_1) & \cdots \\ \vdots & \\ \cdot & \cdots \end{vmatrix} d^3r_1 \cdots$$

$$= \int \begin{vmatrix} \varphi_1^*(r_1)\varphi_1(r_1) & \varphi_1^*(r_1)\varphi_2(r_1) & \cdots \\ \varphi_2^*(r_2)\varphi_1(r_2) & \varphi_2^*(r_2)\varphi_2(r_2) & \\ \vdots & \vdots & \\ \varphi_N^*(r_N)\varphi_1(r_N) & \cdot & \cdots \end{vmatrix} d^3r_1 \cdots \tag{2.73}$$

The second form of "denominator" is obtained by noting that each term in the expansion of the left-hand determinant contributes equally. Using this symmetry, one obtains exactly the same value of the integral by retaining only the principal diagonal and multiplying the result by $N!$ By the standard rules of multiplication of determinants by scalars, each row in the right-hand array can now be multiplied by a function occurring in the principal diagonal of the first array, and this results in the third form of (2.73).

The numerators are handled in an analogous manner, although the problem is somewhat trickier. We may not assume that the various terms in the potential energy are invariant with respect to interchange of the particle coordinates. However the total Hamiltonian *is* invariant under such interchange. We obtain, for one-body potentials

$$\int |\Psi|^2 \sum_i U(r_i)\, d^3r_1 \cdots = \sum_{i,j} \int \varphi_i^*(r_i) U(r_i) \varphi_j(r_i)\, d^3r_i \times \int D_j^i\, d^3r_1 \cdots d^3r_{i-1}\, d^3r_{i+1} \cdots \tag{2.74}$$

and for two-body potentials

$$\int |\Psi|^2 \sum_{i \neq j}\sum V(r_i, r_j)\, d^3r_1 \cdots = \sum_{i,j,l,m} \int \varphi_i^*(r_i)\varphi_j^*(r_j)V(r_i, r_j)\varphi_l(r_i)\varphi_m(r_j)$$
$$\times \int D^{ij}_{lm}\, d^3r_1 \cdots d^3r_{i-1}\, d^3r_{i+1} \cdots d^3r_{j-1}\, d^3r_{j+1} \cdots . \tag{2.75}$$

The sum of these terms when divided by denominator, equals the ferromagnetic energy.

The integrands D^i_j and D^{ij}_{lm} are obtained from the third line of (2.73) by striking out the ith row and jth column to obtain $D^i_j \times (-1)^{i+j}$; and the ith and jth rows, lth and mth columns, to obtain $D^{ij}_{lm} \times (-1)^{i+j+l+m}$. The functions which have been so stricken are explicitly taken into the energy integrals. Now all these determinants can be easily evaluated by expanding, integrating, and re-collecting terms. One finds first

$$\text{Denominator} = \det|L_{ij}| \qquad \text{where} \qquad L_{ij} \equiv \int \varphi_i^*(r)\varphi_j(r)\, d^3r\,. \tag{2.76}$$

The integral of D^i_j is seen to be merely the *cofactor* of L_{ij} in this determinant. And by the usual theory of determinants, we have

$$\frac{\int D^i_j\, d^3r_1 \cdots}{\text{denominator}} = (L^{-1})_{ij}\,. \tag{2.77}$$

And similarly,

$$\frac{\int D^{ij}_{lm}\, d^3r_1 \cdots}{\text{denominator}} = (L^{-1})_{li}(L^{-1})_{mj} - (L^{-1})_{lj}(L^{-1})_{mi}\,. \tag{2.78}$$

Here L^{-1} indicates the matrix inverse to L, the latter being defined as the square matrix array with elements L_{ij}. In solids which possess translational invariance, these are cyclic matrices and can easily be inverted.[6]

For example, consider a one-dimensional chain with nearest-neighbor overlap only. Thus, $L_{ii} = 1$ and $L_{i,i\pm1} \equiv l$, and all other matrix elements vanish. To make the matrix cyclic, we set $N + 1 = 1$, which is equivalent to adopting periodic boundary conditions. The eigenvectors $\boldsymbol{v}$ of $\boldsymbol{L}$ are the plane waves,

$$\boldsymbol{v} = (v_1, \ldots, v_n, \ldots), \quad v_n = e^{ikn} \qquad \text{with} \qquad \boldsymbol{k} = \pm 2\pi\,\frac{\text{integer}}{N} \tag{2.79}$$

and the corresponding eigenvalues are

$$L_k = 1 + 2l\cos \boldsymbol{k}\,. \tag{2.80}$$

[6] The problem of inversion of the overlap matrix is discussed further in [2.18–20].

Thus, it may be verified that $L_{nm} = N^{-1} \sum_k e^{ik(n-m)} L_k$, and more importantly, that the matrix elements of the inverse matrix are

$$
\begin{aligned}
(L^{-1})_{nm} &= \frac{1}{N} \sum_k e^{ik(n-m)} \frac{1}{L_k} \\
&= \frac{1}{2\pi} \int_{-\pi}^{\pi} d\theta \frac{\cos\theta(n-m)}{1 + 2l\cos\theta} = d^{|n-m|} \frac{1+d^2}{1-d^2},
\end{aligned}
\tag{2.81}
$$

where

$$
d = \frac{-1}{2l}\left(1 - \sqrt{1-4l^2}\right),\ d^2 < 1.
$$

In the second line the sum over $\boldsymbol{k}$ was replaced by the integral appropriate in the limit $N \to \infty$. It is seen that the elements of the inverse matrix decrease exponentially with distance $|n-m|$. Therefore if l is sufficiently small, only $|n-m| = 0$ will survive, which corresponds (as we shall see) to a Heisenberg nearest-neighbor interaction.

Let us now illustrate this in three dimensions, in the simple cubic structure. Using (2.72) and those following, we find that the energy of the ferromagnetic stage in the three-dimensional array is

$$
\begin{aligned}
E_{\text{Ferro}} &= \sum_i \sum_j (L^{-1})_{ij} \int \varphi_i^*(\boldsymbol{r}) U(\boldsymbol{r}) \varphi_j(\boldsymbol{r})\, d^3r \\
&+ \sum_i \sum_j \sum_l \sum_m (L^{-1})_{li} (L^{-1})_{mj} \int [\varphi_i^*(\boldsymbol{r}) \varphi_j^*(\boldsymbol{r}') - \varphi_i^*(\boldsymbol{r}') \varphi_j^*(\boldsymbol{r})] \\
&\times V(\boldsymbol{r}, \boldsymbol{r}') \varphi_l(\boldsymbol{r}) \varphi_m(\boldsymbol{r}')\, d^3r\, d^3r'.
\end{aligned}
\tag{2.82}
$$

Let $\boldsymbol{R}_{il} \equiv a\boldsymbol{n} = a(n_1, n_2, n_3)$ and $\boldsymbol{R}_{mj} \equiv a\boldsymbol{n}' = a(n_1', n_2', n_3')$ (2.83)

define

$$
U_k = \sum_i e^{-i\boldsymbol{k}\cdot\boldsymbol{R}_{ij}} \int \varphi_i^*(\boldsymbol{r}) U(\boldsymbol{r}) \varphi_j(\boldsymbol{r})\, d^3r \tag{2.84}
$$

and

$$
L_{\boldsymbol{k}} = \sum_i e^{-i\boldsymbol{k}\cdot\boldsymbol{R}_{ij}} L_{ij} = 1 + 2l(\cos k_x a + \cos k_y a + \cos k_z a). \tag{2.85}
$$

Thus if $l > \frac{1}{6}$ some $L_{\boldsymbol{k}}$ vanishes *and L_{ij} does not have an inverse.* Using the identity $l/f = \int_0^\infty ds\, e^{-sf}$ and the definition of $I_n(z)$, the Bessel function of imaginary argument, $I_n(z) = 1/2\pi \int_{-\pi}^{\pi} d\theta \cos n\theta \exp(z\cos\theta)$, we find

$$
(L^{-1})_{mj} = L^{-1}(n_1', n_2', n_3') = \int_0^\infty ds\, e^{-s} I_{n_1'}(2sl) I_{n_2'}(2sl) I_{n_3'}(2sl) \tag{2.86}
$$

indicating the explicit dependence on distance by the argument (n_1', n_2', n_3') and assuming $l < \frac{1}{6}$. Note that to leading order in powers of l, $L^{-1}(0) = 1 + 6l^2 \doteq 1$; and $L^{-1}(1, 0, 0) = -l$; $L^{-1}(1, 1, 0)$ and $L^{-1}(2, 0, 0) = O(l^2)$; etc. Finally, the ferromagnetic energy is,

$$E_{\text{Ferro}} = \sum_k \frac{U_k}{L_k} + \sum_{l,m,n,n'} L^{-1}(\boldsymbol{n})L^{-1}(\boldsymbol{n}') \int [\varphi_i^*(\boldsymbol{r})\varphi_j^*(\boldsymbol{r}') - \varphi_i^*(\boldsymbol{r}')\varphi_j^*(\boldsymbol{r})] \times V(\boldsymbol{r}, \boldsymbol{r}')\varphi_l(\boldsymbol{r})\varphi_m(\boldsymbol{r}')\, d^3r\, d^3r' \tag{2.87}$$

with $\boldsymbol{R}_i = \boldsymbol{R}_l + a\boldsymbol{n}$, and $\boldsymbol{R}_j = \boldsymbol{R}_m - a\boldsymbol{n}'$. The L^{-1} (0) terms yield the direct and Heisenberg exchange interactions, denoted V and U, respectively, in the notation we used for the hydrogen molecule. The L^{-1} (1, 0, 0) terms include integrals reminiscent of the three-atom molecule

$$L^{-1}(1, 0, 0)L^{-1}(0) \int [\varphi_{l+a(1,0,0)}^*(\boldsymbol{r})\varphi_m^*(\boldsymbol{r}') - \varphi_{l+a(1,0,0)}^*(\boldsymbol{r}')\varphi_m^*(\boldsymbol{r})]V(\boldsymbol{r}, \boldsymbol{r}') \times \varphi_l(\boldsymbol{r})\varphi_m(\boldsymbol{r}')\, d^3r\, d^3r'.$$

These, and successively more complicated terms cannot be associated with simple operators of the form $\boldsymbol{S}_i \cdot \boldsymbol{S}_j$, but require such forms as

$$(\boldsymbol{S}_i \cdot \boldsymbol{S}_j\, \boldsymbol{S}_j \cdot \boldsymbol{S}_m) \qquad \text{or} \qquad (\boldsymbol{S}_i \cdot \boldsymbol{S}_j\, \boldsymbol{S}_n \cdot \boldsymbol{S}_m), \text{ etc.,} \tag{2.88}$$

in their description. Fortunately, they enter with ever-increasing powers of l, and therefore may be presumed negligible if l is sufficiently small. *Note that there is no "catastrophe"*: l must be small compared to $\frac{1}{6}$, but not to $1/N$, in order that a Heisenberg Hamiltonian, with (at worst) corrections of the type indicated in (2.88), describe the energy of the ferromagnetic state.

We turn to another important state, the antiferromagnetic Néel configuratkon, defined as follows: Every electron of spin "up" is surrounded by neighbors of spin "down." We note that because the spin functions of nearest-neighboring electrons are orthogonal, the one-electron wavefunctions of space and spin are also orthogonal, and therefore lack of orthogonality of the spatial parts is irrelevant. There is the no "exchange" whatever as $U(\boldsymbol{r})$ and $V(\boldsymbol{r}, \boldsymbol{r}')$ do not involve the spins, and the energy is straightaway,

$$E_{\text{Néel}} = \sum_i \int \varphi_j^*(\boldsymbol{r})U(\boldsymbol{r})\varphi_i(\boldsymbol{r})\, d^3r + \sum_{i,j} \int |\varphi_i^*(\boldsymbol{r})\varphi_j(\boldsymbol{r}')|^2 V(\boldsymbol{r}, \boldsymbol{r}')\, d^3r\, d^3r'. \tag{2.89}$$

As we shall see in a subsequent chapter, the energy of the Néel state is likely within 25 percent of the true ground-state energy of an antiferromagnet in three dimensions. *Therefore, within the stated accuracy and restricted range of validity of the H-L scheme, one can determine whether the ground state is ferromagnetic or antiferromagnetic by comparing* (2.87) and (2.89), *without in fact any need to calculate exchange constants or to first set up a Heisenberg Hamiltonian!*

One often encounters the following statement: "The Heitler-London scheme and the band structure theory coalesce for *filled* bands." As we have an enlightening example of this phenomenon in the present section, it is appropriate to terminate with a brief remark on the subject. Return to the determinantal wavefunction given in (2.69) and take linear combinations of the columns (this does not affect the value of a determinant) to obtain

$$\Psi = \frac{1}{\sqrt{N!}} \begin{vmatrix} \psi_{k_1}(r_1) & \psi_{k_2}(r_1) & \cdots \\ \psi_{k_1}(r_2) & \psi_{k_2}(r_2) & \cdots \\ \cdot & \cdot & \\ \cdot & \cdot & \\ \psi_{k_1}(r_N) & \cdot & \end{vmatrix} \tag{2.90}$$

where

$$\psi_k(\boldsymbol{r}) = \frac{1}{\sqrt{N}} \sum_i \mathrm{e}^{i\boldsymbol{k}\cdot\boldsymbol{R}_i} \varphi_i(\boldsymbol{r}) \tag{2.91}$$

is the desired linear combination. It is exactly the *Bloch* wavefunction for an electron in the tight-binding limit. The set of functions in (2.91) is an orthogonal set, albeit an unnormalized one. For,

$$\begin{aligned} \int \psi_k^*(\boldsymbol{r}) \psi_{k'}(\boldsymbol{r})\, d^3r &= \frac{1}{N} \sum_i \sum_{\boldsymbol{R}_{ij}} \mathrm{e}^{i(\boldsymbol{k}-\boldsymbol{k}')\cdot\boldsymbol{R}_i} \mathrm{e}^{i\boldsymbol{k}'\cdot\boldsymbol{R}_{ij}} \int \varphi_j^*(\boldsymbol{r}) \varphi_i(\boldsymbol{r})\, d^3r \\ &= \delta_{\boldsymbol{k},\boldsymbol{k}'} \sum_{\boldsymbol{R}_{ij}} \mathrm{e}^{i\boldsymbol{k}\cdot\boldsymbol{R}_{ij}} L_{ij} \end{aligned} \tag{2.92}$$

from which one may deduce (2.80) and other results in the text. This sort of procedure has been used by *Takano* [2.17] to examine the theory of spin waves.

Problem 2.3. Assume a simple cubic structure, quasiconstant electron wavefunctions

$$\phi_j(\boldsymbol{r}) = +(7a^3)^{-1/2}$$

when $\boldsymbol{r}$ is within the unit cube (a^3) centered about $\boldsymbol{R}_j$, or within any of the six nearest-neighbor unit cubes centered about $\boldsymbol{R}_j \pm (a, 0, 0), \ldots, \boldsymbol{R}_j \pm (0, 0, a)$, and

$$\phi_j(\boldsymbol{r}) = 0$$

otherwise. Let the potentials be

$$U(\boldsymbol{r}) = -A$$

that is, constant, and

$$V(\boldsymbol{r}, \boldsymbol{r}') = +B\delta(\boldsymbol{r} - \boldsymbol{r}').$$

Using these functions to calculate the various quantitites in the method of Löwdin and Carr, find the following:

(a) The ferromagnetic and Néel state energies: Determine which lies lower. How does this depend on the magnitude of A? of B?

(b) In the pure Heisenberg model, simple cubic lattice, there are $3N$ bonds and therefore $E_{\text{Néel}} - E_{\text{Ferro}} = \frac{3}{2}NJ_{12}$, where $R_{12} = a$ = near neighbor distance and J_{12} = near neighbor exchange parameter. Using the energies calculated in part (a), obtain J_{12}. Then use the definition, (2.18), to calculate J_{12}. Explain the agreement (or disagreement) between the results of these alternative calculations.

3. Quantum Theory of Angular Momentum

This chapter affords a brief summary of the quantum theory of angular momentum. As the angular momentum of a charged particle is proportional to its magnetization, this subject is at the core of the theory of magnetism. We shall show that motional angular momentum is inadequate, and introduce spin angular momentum. We shall develop operator techniques expressing angular momentum or spin operators in terms of more primitive fermion or boson operators. The topics of spin-one-half and spin-one are treated individually, for use in subsequent chapters on the theory of magnetism.

3.1 Kinetic Angular Momentum

In the absence of any external forces, the classical angular momentum of a point particle has a constant value

$$\boldsymbol{L} = \boldsymbol{r} \times \boldsymbol{p} \tag{3.1}$$

and therefore in quantum theory the analogous quantity, which shall be denoted "kinetic angular momentum" is an operator,

$$\boldsymbol{L} = \boldsymbol{r} \times \frac{\hbar}{\mathrm{i}} \nabla \tag{3.2}$$

according to the usual rule for constructing quantum-mechanical momenta: $p_x = \hbar/\mathrm{i}(\partial/\partial x)$, etc. *The angular momentum associated with a wavefunction* ψ *can be easily determined if this function is spherically symmetric* so that it does not depend on the angular coordinates φ and θ, but only on the magnitude of the radius vector $\boldsymbol{r}$. For in that case we can prove that $\boldsymbol{L} \equiv 0$ by noting that

$$\nabla\psi(r) = \hat{\boldsymbol{u}}_r \frac{d}{dr} \psi(r) \tag{3.3}$$

where $\hat{\boldsymbol{u}}_r$ = unit vector in the radial direction = $\boldsymbol{r}/r$, and therefore,

$$\boldsymbol{L}\psi(r) = \frac{\hbar}{\mathrm{i}} (\boldsymbol{r} \times \hat{\boldsymbol{u}}_r) \frac{d}{dr} \psi(r) \equiv 0. \tag{3.4}$$

A function with *nontrivial* angular dependence has *nonvanishing* kinetic angular momentum, although it is not always so simple to discover its magnitude. For this requires the solution of an eigenvalue problem; and moreover, the three components of $\boldsymbol{L} = (L_x, L_y, L_z)$ do *not* commute with one another and therefore cannot be simultaneously specified. Briefly,

$$\begin{aligned}[L_x, L_y] &= L_xL_y - L_yL_x \\ &= \frac{\hbar}{\mathrm{i}}\frac{\hbar}{\mathrm{i}}\left[\left(y\frac{\partial}{\partial z} - z\frac{\partial}{\partial y}\right)\left(z\frac{\partial}{\partial x} - x\frac{\partial}{\partial z}\right) - \left(z\frac{\partial}{\partial x} - x\frac{\partial}{\partial z}\right)\left(y\frac{\partial}{\partial z} - z\frac{\partial}{\partial y}\right)\right] \\ &= \hbar^2\left(x\frac{\partial}{\partial y} - y\frac{\partial}{\partial x}\right) \\ &= \mathrm{i}\hbar L_z. \end{aligned} \tag{3.5}$$

With the other two components obtained by cyclic permutations of the above, we find a set of relations which can most conveniently be described by the vector cross product, as given by the usual determinantal rule

$$\boldsymbol{L} \times \boldsymbol{L} \equiv \begin{vmatrix} \hat{\boldsymbol{u}}_x & \hat{\boldsymbol{u}}_y & \hat{\boldsymbol{u}}_z \\ L_x & L_y & L_z \\ L_x & L_y & L_z \end{vmatrix} = \mathrm{i}\hbar\boldsymbol{L}. \tag{3.6}$$

While it is true that this rule was obtained by the definition of angular momentum given in (3.2), it is in fact more general. There are operators (identified later) which obey (3.6) but not (3.2). It is therefore important that we define angular momentum as follows: *A vector operator is a genuine angular momentum operator* (*whether kinetic, spin, or generalized angular momentum*) *only if it satisfies* (3.6) *or the equivalent set of equations* (3.10–16).

Example: If $\boldsymbol{L} = (L_x, L_y, L_z)$ is an angular momentum, then $\boldsymbol{L}' = (-L_x, -L_y, -L_z)$ is not. This is related to the pseudovector nature of the classical angular momentum $\boldsymbol{r} \times \boldsymbol{p}$, and is also directly a consequence of (3.6).

Returning to the kinetic angular momentum, we express it in spherical polar coordinates

$$\boldsymbol{r} = (r \sin\theta \cos\phi, r \sin\theta \sin\phi, r \cos\theta)$$

in which the cartesian components of $\boldsymbol{L}$ take the form

$$L_z = \frac{\hbar}{\mathrm{i}}\frac{\partial}{\partial\phi} \tag{3.7}$$

$$L^+ = \hbar e^{\mathrm{i}\phi}\left(\frac{\partial}{\partial\theta} + \mathrm{i}\cot\theta\frac{\partial}{\partial\phi}\right) \tag{3.8}$$

$$L^- = \hbar e^{-i\phi}\left(-\frac{\partial}{\partial\theta} + i\cot\theta\frac{\partial}{\partial\phi}\right). \tag{3.9}$$

Here, for future convenience, we introduced the operators

$$L^\pm \equiv L_x \pm iL_y. \tag{3.10}$$

But if desired, the Cartesian components may be extracted by (what is equivalent to the last equation):

$$L_x = \frac{1}{2}(L^+ + L^-) \quad \text{and} \quad L_y = \frac{1}{2i}(L^+ - L^-). \tag{3.11}$$

The $\pm$ operators are the more useful, though, and have the names of angular momentum *raising* (+) and *lowering* (−) operators, respectively. The reason for this nomenclature is to be sought in the commutation laws for these operators. Because it is possible to derive these commutators using only (3.6), without making use of the actual differential forms for L_z, etc., given above, the following equations are generally valid for *any* angular momentum:

$$[L_z, L^+] = \hbar L^+ \tag{3.12}$$

and

$$[L_z, L^-] = -\hbar L^-. \tag{3.13}$$

Given an eigenfunction Ψ_m of L_z with eigenvalue m, that is,

$$L_z\Psi_m = \hbar m\Psi_m$$

apply both sides of (3.12) to this eigenfunction and after rearranging terms, obtain

$$L_z(L^+\Psi_m) = \hbar(m+1)(L^+\Psi_m).$$

This shows that $L^+\Psi_m$ is an eigenfunction of L_z belonging to eigenvalue $m + 1$. It is similarly shown that $L^-\Psi_m$ belongs to eigenvalue $m - 1$, whence the terminology of "raising" and "lowering" operators. A final commutation law which can be obtained from (3.6) is

$$[L^+, L^-] = 2i[L_y, L_x] = 2\hbar L_z. \tag{3.14}$$

There is only one important operator which has not yet been introduced; it is the scalar associated with the vector angular momentum, its length.

$$L^2 = L_x^2 + L_y^2 + L_z^2 \tag{3.15a}$$

$$= \frac{1}{2}(L^+L^- + L^-L^+) + L_z^2 \tag{3.15b}$$

$$= L^+L^- + L_z(L_z - \hbar) \tag{3.15c}$$

$$= L^-L^+ + L_z(L_z + \hbar) \tag{3.15d}$$

The various alternative forms for L^2 are obtained from each other by means of the commutation relations of (3.14), and the definitions (3.10, 11). Some will be more useful than others when dealing with eigenfunctions of L_z, for example.

It is trivial to prove that L^2 commutes with all the components of $\boldsymbol{L}$, that is,

$$[L^2 , L_z] = [L^2, L_x] = [L^2, L_y] = [L^2, L^\pm] \equiv 0. \tag{3.16}$$

This is seen by inspection for L_z as L^2 does not involve the angle ϕ, and therefore must commute with $\partial/\partial\phi$. It follows by rotatioual symmetry, that it also commutes with both other components of $\boldsymbol{L}$.

Because L^2 and L_z commute, they have a complete set of eigenfunctions in common, which can be labeled by the eigenvalue of each of these two operators. Because L_x and L_y (and also $L^\pm$) do not commute with L_z, they cannot be simultaneously diagonalized with the former two operators, and therefore their eigenvalues cannot *also* be specified. But there is no loss of generality in singling out the component L_z, since the z axis can be picked along any arbitrary direction. In the case of kinetic angular momentum, the desired eigenfunctions will be the well-known spherical harmonics, as we shall now discover.

3.2 Spherical Harmonics

Because the radial dependence of the wavefunction plays no role in the determination of angular momentum, we may dispense with it altogether and consider functional dependence on the angles only. The wavefunctions are normalized on the unit sphere,

$$\int_0^{2\pi} d\phi \int_{-\pi}^{\pi} d\theta \sin\theta \psi^*(\theta, \phi)\psi(\theta, \phi) = 1 \tag{3.17}$$

and the two simultaneous eigenvalue equations are

$$L_z\psi \equiv \frac{\hbar}{\mathrm{i}}\frac{\partial}{\partial\phi}\psi = \hbar m\psi \tag{3.18}$$

and

$$L^2\psi = -\hbar^2\left[\frac{1}{\sin\theta}\frac{\partial}{\partial\theta}\left(\sin\theta\frac{\partial}{\partial\theta}\right) + \frac{1}{\sin^2\theta}\frac{\partial^2}{\partial\phi^2}\right]\psi = \hbar^2\lambda\psi \tag{3.19}$$

with eigenvalues, respectively, of m (the "magnetic quantum number") and λ. We still show the dependence on the quantum-mechanical unit of angular momentum $\hbar$, but eventually it will be most convenient to use units such that $\hbar = 1$, and such units will be assumed throughout most of the book.

The L_z equation can be integrated directly, with the result

$$\psi(\theta, \phi) = e^{im\phi}\psi(\theta, 0). \tag{3.20}$$

A boundary condition must now be invoked to determine the quantum numbers m, an obvious choice being to require $\psi(\theta, \phi)$ to be single-valued on the unit sphere. Therefore we require

$$e^{im(\phi+2\pi)} = e^{im\phi} \tag{3.21}$$

which is satisfied by the choice

$$m = \text{integer} = 0, \pm 1, \pm 2, \ldots \tag{3.22}$$

Although other boundary conditions are possible, they would violate other requirements of quantum theory which we have not yet discussed, which shall be examined in the following section. The same boundaıy condition of (3.21) has also been used by *Dirac* (see *Note* at end of section) to prove that the electric charge is quantized, in units of a fundamental charge q.

The solution, (3.20), may now be introduced into the second differential equation, (3.19), which is the eigenvalue equation for L^2. One finds directly

$$\left(\frac{1}{\sin\theta}\frac{\partial}{\partial\theta}\sin\theta\frac{\partial}{\partial\theta} - \frac{m^2}{\sin^2\theta} + \lambda\right)\psi = 0 \tag{3.23}$$

which is the equation obeyed by the *associated Legendre polynomials*, the properties of which are well established. But if they were not known, the following constructive procedure would solve this eigenvalue equation by elementary means:

Assume that the wavefunction $\psi(\theta, \phi) = e^{im\phi}\psi(\theta, 0)$ is a solution of the following first-order partial differential equation,

$$L^+\psi(\theta, \phi) = 0. \tag{3.24}$$

Therefore, by (3.15d), ψ is an eigenfunction of L^2 with eigenvalue

$$\lambda = m(m + 1). \tag{3.25}$$

The value of m is a special one, by virtue of the previous equation, $L^+\psi = 0$. *It cannot be stepped up.* It is the maximum value of the azimuthal quantum number for the calculated value of λ. Therefore we shall denote it by the symbol l, defined by

$$\lambda = l(l+1) \tag{3.26}$$

and let m continue to represent the azimuthal (or "magnetic") quantum number. For the function which obeys (3.24), $l = m$. By repeated application of L^-, we can generate functions always belonging to the same value of λ or l, but with decreasing magnetic quantum numbers $m = l - 1$, $m = l - 2$, ... The series terminates with $m = -l$, by (3.15c).

The first eigenfunction, with $m = l$, is obtained explicitly by integrating the linear first-order differential equation (3.24). Given

$$\left(\frac{\partial}{\partial\theta} - l\cot\theta\right)\psi = 0 \tag{3.27}$$

divide by $(\cos\theta)\cdot(\psi)$, and integrate to obtain

$$\psi = (\sin^l\theta)(e^{il\phi})\,\text{const.} \tag{3.28}$$

Except for normalization factors, the remaining functions of the set having the common eigenvalue l are found by repeated applications of the operator L^- to this solution. These functions can be normalized and given standard phases, and are denoted the *spherical harmonics*, $Y_{l,m}$.

$$Y_{l,m}(\theta,\phi) = \frac{(-1)^{l+m}}{2^l l!}\sqrt{\frac{(2l+1)(l-m)!}{4\pi(l+m)!}}\cdot(\sin\theta)^m\left(\frac{\partial}{\partial\cos\theta}\right)^{l+m}(\sin\theta)^{2l}e^{im\phi}. \tag{3.29}$$

These are, then, the associated Legendre polynomials of $(\cos\theta)$, multiplied by $\exp(im\phi)$ and suitably normalized over the unit sphere.[1] These functions form a complete, orthonormal set (see Table 3.1).

Note: Dirac assumed the existence, somewhere in the universe, of a magnetic *monopole*. Although such objects have never been found on earth or in astronomical observations, which have so far always indicated that the magnetic dipole is the fundamental source of magnetic fields, the search goes on for this elusive particle,[2] and we may follow Dirac in assuming the existence of at least one such source—of strength μ_0. The magnetic field of such a pole is

$$H = \frac{\mu_0}{r^2}$$

pointing in the radial direction, which implies a vector potential (in spherical polar coordinates)

$$A = (A_r, A_\theta, A_\phi) \qquad \text{with } A_r = A_\theta = 0 \qquad \text{and} \qquad A_\phi = \frac{\mu_0}{r}\tan\left(\frac{1}{2}\theta\right).$$

---

[1] The phase $(-1)^{l+m}$ is chosen in conformity with [3.1,2].

[2] See, e.g. [3.3]

Table 3.1

l	m	$Y_{l,m}$ (normalized on unit sphere)	
0	0	$\frac{1}{2\sqrt{\pi}}$	
1	1	$\frac{1}{r}\left[-\sqrt{\frac{3}{8\pi}}(x+iy)\right]$	$=-\sqrt{\frac{3}{8\pi}}\sin\theta e^{i\varphi}$
	0	$\frac{1}{r}\sqrt{\frac{3}{4\pi}}z$	$=\sqrt{\frac{3}{4\pi}}\cos\theta$
2	2	$\frac{1}{r^2}\left[\sqrt{\frac{15}{32\pi}}(x+iy)^2\right]$	$=\sqrt{\frac{15}{32\pi}}\sin^2\theta e^{2i\varphi}$
	1	$\frac{1}{r^2}\left[-\sqrt{\frac{15}{8\pi}}z(x+iy)\right]$	$=-\sqrt{\frac{15}{8\pi}}\cos\theta\sin\theta e^{i\varphi}$
	0	$\frac{1}{r^2}\left[\sqrt{\frac{5}{16\pi}}(3z^2-r^2)\right]$	$=\sqrt{\frac{5}{16\pi}}(2\cos^2\theta-\sin^2\theta)$
3	3	$\frac{1}{r^3}\left[-\sqrt{\frac{35}{64\pi}}(x+iy)^3\right]$	$=-\sqrt{\frac{35}{64\pi}}\sin^3\theta e^{3i\varphi}$
	2	$\frac{1}{r^3}\left[\sqrt{\frac{105}{32\pi}}z(x+iy)^2\right]$	$=\sqrt{\frac{105}{32\pi}}\cos\theta\sin^2\theta e^{2i\varphi}$
	1	$\frac{1}{r^3}\left[-\sqrt{\frac{21}{64\pi}}(5z^2-r^2)(x+iy)\right]$	$=-\sqrt{\frac{21}{64\pi}}(4\cos^2\theta\sin\theta-\sin^3\theta)e^{i\varphi}$
	0	$\frac{1}{r^3}\left[\sqrt{\frac{7}{16\pi}}(5z^2-3r^2)z\right]$	$=\sqrt{\frac{7}{16\pi}}(2\cos^3\theta-3\cos\theta\sin^2\theta)$
4	4	$\frac{1}{r^4}\left[\frac{3}{16}\sqrt{\frac{35}{2\pi}}(x+iy)^4\right]$	$=\frac{3}{16}\sqrt{\frac{35}{2\pi}}\sin^4\theta e^{4i\varphi}$
	3	$\frac{1}{r^4}\left[-\frac{3}{8}\sqrt{\frac{35}{\pi}}z(x+iy)^3\right]$	$=-\frac{3}{8}\sqrt{\frac{35}{\pi}}\sin^3\theta\cos\theta e^{3i\varphi}$
	2	$\frac{1}{r^4}\left[\frac{3}{8}\sqrt{\frac{5}{2\pi}}(7z^2-r^2)(x+iy)^2\right]$	$=\frac{3}{8}\sqrt{\frac{5}{2\pi}}(6\sin^2\theta\cos^2\theta-\sin^4\theta)e^{2i\varphi}$
	1	$\frac{1}{r^4}\left[-\sqrt{\frac{45}{64\pi}}(7z^2-3r^2)z(x+iy)\right]$	$=-\sqrt{\frac{45}{64\pi}}(4\sin\theta\cos^3\theta-3\cos\theta\sin^3\theta)e^{i\varphi}$
	0	$\frac{1}{r^4}\left[\frac{3}{16\sqrt{\pi}}(35z^4-30r^2z^2+3r^4)\right]$	$=\frac{3}{16\sqrt{\pi}}(8\cos^4\theta+3\sin^4\theta-24\sin^2\theta\cos^2\theta)$

Note: $\frac{1}{r^4}\left(x^4+y^4+z^4-\frac{3}{5}r^4\right)=\frac{4\sqrt{\pi}}{15}\left[Y_{4,0}+\sqrt{\frac{5}{14}}\left(Y_{4,4}+Y_{4,-4}\right)\right]$

$Y_{l,-m}=(-1)^m Y^*_{l,m}$

This may be verified by the relation $\nabla\times\boldsymbol{A}=\boldsymbol{H}$, defining $\boldsymbol{A}$. As we know, the Hamiltonian, as well as any other dynamical operator of a $\pm$ charged particle, is modified in a magnetic field because of a change in the momentum of the particle

$$\boldsymbol{p}\to\boldsymbol{p}\pm\frac{e}{c}\boldsymbol{A}$$

with c = speed of light. But notice that $\boldsymbol{A}\to\infty$ along the line $\theta=\pi$, in just such a way that the integral along a path enclosing this line has a definite value, namely,

$$\int_{\theta=\pi} \boldsymbol{A}\cdot d\boldsymbol{s} = \mu_0 4\pi .$$

A particle which in the absence of the monopole had magnetic quantum number m, now has magnetic quantum number

$$m' = m \pm \left(e \int \boldsymbol{A} \frac{d\boldsymbol{s}}{2\pi\hbar c}\right) = m \pm \frac{e}{2\pi\hbar c}\mu_0 4\pi .$$

The wavefunction must still be single-valued, hence m' must be an integer, from which it follows that $\pm e$ must be an integer multiple of the "fundamental charge"

$$q = \frac{\hbar c}{2\mu_0} \qquad \text{Q.E.D.}$$

By showing moreover that there is no fundamental unit of length associated with the magnetic monopole, Dirac proved that there could not be any bound states of a charged particle in the field $\boldsymbol{H}$, so that the main physical effect of the monopole is the quantization of charge [3.4].

3.3 Reason for Integer l and m

It is a straightforward matter to show that l and m are all integers, or all half-odd integers. The choice between these two sets is, however, not so simple.

Let us prove the first statement by recalling the boundary condition, (3.21), by which we required the wavefunctions to be single-valued on the unit sphere. This is too strict; in fact, the only physical requirement is that the probability density—a physical *observable*—be single-valued in an arbitrary state. That is, if we take a wave function which depends on ϕ as follows:

$$\psi(\phi) = \sum_m A_m(\mathrm{e}^{\mathrm{i}m\phi} + \mathrm{e}^{\mathrm{i}(m+1)\phi} + \mathrm{e}^{\mathrm{i}(m-1)\phi} + \cdots)$$

and require: (3.30)

$$|\psi(\phi + 2\pi)|^2 = |\psi(\phi)|^2$$

then the only way to satisfy this in general is to have all the m's integers, *or* all the m's half-odd integers. It is a simple exercise to verify that if *both* are admitted, or other choices of m are made, then some probability densities can be constructed which are not single-valued.

The choice between the integers and half-integers is now made on the basis of the following reasoning. Physical quantities must be independent of the choice of coordinate system (whether Cartesian, spherical, or whatever) and of the the origin of that coordinate system. For example, even for particles not undergoing circular motion, the complete set of eigenfunctions of angular momentum must be adequate to describe the angular part of the motion. And conversely, purely spherical motion must be describable in terms of other complete sets of wavefunctions, such as the plane-wave states. Let us take this for an example. The well-known formula for the expansion of plane waves in spherical harmonics is

$$e^{i\boldsymbol{k}\cdot\boldsymbol{r}} = 4\pi \sum_{l=0}^{\infty} \sum_{m=-l}^{+l} i^l j_l(kr) Y_{l,m}(\theta, \phi) Y_{l,m}^*(\Theta, \Phi) \tag{3.31}$$

where the coefficients

$$j_l(kr) = \sqrt{\frac{\pi}{2kr}} J_{l+\frac{1}{2}}(kr) \tag{3.32}$$

are spherical Bessel functions, here defined in terms of the ordinary Bessel functions $J_p(z)$. If Θ and Φ are the angles of the $\boldsymbol{k}$ vector and θ and ϕ are the angles of the coordinate vector $\boldsymbol{r}$, the expansion formula is further reduced by expressing it in terms of the relative angle ω,

$$\cos \omega = \frac{\boldsymbol{k}\cdot\boldsymbol{r}}{kr} = \cos\theta \cos\Theta + \sin\theta \sin\Theta \cos(\phi - \Phi) \tag{3.33}$$

using the so-called *addition* formula,

$$P_n(\cos\omega) = \frac{4\pi}{2n+1} \sum_{m=-n}^{+n} Y_{n,m}^*(\theta, \phi) Y_{n,m}(\Theta, \Phi). \tag{3.34}$$

While we make no attempt to prove (3.31–34), they are well established in elementary texts in electromagnetic and quantum theory. The Legendre polynomials, for example,

$$P_0(x) = 1 \qquad P_1(x) = x \qquad P_2(x) = \frac{1}{2}(3x^2 - 1) \tag{3.35}$$

are particularly well known (almost elementary) functions, constituting a complete set of orthonormal polynomials on the interval $-1 \leq x \leq +1$.

The expansion of the plane waves given above or of any other useful complete set of nonspherical wavefunctions in a complete set of spherical wavefunctions requires only the *integer* values of l and m, and this is what finally fixes our choice.

Note that the plane waves are a complete set of states of the infinitesimal translation operator. The half-odd integer states must therefore be reserved for a space in which *only* rotations are permitted, the so-called *spin space.*

3.4 Matrices of Angular Momentum

The spherical harmonics are the orthonormal eigenfunctions of L^2 and L_z, and by using them we can calculate the matrix structure of such operators as L_x and L_y, or $L^\pm$. For example, supplementing the relationship we had already derived,

$$L^+ Y_{l,l} = 0,$$

we can obtain

$$L^+ Y_{l,m} = \hbar\sqrt{(l - m)(l + m + 1)}\, Y_{l,m+1} \tag{3.36}$$

by operating directly on the spherical harmonics given in (3.29). The following relations are also very useful:

$$L^- Y_{l,m} = \hbar\sqrt{(l - m + 1)(l + m)}\, Y_{l,m-1} \tag{3.37}$$

$$L_z Y_{l,m} = \hbar m Y_{l,m} \tag{3.38}$$

and

$$\begin{aligned} L^2 Y_{l,m} &= [L_z^2 + \frac{1}{2}(L^-L^+ + L^+L^-)Y_{l,m} \\ &= \hbar^2[m^2 + \frac{1}{2}(l - m)(l + m + 1) + \frac{1}{2}(l - m + 1)(l + m)]Y_{l,m} \\ &= \hbar^2 l(l + l)Y_{l,m}. \end{aligned} \tag{3.39}$$

Note that in this last equation, no special property of the spherical harmonics was used. The eigenvalue of L^2 can be calculated using the previous three equations, that is, by knowing the matrix structure of $L^\pm$ and L_z. But this matrix structure, also, can be derived without explicit recourse to the spherical harmonics, by using the commutation relations, (3.12–14). Thus it would apply also to the half-odd-integer solutions in which we shall soon be interested. To distiuguish these, we shall want to extend our notation.

The conventional notation of $L^\pm$, L_z, and L^2, l, and m, will continue to be used when we are dealing with kinetic angular momentum. For general or arbitrary angular momentum, when only the basic commutation relations are invoked but the arguments are not restricted to integer values of l and m, $\boldsymbol{J}$ will replace $\boldsymbol{L}$ and j will replace l, although the eigenvalue associated with J will still be denoted m. We shall also use $\boldsymbol{S}$ interchangeably with $\boldsymbol{J}$. The general matrix structure of the general angular momentum operators can be obtained using only

(3.12–16) and the construction given in the first section. One obtains the following results:

$$(j'm'|J^+|jm) = \delta_{j,j'}\delta_{m+1,m'}\hbar\sqrt{(j-m)(j+m+1)} \tag{3.40}$$

$$(j'm'|J^-|jm) = \delta_{j,j'}\delta_{m-1,m'}\hbar\sqrt{(j-m+1)(j+m)} \tag{3.41}$$

$$(j'm'|J_z|jm) = \delta_{j,j'}\delta_{m,m'}\hbar m \tag{3.42}$$

$$(j'm'|J^2|jm) = \delta_{j,j'}\delta_{m,m'}\hbar^2 j(j+1). \tag{3.43}$$

Because of their common factor $\delta_{j,j'}$, the general angular momentum operators can be represented by $(2j+1) \times (2j+1)$ square matrices (*irreducible representations*) which act in a subspace of the $2j+1$ linearly independent eigenfunctions of J^2 belonging to a given j value. Such matrices satisfy the basic commutation relations (3.6) or (3.12–16), *by construction*, and are therefore genuine angular momenta in their own right.

Larger matrices can also be found which are angular momenta, but then they have sub-blocks with the above matrix structure, and are called *reducible representations*. The most compact matrices having the properties of angular momentum are *irreducible representations* of the angular momentum operators. For example, the irreducible representation of angular momentum $l = 1$ is

$$\begin{aligned}
\mathscr{L}^+ &= \hbar\begin{bmatrix} 0 & \sqrt{2} & 0 \\ 0 & 0 & \sqrt{2} \\ 0 & 0 & 0 \end{bmatrix} \\
\mathscr{L}^- &= \hbar\begin{bmatrix} 0 & 0 & 0 \\ \sqrt{2} & 0 & 0 \\ 0 & \sqrt{2} & 0 \end{bmatrix} \\
\mathscr{L}_z &= \hbar\begin{bmatrix} 1 & 0 & 0 \\ 0 & 0 & 0 \\ 0 & 0 & -1 \end{bmatrix} \\
\mathscr{L}^2 &= 2\hbar^2\begin{bmatrix} 1 & 0 & 0 \\ 0 & 1 & 0 \\ 0 & 0 & 1 \end{bmatrix}
\end{aligned} \tag{3.44}$$

For angular momentum $l = 2$ there are 5×5 matrices, for angular momentum $l = 3$ there are 7×7, etc. It is a striking fact, however, that all the *even*-dimensional representations are missing from the list if we insist on the integer values of angular momentum possessed by the spherical harmonics. The missing matrices, the even-dimensional ones, are the spin angular momentum operators for j,m = half-odd integers. Upon including these, the series is made complete.

3.5 Pauli Spin Matrices

Here we shall discuss the smallest of the even-dimensional irreducible representations of angular momentum, the matrices of *spin one-half*. They are very similar to the 2×2 matrices, which were invented by Pauli precisely for the purpose of describing the intrinsic spin angular momentum of the electron.

In this smallest subspace, the spin operators are, according to (3.40–43),

$$S^+ = \hbar \begin{bmatrix} 0 & 1 \\ 0 & 0 \end{bmatrix} \qquad S^- = \hbar \begin{bmatrix} 0 & 0 \\ 1 & 0 \end{bmatrix}$$

and

$$S_z = \hbar \begin{bmatrix} \frac{1}{2} & 0 \\ 0 & -\frac{1}{2} \end{bmatrix} \tag{3.45}$$

and

$$S^2 = \hbar^2 \frac{3}{4} \begin{bmatrix} 1 & 0 \\ 0 & 1 \end{bmatrix}. \tag{3.46}$$

The Pauli spin matrices are obtained from S_z, $S_x = \frac{1}{2}(S^+ + S^-)$, and $S_y = (S^+ - S^-)/2\mathrm{i}$ by the relation

$$\boldsymbol{S} = \frac{1}{2}\hbar\boldsymbol{\sigma} = \frac{1}{2}\hbar(\sigma_x, \sigma_y, \sigma_z) \tag{3.47}$$

and so are explicitly

$$\sigma_x = \begin{bmatrix} 0 & 1 \\ 1 & 0 \end{bmatrix} \qquad \sigma_y = \mathrm{i} \begin{bmatrix} 0 & -1 \\ 1 & 0 \end{bmatrix} \qquad \sigma_z = \begin{bmatrix} 1 & 0 \\ 0 & -1 \end{bmatrix}$$

$$\text{and the unit operator:} \quad \mathbf{1} = \begin{bmatrix} 1 & 0 \\ 0 & 1 \end{bmatrix}. \tag{3.48}$$

The eigenvectors of S_z, and σ_z are the two-component *spinors*

$$\chi_+ = \begin{bmatrix} 1 \\ 0 \end{bmatrix} \quad \text{and} \quad \chi_- = \begin{bmatrix} 0 \\ 1 \end{bmatrix} \tag{3.49}$$

corresponding to $m = \pm \frac{1}{2}$, respectively. Conventionally, $\sigma^{\pm}$ are defined by $\sigma^{\pm} = S^{\pm}$.

The higher-dimensional even matrices, 4×4, 6×6, etc., can be similarly constructed using the general set of rules for the matrix elements given in (3.40–

43) and will represent the higher spins 3/2, 5/2, etc. There is no need to give these explicitly, particularly because matrices become unwieldy with increasing size, and because we shall find far more convenient operator representations.

3.6 Compounding Angular Momentum

It is frequently necessary to compound constituent angular momenta into a total angular momentum operator and, conversely, to decompose and simplify operators with complex structure. Let us go into this matter briefly and directly, even at the expense of omitting formal proofs.

Given two angular momenta $\boldsymbol{J}_1$ and $\boldsymbol{J}_2$, it seems certain (intuitively) that $\boldsymbol{J} = \boldsymbol{J}_1 + \boldsymbol{J}_2$ will be an allowable angular momentum, obeying the commutation laws of Eqs. (3.6) and those following. However, no other linear combination of $\boldsymbol{J}_1$ and $\boldsymbol{J}_2$ will do (cf. Problem 3.1).

Problem 3.1. Given angular momenta $\boldsymbol{J}_1$ and $\boldsymbol{J}_2$, show that $a\boldsymbol{J}_1 + b\boldsymbol{J}_2$ is itself an angular momentum only if $a = b = 1$; or $a = 1$, $b = 0$; or $a = 0$, $b = 1$.

How does one express the eigenfunctions of $\boldsymbol{J}_1$ and $\boldsymbol{J}_2$ in terms of the eigenfunctions of $\boldsymbol{J}$ and vice versa?

Supposing we start by knowing a complete set of angular momenta eigenfunctions of $\boldsymbol{J}_1$ and $\boldsymbol{J}_2$, whether spinors or spherical harmonics, or whatever, which we label by the two sets of quantum numbers

$$|j_1 m_1 j_2 m_2). \tag{3.50}$$

For fixed j_1 and j_2 there are $(2j_1 + 1)\cdot(2j_2 + 1)$ orthonormal eigenfunctions of this type corresponding to the various choices of m_1 and m_2. Each one is, therefore, also an eigenfunction of J^z, with eigenvalue $m = m_1 + m_2$. (Henceforth, subscripts will label individual angular momenta, and superscripts their components; for example, J_i^x.) This has a maximum value $j_1 + j_2$, therefore it is the maximum value of m, which is by definition j, the quantum number of J^2 [recall (3.24–26)]. By repeated applications of the operator $J^- = J_1^- + J_2^-$ to this state of maximal m, we generate a total of $2(j_1 + j_2) + 1$ orthogonal states all belonging to the same j value, but with $m = j_1 + j_2, j_1 + j_2 - 1, \ldots, -j_1 - j_2$. (Note that j is not changed, because J^- commutes with J^2.) Only *one* of these states has $m = j_1 + j_2 - 1$, for indeed, every m value occurs only once in this list. But there are *two* of the original states which belong to this eigenvalue of J^z: $|j_1 j_1 - 1 j_2 j_2)$ and $|j_1 j_1 j_2 j_2 - 1)$. The function which has been included is in fact the *sum* of these two functions,

$$|j_1 j_1 - 1 j_2 j_2) + |j_1 j_1 j_2 j_2 - 1) \tag{3.51}$$

because $J^- = J_1^- + J_2^-$. This leaves us free to consider the difference of the two

$$|j_1 j_1 - 1 j_2 j_2) - |j_1 j_1 j_2 j_2 - 1). \tag{3.52}$$

This function is orthogonal to the previous one, it belongs to the same m value, and when $J^+ = J_1^+ + J_2^+$ is a applied to it the result vanishes. Therefore the m value to which it belongs must be maximal, and (3.52) must be an eigenfunction belonging to $j = j_1 + j_2 - 1$. Repeated applications of J^- to this function, (3.52), results indeed in $2(j_1 + j_2 - 1) + 1$ orthogonal functions, corresponding to all the attainable m values.

Two of the three states belonging to $m = j_1 + j_2 - 2$ are thus accounted for, and the third can now be used to construct the set of functions belonging to $j = j_1 + j_2 - 2$. This procedure may be continued until all $(2j_1 + 1)\cdot(2j_2 + 1)$ initial states are exhausted. Such a stage is reached when $j = |j_1 - j_0|$, which is therefore the *minimum* magnitude of the compound angular momentum, just as $j_1 + j_2$ is the *maximum* magnitude. The proof of this so-called "triangle inequality" is given in Problem 3.2 and by an operator method in (3.75, 76).

Problem 3.2. Show that the procedure indicated in the text does in fact use every one of the initial functions, by proving the identity

$$\sum_{j=|j_1-j_2|}^{j_1+j_2} (2j + 1) = (2j_1 + 1)\cdot(2j_2 + 1).$$

Having constructed the complete set of eigenfunctions of J^2 and J^z from the set of eigenfunctions of J_1^z and J_2^z, one may formalize this procedure somewhat. The two sets of orthonormal states are related by a canonical (unitary) transformation, that is, we can express the new wavefunctions $|j_1, j_2 jm)$ in terms of the old by

$$|j_1 j_2 jm) = \sum_{m_1 m_2} |j_1 m_1 j_2 m_2)(j_1 m_1 j_2 m_2 | j_1 j_2 jm)$$

where the matrix elements of the transformation operator

$$(j_1 m_1 j_2 m_2 | j_1 j_2 jm) \tag{3.53}$$

are the vector-coupling, or *Clebsch-Gordan coefficients*. A somewhat symmetric form for these coefficients was derived first by *Wigner* using group-theoretical methods, and later by *Racah* and by *Schwinger*.[4] (Later in this chapter there is an introduction to Schwinger's operator technique in angular momentum.) We quote the result:

$$(j_1 m_1 j_2 m_2 | j_1 j_2 jm) = \delta_{m_1+m_2, m}$$
$$\cdot \left[\frac{(2j+1)(j_1+j_2-j)!(j_1-j_2+j)!(-j_1+j_2+j)!(j_1+m_1)!(j_1-m_1)!(j_2+m_2)!}{(j_1+j_2+j+1)!}\right.$$

[4] For references to Wigner's and other work, and for detailed formulas see [3.1].

$$\cdot (j_2 - m_2)!(j + m)!(j - m)!\Big]^{1/2}$$
$$\cdot \sum_z (-1)^z [z!(j_1 + j_2 - j - z)!(j_1 - m_1 - z)!(j_2 + m_2 - z)!$$
$$(j - j_2 + m_1 + z)!(j - j_1 - m_2 + z)!]^{-1}. \tag{3.54}$$

This unwieldy formula is evaluated in a form useful in problems involving angular momenta $\frac{1}{2}$ or 1 in Table 3.2.

Table 3.2. Nonvanishing Clebsch-Gordan coefficients for $j_2 = \frac{1}{2}$ and 1

$(j_1 m_1 \frac{1}{2} m_2 \| j_1 \frac{1}{2} jm)$		
	$m_2 = \frac{1}{2}$	$m_2 = -\frac{1}{2}$
$j = j_1 + \frac{1}{2}$	$+\sqrt{\frac{j_1 + m + \frac{1}{2}}{2j_1 + 1}}$	$+\sqrt{\frac{j_1 - m + \frac{1}{2}}{2j_1 + 1}}$
$j = j_1 - \frac{1}{2}$	$-\sqrt{\frac{j_1 - m + \frac{1}{2}}{2j_1 + 1}}$	$+\sqrt{\frac{j_1 + m + \frac{1}{2}}{2j_1 + 1}}$

$(j_1 m_1 1 m_2 \| j_1 1 jm)$			
j	$m_2 = 1$	$m_2 = 0$	$m_2 = -1$
$j_1 + 1$	$+\sqrt{\frac{(j_1 + m)(j_1 + m + 1)}{(2j_1 + 1)(2j_1 + 1)}}$	$+\sqrt{\frac{(j_1 - m + 1)(j_1 + m + 1)}{(2j_1 + 1)(j_1 + 1)}}$	$+\sqrt{\frac{(j_1 - m)(j_1 - m + 1)}{(2j_1 + 1)(2j_1 + 1)}}$
j_1	$-\sqrt{\frac{(j_1 + m)(j_1 - m + 1)}{2j_1(j_1 + 1)}}$	$+\frac{m}{\sqrt{j_1(j_1 + 1)}}$	$+\sqrt{\frac{(j_1 - m)(j_1 + m + 1)}{2j_1(j_1 + 1)}}$
$j_1 - 1$	$+\sqrt{\frac{(j_1 - m)(j_1 - m + 1)}{2j_1(2j_1 + 1)}}$	$-\sqrt{\frac{(j_1 - m)(j_1 + m)}{j_1(2j_1 + 1)}}$	$+\sqrt{\frac{(j_1 + m + 1)(j_1 + m)}{2j_1(2j_1 + 1)}}$

For the purposes of nuclear and atomic physics, a symmetrized form of these coefficients, the *Wigner 3-j symbols*, is simpler to manipulate. However, it does not serve any purpose to introduce them here, and we refer the reader to Edmonds' book. [3.1].

We shall return to the subject of compound angular momentum, after introducing an operator technique for expressing the angular momenta.

3.7 Equations of Motion of Interacting Angular Momenta

An interaction Hamiltonian involving two or more angular momenta (e.g., spin-orbit coupling of a valence electron in an atom) must be free of unwarranted dependence on an arbitrary coordinate system, so must be a scalar. For two angular momenta, the Hamiltonian must therefore be a function only of $\boldsymbol{J}_1 \cdot \boldsymbol{J}_2$. The Heisenberg exchange operator, $-J(\boldsymbol{S}_1 \cdot \boldsymbol{S}_2)$, is the simplest example of this. The time development of all other operators are determined by the Hamiltonian, through the equations of motion (Poisson bracket equations of motion classically, commutator bracket equations, quantum mechanically), therefore we are interested in the commutation relations of the components $J_{1x}, J_{2y}, \ldots, J_{2z}$ with the bilinear scalar. They are, compactly

$$\frac{1}{\mathrm{i}\hbar}[\boldsymbol{J}_1 \cdot \boldsymbol{J}_2, \boldsymbol{J}_1] = \boldsymbol{J}_1 \times \boldsymbol{J}_2. \tag{3.55}$$

Permutation of the two operators yields a similar equation for $\boldsymbol{J}_2$, and the antisymmetry of the vector product then establishes that the total angular momentum operator, $\boldsymbol{J}_1 + \boldsymbol{J}_2$, commutes with $\boldsymbol{J}_1 \cdot \boldsymbol{J}_2$. This motivates representations such as the one introduced in the preceding section.

3.8 Coupled Boson Representation

In semiclassical theories of magnetism it is common to approximate the unwieldy spin operators or matrices by harmonic-oscillator operators, for the general matrix structure of the spins, (3.40–43), resembles in many respects the matrix structure of harmonic oscillator operators. That this is no coincidence has been proved by Schwinger in his theory of angular momentum [3.5] based on coupled harmonic-oscillator fields.

As is well known, one may take linear combinations of momentum and coordinate operators, to obtain harmonic-oscillator "raising" and "lowering" operators $\boldsymbol{a}^*$ and $\boldsymbol{a}$; for example,

$$\boldsymbol{a}^* = \frac{1}{\sqrt{2\hbar}}\left(\frac{\hbar}{\mathrm{i}}\frac{\partial}{\partial x} + \mathrm{i}x\right) \qquad \boldsymbol{a} = \frac{1}{\sqrt{2\hbar}}\left(\frac{\hbar}{\mathrm{i}}\frac{\partial}{\partial x} - \mathrm{i}x\right).$$

Let us make clear the reason for this terminology. The commutation relations which may be derived for the operators $\boldsymbol{a}, \boldsymbol{a}^*$ are those of *Bosons*

$$a_i a_j - a_j a_i \equiv [a_i, a_j] = [a_i^*, a_j^*] = 0 \quad \text{and} \quad [a_i, a_j^*] = \delta_{ij} \tag{3.56}$$

with i, j referring to different particles, from which we deduce also

$$[\mathfrak{n}_i, a_j^*] = \delta_{ij} a_j^* \quad \text{and} \quad [\mathfrak{n}_i, a_j] = -\delta_{ij} a_i. \tag{3.57}$$

introducing $\mathfrak{n}_i = a_i^* a_i =$ *occupation-number operator*, referring to the degree of excitation (occupation) of the ith harmonic oscillator. One may construct a complete set of states, labeled by occupation numbers, first introducing a "vacuum." This is the ground state of all the harmonic oscillators, the no-particle state, denoted $|0)$, which is annihilated by every lowering operator

$$a_i|0) \equiv 0 \qquad \text{for all } i. \tag{3.58}$$

The one-particle states are clearly

$$a_i^*|0)$$

and the (normalized) two-particle states are

$$a_1^* a_2^*|0) \qquad \text{or} \qquad \frac{(a_1^*)^2}{\sqrt{2}}|0),$$

etc. The general formula for the many-particle, normalized state is

$$\frac{(a_1^*)^{n_1}(a_2^*)^{n_2}\ldots}{\sqrt{n_1!n_2!\ldots}}|0) \tag{3.59}$$

as proved in Problem 3.3. In these states, the occupation-number operators $\mathbf{n}_i$ are diagonal, with non-negative integer eigenvalues $n_i = 0, 1, 2, \ldots$. The total occupation of a state is given by the sum of the eigenvalues,

$$\sum n_i.$$

Problem 3.3. Prove that the state in Eq. (3.59) is normalized by evaluating $(0|\ldots(a_2)^{n_2}(a_1)^{n_1}(a_1^*)^{n_1}(a_2^*)^{n_2}\ldots|0)$, using only the definition of the vacuum, (3.58), and the commutation relations of (3.56, 57). Show that n_i is the eigenvalue of $\mathbf{n}_i$.

Schwinger showed that with the aid of only two harmonic oscillators the entire matrix structure of a single angular momentum, as summarized in (3.40–43), could be exactly reproduced. The advantages are great, even if no new results were obtained, most especially for large values of j for which the angular-momentum matrices are large and unwieldy. New features can also be studied with Schwinger's approach, due to the great understanding which has been achieved concerning the harmonic-oscillator fields as compared to the more obscure angular momentum-operators. Labeling the two oscillators by subscripts 1 and 2, let us introduce the *spinor operators*

$$\boldsymbol{a}^+ = (a_1^*, a_2^*) \qquad \text{and} \qquad \boldsymbol{a} = \begin{pmatrix} a_1 \\ a_2 \end{pmatrix}, \tag{3.60}$$

which are merely two-component vectors with operator components. If, moreover, we contract these operators with the Pauli spin matrices, we obtain the desired representation. That is, let

$$J^z = \frac{\hbar}{2}\boldsymbol{a}^+ \cdot \boldsymbol{\sigma}_z \cdot \boldsymbol{a} = \frac{\hbar}{2}(a_1^* a_1 - a_2^* a_2) = \frac{\hbar}{2}(\mathfrak{n}_1 + \mathfrak{n}_2)$$

and similarly for the other components, with the following compact result:

$$\boldsymbol{J} = \frac{\hbar}{2}\boldsymbol{a}^+ \cdot \boldsymbol{\sigma} \cdot \boldsymbol{a}\,. \tag{3.61}$$

One-half times the contraction with the unit operator will be denoted the $\boldsymbol{j}$ operator

$$\boldsymbol{j} = \frac{1}{2}\boldsymbol{a}^+ \cdot \boldsymbol{a} = \frac{1}{2}(a_1^* a_1 + a_2^* a_2) = \frac{1}{2}\mathfrak{n}_1 + \frac{1}{2}\mathfrak{n}_2. \tag{3.62}$$

It may be verified that the eigenvalues of this operator are indeed $j = 0, \frac{1}{2}, 1, \frac{3}{2}, \ldots,$ where $\boldsymbol{J}^2$ has eigenvalue $\hbar^2 j(j+1)$.

Problem 3.4. Show that $J^+ = \hbar a_1^* a_2$ and $J^- = \hbar a_2^* a_1$. (These formulas and the ones in the text may be remembered by associating a change $\Delta m = +\frac{1}{2}$ with each type-1 particle, and $\Delta m = -\frac{1}{2}$ with each type-2 particle.)

Problem 3.5. Prove the operator identity

$$\boldsymbol{J}^2 = \hbar^2 \boldsymbol{j}(\boldsymbol{j} + 1).$$

Problem 3.6. Quantization of the electric and magnetic fields of light waves is the first step in quantum electrodynamic theory. Assuming these are harmonic-oscillator fields, let subscript 1 refer to the electric field and subscript 2 to the magnetic field in operators a_1, a_1^* and a_2, a_2^* (e.g., for a particular plane-wave mode). Use the fact that half the total energy must be contained in each field to prove that the eigenvalues j can only take on integer values, thus showing that the photon has quantized spin of unity. Compare with (3.103, 104).

Normalized eigenfunctions of a single angular momentum, possessing definite m and j eigenvalues shall be denoted $|jm)$, and are simply

$$|jm) = \frac{(a_1^*)^{j+m}(a_2^*)^{j-m}}{\sqrt{(j+m)!(j-m)!}}\,|0). \tag{3.63}$$

The coupled-Boson operators have more flexibility than the original angular-momentum operators. For example, we can make use of the extra degrees of freedom to construct the so-called *hyperbolic operators* which conserve m, but *raise* or *lower* j

$$K^+ = \hbar a_1^* a_2^* \qquad K^- = \hbar a_2 a_1 \qquad \text{and} \qquad K^z = \frac{\hbar}{2}(\mathfrak{n}_1 + \mathfrak{n}_2 + 1). \tag{3.64}$$

They obey the commutation relations,

$$[K^z, K^+] = \hbar K^+ \qquad [K^z, K^-] = -\hbar K^- \qquad \text{and}$$
$$[K^+, K^-] = -2\hbar K^z. \tag{3.65}$$

Only the last of these differs, by a sign, from the commutation relations of the angularmomentum operators. The following equations may also be verified:

$$(J^z)^2 - \frac{1}{4}\hbar^2 = (K^z)^2 - \frac{1}{2}(K^+K^- + K^-K^+) \tag{3.66a}$$

$$= K^z(K^z - \hbar) - K^+K^- \tag{3.66b}$$

$$= K^z(K^z + \hbar) - K^-K^+. \tag{3.66c}$$

3.9 Rotations

The study of rotations of coordinate systems and of particles is intimately tied in with the theory of angular momentum. For example, given an arbitrary function $f(\phi)$ of the azimuthal angle ϕ, its argument may be rotated through an angle α by applying a differential operator

$$e^{\alpha(d/d\phi)}f(\phi) = f(\phi + \alpha).$$

This formula may be checked by a Taylor series expansion of both sides, in powers of α. The operator which is exponentiated will be recognized as proportional to L^z, and is just one of the three differential operators, or linear combinations thereof, which may be exponentiated to effect desired rotations in the coordinate system. The most general rotation is expressible in terms of the three Euler angles α, β, and γ, and takes the form of the unitary operator

$$D(\alpha\beta\gamma) = e^{i(\alpha/\hbar)J^z}e^{i(\beta/\hbar)J^y}e^{i(\gamma/\hbar)J^z} = (D^\dagger)^{-1}. \tag{3.67}$$

One writes $\boldsymbol{J}$ instead of $\boldsymbol{L}$ because the rotations are not limited to the integer angular momenta. Of course, the three-Euler-angle formula is equivalent to a single rotation about a suitably chosen axis, with its direction along a unit vector $\hat{\boldsymbol{u}}$, and so

$$D(\alpha\beta\gamma) = e^{i\alpha'\hat{\boldsymbol{u}}\cdot\boldsymbol{J}} \tag{3.68}$$

for a suitable angle α'. Thus, two or more successive rotations can be expressed by a single one. One of the practical consequences of this equality, after the rotations are expressed in terms of a complete set of spherical harmonics (for integer angular momentum), is a very useful formula for expressing *products* of spherical harmonics as a *linear* combination of spherical harmonics, viz.,

$$Y_{l_1m_1}(\theta, \phi)Y_{l_2m_2}(\theta, \phi) = \sum_{l,m} Y_{l,m}(\theta, \phi)\sqrt{\frac{(2l_1+1)(2l_2+1)}{4\pi(2l+1)}} \cdot (l_1m_1l_2m_2 \,|\, l_1l_2lm)(l_10l_20 \,|\, l_1l_2l0). \tag{3.69}$$

We shall not prove this formula here, but proceed instead to show in what way the rotations are related to the unitary transformations of Boson operators. The transformation which takes a operators into a linear combination of each other is a unitary canonical transformation if it preserves the Boson commutation relations, and the Hermitean nature of Hermitean operators. If it also preserves the eigenvalue j (that is, commutes with $\boldsymbol{j}$), it corresponds precisely to a rotation. In particular,

$$\begin{aligned}
a_1^* &\to [e^{+(i/2)(\alpha+\gamma)} \cos \tfrac{1}{2}\beta]a_1^* + [e^{(i/2)(\gamma-\alpha)} \sin \tfrac{1}{2}\beta]a_2^* \\
a_2^* &\to -[e^{(i/2)(\alpha-\gamma)} \sin \tfrac{1}{2}\beta]a_1^* + [e^{(i/2)(\alpha+\gamma)} \cos \tfrac{1}{2}\beta]a_2^* \\
a_1 &\to [e^{-(i/2)(\alpha+\gamma)} \cos \tfrac{1}{2}\beta]a_1 + [e^{-(i/2)(\gamma-\alpha)} \sin \tfrac{1}{2}\beta]a_2 \\
a_2 &\to -[e^{-(i/2)(\alpha-\gamma)} \sin \tfrac{1}{2}\beta]a_1 + [e^{-(i/2)(\alpha+\gamma)} \cos \tfrac{1}{2}\beta]a_2
\end{aligned} \tag{3.70}$$

is the transformation which results if in (3.67) we express the components of $\boldsymbol{J}$ in the coupled-Boson representation, and calculate

$$a_i \to D(\alpha\beta\gamma)a_iD^{-1}(\alpha\beta\gamma).$$

For complex angles, this is no longer a unitary but a *similarity* transformation, which preserves the commutation relations but not Hermiticity. One may also imagine unitary or similarity transformations which do not conserve j, and which mix the $\boldsymbol{a}$ operators with $\boldsymbol{a}^*$ operators. Such transformations mix the angular-momentum operators with the hyperbolic ones and are generated by including the components of $\boldsymbol{K}$ as well as of $\boldsymbol{J}$ in D.

3.10 More on Compound Angular Momentum

The addition of the spin of an electron to its mechanical angular momentum, the addition of the angular momenta or spins of two distinct electrons, and in general the coupling of two or more angular momenta of various origins, require mathematical techniques beyond the simple formulation given so far. Here we shall show some of the machinery established by Schwinger for handling these problems, without, however, entering into the details. We shall study the coupling of two angular momenta, and it might appear that by induction one may use this theory to couple an arbitrary number of angular momenta. However, in practice this is not quite so, and even for three angular momenta certain additional simplifications must be sought if the problem is to remain manageable.

Two angular momenta require four Bose particles for their description. But with four Boson operators, far more than the components of $\boldsymbol{J}_1$ and $\boldsymbol{J}_2$ can be constructed: we can obtain $\boldsymbol{J} = \boldsymbol{J}_1 + \boldsymbol{J}_2$, and also all the relevant hyperbolic operators. The Clebsch-Gordan coefficients, rotations in one or the other angular-momentum space, and all the other possible subjects of interest also may be investigated by studying the Bose operators and their eigenfunctions.

Let the Bose operators referring to angular momentum 1 be labeled a_1 and a_2, and those referring to angular momentum 2 be labeled b_1 and b_2. Any $\boldsymbol{a}$ operator commutes with any $\boldsymbol{b}$ operator. The notation and terminology will be that introduced in (3.56–63). This takes care of the usual operators, the components of J_i, and J_j^2. In addition, the following new ones are required:

$$I^+ = \hbar \boldsymbol{a}^+\boldsymbol{b} \qquad I^- = \hbar \boldsymbol{b}^+\boldsymbol{a} \qquad I^z = \hbar(\boldsymbol{j}_1 - \boldsymbol{j}_2), \qquad \text{where} \tag{3.71}$$

$$\boldsymbol{a}^+\boldsymbol{b} = a_1^* b_1 + a_2^* b_2, \tag{3.72}$$

etc., in the spinor notation, and

$$K^+ = \hbar(a_1^* b_2^* - a_2^* b_1^*) \quad K^- = \hbar(a_1 b_2 - a_2 b_1) \quad K^z = \hbar(\boldsymbol{j}_1 + \boldsymbol{j}_2 + 1). \tag{3.73}$$

The following relations may be verified by substituting the definitions given above:

$$J^2 = J_x^2 + J_y^2 + J_z^2$$
$$= I^z(I^z - \hbar) + I^+I^- \tag{3.74a}$$
$$= I^z(I^z + \hbar) + I^-I^+ \tag{3.74b}$$
$$= K^z(K^z - \hbar) - K^+K^- \tag{3.74c}$$
$$= K^z(K^z + \hbar) - K^-K^+. \tag{3.74d}$$

Using the definition of I^z (3.71), and the fact that I^+I^- in (3.74a) is a positive semidefinite operator (so is I^-I^+), one proves directly that $j(j+1) \leq (j_1 - j_2)(j_1 - j_2 - 1)$. The choice $j_1 \leq j_2$ [if $j_2 < j_1$, then use (3.74b)] establishes

$$j \geq |j_2 - j_1|. \tag{3.75}$$

Similarly, the two equations involving K can be used to prove

$$j \leq (j_1 + j_2) \tag{3.76}$$

and thus to establish the "triangle inequality" first introduced in the discussion following (3.52) and in Problem 3.2.

A state of definite eigenvalues $j_1 m_1$ and $j_2 m_2$ is

$$|j_1 m_1 j_2 m_2) = \frac{(a_1^*)^{j_1+m_1}(a_2^*)^{j_1-m_1}(b_1^*)^{j_2+m_2}(b_2^*)^{j_2-m_2}}{\sqrt{(j_1+m_1)!(j-m_1)!(j_2+m_2)!(j_2-m_2)!}}\,|0). \tag{3.77}$$

A state of definite j, m, j_1 and j_2 may equally well be labeled j, m, μ, and v, where

$$\mu = j_1 - j_2 \qquad \text{and} \qquad v = j_1 + j_2 + 1 \tag{3.78}$$

and denoted

$$|jm\mu v\rangle \tag{3.79}$$

where $\mu \leq j$ and $v \geq j + 1$. For $m = j$ and $v = j + 1$, the appropriate state is

$$|jj\mu j + 1\rangle = \frac{(a_1^*)^{j+\mu}(b_1^*)^{j-\mu}}{\sqrt{(j+\mu)!(j-\mu)!}}|0\rangle. \tag{3.80}$$

Application of the spin-lowering operator yields the states $|jm\mu j + 1\rangle$, and finally, arbitrary states are given by

$$|jm\mu v\rangle = \sqrt{\frac{(2j+1)!}{(v+j)!(v-j-1)!}}(K^+)^{v-j-1}|jm\mu j + 1\rangle. \tag{3.81}$$

The inner product of these wavefunctions with the $|j_1m_1j_2m_2)$ of (3.77) yields the Clebsch-Gordan coefficients by Schwinger's method.

The Bose operator method may be extended in various directions to recover well-known results (e.g., for the spherical harmonics) or to discover new ones. A connection with the continuous representations may be made using the operator identity

$$a_i F(a_i^*)|0) = \frac{\partial}{\partial a_i^*}F(a_i^*)|0) \tag{3.82}$$

which may be used to construct generating functions for various sets of orthogonal polynomials. But this subject is properly outside the scope of the present elementary treatment.

3.11 Other Representations

The first formal justification of the Bloch theory of spin waves may be found in the work of *Holstein* and *Primakoff* [3.6]. Bloch had naturally assumed that spin waves obey Bose-Einstein statistics, but these authors showed how spin operators could be expressed in terms of true Bose fields, much as we shall show it in the chapter on spinwave theory. The Holstein-Primakoff representation is best understood as a special case of the Schwinger coupled-Boson representation, that is, as an irreducible representation of the latter in a subspace of fixed j. Recall (3.62), which we rewrite as

$$a_2^* a_2 = \sqrt{2j - \mathfrak{n}_1} \cdot \sqrt{2j - \mathfrak{n}_1} \, . \tag{3.83}$$

In a subspace of fixed eigenvalue j, this equation is solved by treating a_2 and its conjugate as two diagonal operators,

$$a_2 = a_2^* = (2j)^{1/2} \sqrt{1 - \frac{\mathfrak{n}}{2j}} \qquad \text{hence,} \tag{3.84}$$

$$J^+ = \hbar a^* (2j)^{1/2} \sqrt{1 - \frac{\mathfrak{n}}{2j}} \qquad J^- = \hbar (2j)^{1/2} \sqrt{1 - \frac{\mathfrak{n}}{2j}}\, \mathrm{a} \qquad \text{and}$$

$$J^z = \hbar(\mathfrak{n} - j) \tag{3.85}$$

omitting the subscript ($_1$) on a and $\mathfrak{n}$. The formalism is incorrect whenever n exceeds $2j$, but within the allowed range, it may be verified that the basic commutation relations (3.12–16) are correctly obeyed. (The rationalization of the square root is discussed, following (3.101) in the section on "spins one.")

If all that one requires is the satisfaction of these commutation laws, and if one relaxes the requirement that J^+ and J^- be Hermitean conjugate operators, then he may perform the so-called Maléev similarity transformation to a new set of operators

$$J^+ = a^* (2j)^{1/2} \left(1 - \frac{\mathfrak{n}}{2j}\right) \hbar \qquad J^- = (2j)^{1/2} a \hbar \qquad \text{and} \qquad J^z = (\mathfrak{n} - j)\hbar. \tag{3.86}$$

The obvious advantage, the rationalization of the square root, is somewhat offset by the complications introduced by the nonunitary transformation.

A variant of the above is due to *Villain* [3.7]. Assuming $\boldsymbol{J} \leq |J_z|$, he writes $J^\pm$ as

$$J^+ = \mathrm{e}^{\mathrm{i}\varphi} \sqrt{\left(J + \frac{1}{2}\hbar\right)^2 - \left(J_z - \frac{1}{2}\hbar\right)^2}, \qquad \text{and} \qquad J^- = (J^+)^\dagger \tag{3.87}$$

from which (3.15d, 40–42) flow easily. The correct commutation relations are ensured by requiring, as in (3.7), or (3.42),

$$[\varphi, J_z] = \mathrm{i}\hbar, \quad \text{i.e.,} \quad \varphi = \hbar \mathrm{i}\, \partial/\partial J_z \quad \text{or} \quad \mathrm{J}_z = -\mathrm{i}\hbar \partial/\partial\varphi. \tag{3.88}$$

Thus, $\exp(-\mathrm{i}x\varphi)$ is a translation operator for J_z and

$$[J_z, \mathrm{e}^{\pm \mathrm{i}m\varphi}] = \pm \hbar m\, \mathrm{e}^{\pm \mathrm{i}m\varphi}. \tag{3.89}$$

In the semi-classical limit, one just drops the terms $\pm\frac{1}{2}\hbar$ in (3.87). The extreme quantum limit $j = \frac{1}{2}$ or 1 deserves separate investigation, and will be discussed in the following sections.

3.12 Spins One-Half

For spins one-half, $n = 0, 1$ in the preceding and the Holstein-Primakoff and Maléev representations are identical within the physically allowed subspace.

Problems involving N interacting spins one-half may be more easily formulated with the aid of one of several Fermion representations. This takes us back a number of years, to the historic paper of *Jordan* and *Wigner* on second quantization [3.8] in which the Fermion anticommuting operators were explicitly constructed out of Pauli spin matrices. It is, however, the converse which is of present interest.

Fermion operators, which we denote by the letter c, are a set of *anticommuting* operators; for example, for any state $|\Phi)$,

$$c_1^* c_2^* |\Phi) = -c_2^* c_1^* |\Phi) \tag{3.90}$$

and therefore, care must be taken of the order in which the operators are written, see Chap. 4 for further discussion. The vacuum is defined, just as for spins, as the state annihilated by all c's,

$$c_i|0) = 0 \qquad \mathfrak{n}_i|0) = 0 \qquad (i = 1, \ldots, N) \tag{3.91}$$

and is the state in which the particle number operators

$$\mathfrak{n}_i = c_i^* c_i \tag{3.92}$$

all have zero eigenvalue.

The anticommutation relations are indicated by curly brackets and are

$$c_i c_j + c_j c_i \equiv \{c_i, c_j\} = 0 \qquad \{c_i^*, c_j^*\} = 0 \qquad \{c_i, c_j^*\} = \delta_{ij}. \tag{3.93}$$

Setting $i = j$ in the above equations yields the relations

$$c_i^2 = (c_i^*)^2 = 0 \qquad c_i^* c_i + c_i c_i^* = 1 \tag{3.94}$$

which are identically the equations obeyed by the Pauli spin matrices $\sigma^{\pm}$. It is only the fact that Pauli spin matrices referring to different particles ($i \neq j$) *commute* with one another, which differentiates them from the anticommuting Fermion operators above. This may be remedied by introducing a set of drone operators d_i and d_i^*, equal in number to the original set of c's, which anticommute with the latter and obey amongst themselves anticommutation relations entirely analogous to (3.93). As a consequence,

$$(d_i + d_i^*)^2 \equiv 1 \quad \text{and} \quad \{d_i + d_i^*, d_j + d_j^*\} = 0 \quad \text{for} \quad i \neq j \tag{3.95}$$

so that finally

$$S_i^+ = \hbar c_i^*(d_i + d_i^*) \quad S_i^- = \hbar(d_i + d_i^*)c_i \quad \text{and} \quad S_i^z = \hbar\left(c_i^* c_i - \frac{1}{2}\right) \tag{3.96}$$

is the desired set of spin one-half operators, which *commute* when $i \neq j$.

Problem 3.7. Prove that the various representations in this section obey (3.12–16), and also that different spins *commute*:

$$[S_i^k, S_j^{k'}] = 0 \qquad \text{for} \quad i \neq j$$

for all components $k, k' = x, y, z$ of the spin vectors.

A second useful representation is the exact analogue of the coupled-Boson picture, but replaces the Bosons by Fermions. The paired Fermions, instead of being labeled 1,2, will be labeled by ↑ and ↓ to make more explicit the role of each operator. Thus,

$$S_i^+ = \hbar c_{i\uparrow}^* c_{i\downarrow} \quad S_i^- = \hbar c_{i\downarrow}^* c_{i\uparrow}, \quad \text{and} \quad S_i^z = \frac{\hbar}{2}(c_{i\uparrow}^* c_{i\uparrow} - c_{i\downarrow}^* c_{i\downarrow}), \tag{3.97}$$

where the anticommutation relations are

$$\{c_{i,m}, c_{j,m'}\} = \{c_{i,m}^*, c_{j,m'}^*\} = 0, \ \{c_{i,m}, c_{j,m'}^*\} = \delta_{i,j}\delta_{m,m'} \quad (m = \uparrow \text{ or } \downarrow). \tag{3.98}$$

This is a very important representation, corresponding to the second quantization of electrons *cum* spin and we shall return to it in Chap. 6.

A final representation of the spins one-half brings us closest to the work of Jordan and Wigner. The technique which we shall discuss has been useful in the solution of one-dimensional problems, and also in the two-dimensional Ising model, an important subject in the statistical mechanics of magnetism. Therefore, last but not least, set

$$S_i^+ = \hbar c_i^* Q_i \qquad S_i^- = \hbar Q_i c_i \qquad \text{and} \qquad S_i^z = \frac{\hbar}{2}(2c_i^* c_i - 1), \tag{3.99}$$

where

$$Q_i = Q_i^* = Q_i^{-1} = \exp\,(i\pi \sum_{j<i} c_j^* c_j) \tag{3.100a}$$

$$= \exp(i\pi \sum_{j<i} s_j^+ s_j^-) \tag{3.100b}$$

$$= \prod_{j<i} (c_j^* + c_j)(c_j^* - c_j), \tag{3.100c}$$

etc. In actual problems, the choice of ordering $i = 1, \ldots, N$ is crucial, because the phase factors Q_i introduce great complexity into a problem unless means are found to eliminate them. When this is possible, however, then the representation above is the simplest of all, because it is the only one to establish a one-to-one correspondence between spins and Fermions, and their respective eigenstates.

3.13 Spins One

There is no special representation for spins one, except for that which deals specifically with vector fields, such as the electromagnetic field. However, the Holstein-Primakoff representation may be used after rationalizing the square root. For, in the physically admissible range $n = 0, 1, 2$, the equation

$$\sqrt{1 - \mathfrak{n}/2} \equiv 1 - \left(\frac{3}{2} - \sqrt{2}\right) \mathfrak{n} - \frac{1}{2}(\sqrt{2} - 1)\, \mathfrak{n}^2 \tag{3.101}$$

is exact, and only fails for $n \geq 3$, which is outside the domain of validity of this particular representation. In fact for any j, a polynomial of order $2j$ can be found which has the same structure as the Holstein-Primakoff root, but is quite tedious to construct for large j. Because the Taylor series expansion of the root,

$$\sqrt{1 - \frac{\mathfrak{n}}{2j}} \doteq 1 - \frac{\mathfrak{n}}{4j} + \cdots \quad ,$$

becomes asymptotically correct in the limit $j = \infty$, it is commonly used in approximate theories.

We have not dealt with vector fields, except in Problem 3.6, where it was suggested that the "intrinsic spin" of the photon was unity. There are good reasons why this should be so. When angular momentum generates infinitesimal rotations of the coordinates, it changes not only the arguments of a vector field, as it does a scalar field (e.g., ordinary wavefunctions), but also mixes the various components of the field amongst themselves. The total effect might be described by an operator

$$J^z = \frac{\hbar}{\mathrm{i}}\left(\frac{\partial}{\partial \phi} - \hat{\boldsymbol{u}}_z \times\right) = L^z + S^z \tag{3.102}$$

of which the first term takes care of arguments, and the second of the rotations of the vector field components. Similar expression for the two other Cartesian components results in an operator $\boldsymbol{J} = \boldsymbol{L} + \boldsymbol{S}$. Recall that $\boldsymbol{J}$ can be a true angular momentum only if $\boldsymbol{L}$ and $\boldsymbol{S}$ each are, in their own right. Therefore, let us investigate this new angular-momentum operator,

$$\boldsymbol{S} = \mathrm{i}\hbar(\hat{\boldsymbol{u}}_x \times\, , \hat{\boldsymbol{u}}_y \times\, , \hat{\boldsymbol{u}}_z \times) \tag{3.103}$$

specifically constructed to operate on vector fields, such as the vector potential $\boldsymbol{A}(\boldsymbol{r})$. Calculating $S^2 = S_x^2 + S_y^2 + S_z^2$, we readily find

$$S^2 = \hbar^2(2) \tag{3.104}$$

and therefore the spin magnitude $s = 1$.

Field amplitudes may then be expanded in eigenfunctions of S^z, the spherical unit eigenvectors,

$$\boldsymbol{e}_{\pm 1} = \frac{-1}{2}(\mathrm{i}\hat{\boldsymbol{u}}_y \pm \hat{\boldsymbol{u}}_z) \qquad \text{and} \qquad \boldsymbol{e}_0 = \hat{\boldsymbol{u}}_z \tag{3.105}$$

with eigenvalues,

$$S^z \boldsymbol{e}_r = \hbar r \boldsymbol{e}_r, \qquad r = -1, 0, 1. \tag{3.106}$$

These are particularly useful in problems with some spherical symmetry, e. g., the field of a radiating atom. Linear combinations of functions having definite J and total azimuthal quantum number M are called *vector* spherical harmonics

$$\mathscr{Y}_{JlM}(\theta, \phi) = \sum_{mr} Y_{lm}(\theta, \phi)\boldsymbol{e}_r(lm1r \,|\, l1JM). \tag{3.107}$$

In addition to J and M, only the l of the spherical harmonic need be indicated, for the spherical unit vectors invariably have unit angular momentum, as we have shown.

The vector spherical harmonics form a complete, orthonormal set for the description of vector fields. The normalization integral, including scalar product, is

$$\int_0^{2\pi} d\phi \int_{-\pi}^{\pi} d\theta \sin\theta \mathscr{Y}^*_{Jlm}(\theta, \phi) \cdot \mathscr{Y}_{J'l'M'}(\theta, \phi) = \delta_{JJ'}\delta_{ll'}\delta_{MM'}. \tag{3.108}$$

3.14 Quadratic Forms

We have already discussed the scalar product of two angular momentum operators, say $\boldsymbol{S}_1 \cdot \boldsymbol{S}_2$. We now obtain the eigenvalues of this operator, and find them to be parameterized only by invariants.

We start by completing the square

$$\boldsymbol{S}_1 \cdot \boldsymbol{S}_2 = \frac{1}{2}(\boldsymbol{S}_1 + \boldsymbol{S}_2)^2 - \frac{1}{2}(\boldsymbol{S}_1^2 + \boldsymbol{S}_2^2) \tag{3.109}$$

and replace $\boldsymbol{S}_1^2$ by its magnitude $s_1(s_1 + 1)$, $\boldsymbol{S}_2^2$ by $s_2(s_2 + 1)$, and $\boldsymbol{S}_1 + \boldsymbol{S}_2$ by the compound operator $\boldsymbol{S}_{\text{tot}}$, obtaining

$$\boldsymbol{S}_1 \cdot \boldsymbol{S}_2 = \frac{1}{2}[s_{\text{tot}}(s_{\text{tot}} + 1) - s_1(s_1 + 1) - s_2(s_2 + 1)] \tag{3.110}$$

with s_{tot} taking on any possible integer value ranging from $|s_1 - s_2|$ to a maximum $(s_1 + s_2)$. Substituting in the above we find for the maximum and minimum values of $\boldsymbol{S}_1 \cdot \boldsymbol{S}_2$

$$-s_2(s_1+1) \le \boldsymbol{S}_1 \cdot \boldsymbol{S}_2 \le s_1 s_2 \tag{3.111}$$

assuming $s_2 \le s_1$. The maximum, occurring when both spins are parallel, could have been predicted classically. The minimum, corresponding to antiparallel orientation, is lower by a fraction $1/s_1$ than the classical estimate; this foreshadows the complications of antiferromagnetism compared to ferromagnetism.

For the special case of spins one-half only, the quadratic form can be diagonalized by a nonlinear transformation [3.9]. We introduce two new spin one-half operators denoted $\boldsymbol{J}$ and $\boldsymbol{P}$, constructed out of the components of $\boldsymbol{S}_1$ and $\boldsymbol{S}_2$ in the following manner ($\hbar = 1$):

$$\begin{aligned}(J_x, J_y, J_z) &= (S_{1x}, 2S_{1y}S_{2z}, 2S_{1z}S_{2z}), \qquad \text{and}\\ (P_x, P_y, P_z) &= (S_{2z}, 2S_{1x}S_{2y}, -2S_{1x}S_{2x}).\end{aligned} \tag{3.112}$$

The inverse of these relations is

$$\begin{aligned}(S_{1x}, S_{1y}, S_{1z}) &= (J_x, 2J_yP_x, 2J_zP_x) \qquad \text{and}\\ (S_{2x}, S_{2y}, S_{2z}) &= (-2P_zJ_x, 2P_yJ_x, P_x).\end{aligned} \tag{3.113}$$

In the new representation, the quadratic form is diagonal

$$\boldsymbol{S}_1 \cdot \boldsymbol{S}_2 = \frac{1}{2}(J_z - P_z) + J_zP_z \tag{3.114}$$

where $J_z = \pm\frac{1}{2}$ and $P_z = \pm\frac{1}{2}$ in the units $\hbar = 1$. For some problems this formulation may be superior to (3.110).

4. Many-Electron Wavefunctions

Although the old quantum theory of Niels Bohr had great success in the interpretation of the line spectra of the hydrogen atom, it did not provide an adequate framework for the interpretation of the complex spectra of the many-electron atoms and molecules. Nevertheless the old theory had many clever proponents, who with a series of particularly shrewd guesses nailed down the structure of the periodic table and the theory of line spectra some time before the new wave mechanics. Notable was the exclusion principle of *Pauli*, which included a fourth quantum number in addition to the three which are associated with the orbital motion of each electron [4.1]. This was the key to explaining why there are two electrons in an s shell, 6 in p shell, etc., and toward a correct theory of the building-up principle of the periodic table. Almost immediately thereafter, *Uhlenbeck* and *Goudsmit* [4.2] published their evidence that the fourth quantum number referred to the spin of the electron, a concept that helped clarify much of the subsequent thinking on the subject. Indeed so much was understood on the basis of the old theory alone, that it might even appear that the 1926 theories of Schrödinger and Heisenberg, far from solving old problems, only raised new difficulties. The reason is that it proved so awkward and difficult to introduce *spin* into the new quantum mechanics.

Once it was established that the probability density is equal to $|\Psi|^2$ the square of the wavefunction, then the following paradox imposed itself: The probability function for N indistinguishable particles must be invariant under any permutation of the particles. Moreover, as any permutation can be achieved by a succession of transpositions (permutations of two particles at a time), one need only investigate the effects of transpositions. The wavefunction, it is then found, can be allowed to be *even* or *odd* under a transpotion, but no other choice leads to the correct, totally symmetric probability density. (For two electrons, hydrogen molecule or helium atom, it was recognized from the first that odd functions describe triplet states, and even functions the singlet states; the agreement of the calculated energy levels with experiment was extraordinary. On the other hand, the energy levels of the even or odd functions of *three* or more electrons did not agree with spectroscopic data, or even with elementary notions of how atoms were constituted. Surely doubts must have been raised regarding whether quantum mechanics was applicable to more than two electrons even in principle.) But the experimental spectra of systems of more than two electrons could only be understood in terms of wavefunctions which were *neither* odd *nor* even under arbitrary transpositions, but, rather, which had complicated properties under the various permutations. This was totally incompatible with the notion of a probability density symmetric under the interchange of identical particles. Nor was it obvious to what values of the total spin, i.e., to what multiplicity these complicated space functions were associated. One can readily guess that the totally antisymmetric function corresponds to the state of maximum multiplicity, all spins parallel; but before the invention by *Pauli* of the spin matrix operators [4.3] the calculation of general symmetry properties of the space functions corresponding to arbitrary multiplicity posed many such logical and practical problems.

A remedy was provided by *Slater*, with his invention of the determinantal wavefunction of space and spin [4.4]. Spin was an observable; therefore, the spin coordinate belonged in the wavefunction on the same footing as the space coordinate. The exclusion principle could now be correctly stated as follows: "A wavefunction for a set of identical electrons must be antisymmetric under the interchange of any two of them; by which is meant, the interchange of the space + spin coordinates of the two electrons."

We shall examine the Slater determinants, insofar as they shed further light on the molecules of two and three hydrogen atoms we have studied so far. We shall then progress in reverse, it might seem, and ask to which functions of space alone, they correspond for the various multiplicities. We shall not now have any difficulty in finding nor in reconciling these with the antisymmetrization principle (*supra*). This is in contrast with the elaborate theory which was developed for this purpose before spin was finally understood.

We then put this knowledge to use, proving a theorem due to *Lieb* and *Mattis* [4.5] which seemingly contradicts Hund's rule that states of highest multiplicity lie lowest. Fortunately this paradox is also soon removed, for the theorem and the rule are found to have distinct domains of applicability. In the reconciliation of opposites, new light is shed on the theory of ferromagnetism. Naturally, this eventually leads us back to the interesting and central subject of "exchange," thence to manybody theory.

It is essential for an understanding of this chapter, that the reader have familiarity with some notation and theorems of angular momentum provided by the previous chapter. Beyond this, we develop the many-electron theory up to, and including second quantization in this rather far-ranging chapter.

4.1 Slater Determinants

It might be well to clarify the notion of "spin function" by analyzing the Hilbert space of ordinary space functions and bringing out the similarities. The student will not be surprised, then, to see space and spin treated on equal footing in the determinantal wavefunction, and will obtain insight into the possibilities and limitations of the form. For a more precise notion of meaning of spin and spin coordinates, the reader may wish to consult the preceding chapter on angular momentum, particularly the first five sections. (See also [4.3]).

Consider a complete set of functions, assumed orthonormal,

$$f_1(r), f_2(r), \ldots, f_n(r), \ldots \tag{4.1}$$

in which an arbitrary function $g(r)$ can be expanded,

$$g(r) = \sum_n g_n f_n(r) \tag{4.2}$$

by analogy with a Fourier expansion. The expansion coefficients g_n play the role of Fourier coefficients (or Fourier transform). The function $g(r)$ can be represented as a vector in Hilbert space,

$$\boldsymbol{g} = (g_1, \ldots, g_n \ldots). \tag{4.3}$$

The overlap integral,

$$C \equiv \int g^*(r)h(r)dr \tag{4.4}$$

is then the scalar dot product between two vectors

$$C = \sum_n g_n^* h_n = \boldsymbol{g} \cdot \boldsymbol{h}. \tag{4.5}$$

Further discussion is best left to mathematical treatises. However, recall that it is just this strong analogy which unites the matrix mechanics of Heisenberg and the wave mechanics of Schrödinger.

In the space of the Pauli spin matrices, the basis vectors are the spinors

$$(1, 0) \qquad \text{and} \qquad (0, 1). \tag{4.6}$$

Similarly, in the functional Hilbert space, the basis vectors are

$$(1, 0, \ldots, 0, \ldots, 0, \ldots), \qquad (0, 1, 0, \ldots, 0, \ldots), \ldots \tag{4.7}$$

etc., as we discover by writing the functions of (4.1) in the vector notation of (4.3). The spaces spanned by (4.6, 7) may have different dimensionality, but the vector analysis is similar. The apparent difference between continuous and discrete variables as in fact been removed.

But one can equally well reverse the process and associate two orthonormal functions to the two basis vectors of (4.6), for example,

$$\chi_+(\xi) \qquad \text{to} \qquad (1, 0) \qquad \text{spin "up"} \tag{4.8}$$

and

$$\chi_{-}(\xi) \quad \text{to} \quad (0, 1) \quad \text{spin "down"}. \tag{4.9}$$

The χ's have properties of ordinary functions. If we wished ξ to be a discrete two-valued variable, we could allow the two functions to become delta functions in some appropriate way, and still formally preserve the analogy with the functions $f_n(r)$.

If now the $f_n(r)$ functions describe electronic states, a product function describing three particles for instance, might be

$$\Psi(1, 2, 3) = [f_j(r_1)\chi_r(\xi_1)][f_k(r_2)\chi_s(\xi_2)][f_n(r_3)\chi_t(\xi_3)] \tag{4.10}$$

with each of r, s, t, assuming the value $+$ or $-$. This is not an allowable wavefunction for identical particles since the probability density

$$P(1, 2, 3) \equiv \Psi^*\Psi \tag{4.11}$$

is not invariant under permutations of the three particles, except if

$$j = k = n \quad \text{and} \quad r = s = t \tag{4.12}$$

an equality which would not have been allowed for electrons according to the Pauli exclusion principle, in the old quantum theory. In the new quantum theory, moreover, we are supposed to consider not $\Psi(1, 2, 3)$, but the antisymmetrized version,

$$\begin{aligned} \Psi_A(1, 2, 3) = {} & \Psi(1, 2, 3) + \Psi(2, 3, 1) + \Psi(3, 1, 2) \\ & - \Psi(2, 1, 3) - \Psi(3, 2, 1) - \Psi(1, 3, 2). \end{aligned} \tag{4.13}$$

If a product function violated the old Pauli principle, such as in (4.12), it would also automatically be projected out by the antisymmetrization procedure, for as we shall see, Ψ_A would vanish identically. This shows that in the new mechanics, electrons cannot violate the Pauli principle even in an arbitrary state—let alone in an eigenstate described by good quantum numbers.

To extend the procedure to N particles would require writing out $N!$ terms. But following Slater, we recognize in (4.13) the rules for writing out a determinant, i.e.,

$$\Psi_A^{\text{Slater}} = \frac{1}{\sqrt{3!}} \det \begin{vmatrix} f_j(r_1)\chi_r(\xi_1) & \cdots & f_j(r_3)\chi_r(\xi_3) \\ f_k(r_1)\chi_s(\xi_1) & \cdots & \vdots \\ f_n(r_1)\chi_t(\xi_1) & \cdots & f_n(r_3)\chi_t(\xi_3) \end{vmatrix}. \tag{4.14}$$

The determinant is conventionally divided by $\sqrt{3!}$ for purposes of normalization. This ensures that in addition to being totally symmetric, the probability distribution of (4.11) calculated with Slater determinants of (4.14) is normalized to unity, if the $f(r)$ and $\chi(\xi)$ functions are members of an orthonormal set.

It is by means of such determinants as (4.14) that one forms a complete, orthonormal set of states for three electrons, and *not* by means of the product functions of (4.10). The number of such linearly independent determinants is always far less than the number of product functions, as all states which cannot be antisymmetrized are "excluded."

4.2 Antisymmetrization

A simple rule may be helpful to understand the process of antisymmetrization called for by the Pauli principle.

Assume a function of N variables and all $N!$ of its permutations:

$$f(1, 2, 3, \ldots, N), f(2, 1, \ldots, N), f(N, 3, 1, \ldots, 2), \ldots \tag{4.15}$$

where in the self-evident notation, j is allowed to stand for the jth variable. To construct a totally antisymmetric function using the f's, there is a systematic procedure:

Define a function antisymmetric in variables 1 and 2,

$$f(1, 2/3, \ldots) \equiv f(1, 2, 3, \ldots) - f(2, 1, 3, \ldots) \tag{4.16}$$

with the bar (/) to signify that the function is antisymmetrized with respect to the variables preceding it. Then by induction, we construct a function antisymmetric in variables 1, 2, and 3,

$$f(1, 2, 3/4, \ldots) \equiv f(1, 2/3, \ldots) - f(1, 3/2, \ldots) - f(3, 2/1, \ldots) \tag{4.17}$$

In general,

$$f(1, \ldots, p/p+1, \ldots) = f' - \sum_{j=1}^{p-1} \mathscr{P}_{j,p} f' \tag{4.18}$$

where

$$f' \equiv f(1, \ldots, p-1/p, \ldots)$$

and $\mathscr{P}_{j,p}$ transposes the jth and pth variables. The procedure can be continued until $p = N$ if desired.

By similar procedures totally symmetric functions, suitable for Bosons, can be constructed. It is merely necessary to replace all the minus (—) signs above by plus (+).

Problem 4.1. Prove by repeated use of (4.18) that the totally antisymmetric function is given by

$$\sum (-1)^P \mathscr{P} f(1, 2, \ldots, N) \equiv f(1, 2, \ldots, N/)$$

the sum being over all $N!$ distinct permutations of the N variables, and the exponent $P = 0$ for even permutations, $+1$ for odd permutations.

By taking other combinations of permutations, it is possible to generate functions with more complicated transformation properties under permutations of the various variables. But fortunately we do not need these; as we have already mentioned, wavefunctions of Fermions must be totally antisymmetric, and of Bosons totally symmetric, provided, of course, that we are sure to include the spin variable as well as any other coordinates (e.g., isotopic spin, band index, or whatever else is required in the complete description of the particle) in the wavefunction.

Starting with a product function of space and spin coordinates of N electrons, say $f(1)g(2) \ldots h(N)$, the antisymmetrization procedure gnerates $N!$ terms, which have the same sign (+ or −) as the corresponding terms in the expansion of a determinant in which $f(1)g(2) \ldots$ is the product along the principal diagonal of the array. If the original product function is normalized, then the determinant is conventionally normalized by dividing it by $\sqrt{N!}$. But this does not in fact normalize it *unless* the various permutations are all orthogonal, as when the one-electron functions in the original product are orthogonal. (We previously encountered problems in nonorthogonality in attempting to extend the Heitler-London scheme to the N-body problem.)

4.3 States of Three Electrons

Let us postpone the trivial problem of two electrons, and study three electrons by means of Slater determinants. The most general single determinant is

$$\begin{vmatrix} \varphi_a(r_1)\chi_r(\xi_1) & \varphi_a(r_2)\chi_r(\xi_2) & \varphi_a(r_3)\chi_r(\xi_3) \\ \varphi_b(r_1)\chi_s(\xi_1) & \varphi_b(r_2)\chi_s(\xi_2) & \varphi_b(r_3)\chi_s(\xi_3) \\ \varphi_c(r_1)\chi_t(\xi_1) & \varphi_c(r_2)\chi_t(\xi_2) & \varphi_c(r_3)\chi_t(\xi_3) \end{vmatrix} \tag{4.19}$$

aside from a normalization constant. For fixed $\varphi_a, \varphi_b, \varphi_c$, the two possibilities for *each* spin function result in eight distinct determinantal functions in total. If some of the space functions are not linearly independent, then the number of independent determinantal functions could be less than eight, and in the example of (4.12), $j = k = n, r = s = t$, the number of possible determinantal functions was zero. But in no case does it exceed eight for three particles. Note that if there were no exclusion principle, and no requirement of antisymmetrization, the number of allowed product configurations of the type in (4.10) would be $6 \times 6 \times 6 = 216$ in the present case. So, a considerable number of states have been pruned.

There are two instances of all spins "up" or "down" simultaneously,

$$\begin{vmatrix} \varphi_a(r_1)\chi_r(\xi_1) & \varphi_a(r_2)\chi_r(\xi_2) & \varphi_a(r_3)\chi_r(\xi_3) \\ \varphi_b(r_1)\chi_r(\xi_1) & \varphi_b(r_2)\chi_r(\xi_2) & \varphi_b(r_3)\chi_r(\xi_3) \\ \varphi_c(r_1)\chi_r(\xi_1) & \varphi_c(r_2)\chi_r(\xi_2) & \varphi_c(r_3)\chi_r(\xi_3) \end{vmatrix} \tag{4.20}$$

$r = +$ or $-$. The remaining six determinants are, in turn,

$$\begin{vmatrix} \varphi_a(1)\chi_r(1) & \varphi_a(2)\chi_r(2) & \varphi_a(3)\chi_r(3) \\ \varphi_b(1)\chi_s(1) & \varphi_b(2)\chi_s(2) & \varphi_b(3)\chi_s(3) \\ \varphi_c(1)\chi_s(1) & \varphi_c(2)\chi_s(2) & \varphi_c(3)\chi_s(3) \end{vmatrix} \tag{4.21}$$

with $r = +$ and $s = -$, or the converse; and

$$\begin{vmatrix} \varphi_a(1)\chi_s(1) & \varphi_a(2)\chi_s(2) & \varphi_a(3)\chi_s(3) \\ \varphi_b(1)\chi_r(1) & \varphi_b(2)\chi_r(2) & \varphi_b(3)\chi_r(3) \\ \varphi_c(1)\chi_s(1) & \varphi_c(2)\chi_s(2) & \varphi_c(3)\chi_s(3) \end{vmatrix} \tag{4.22}$$

and finally

$$\begin{vmatrix} \varphi_a(1)\chi_s(1) & \varphi_a(2)\chi_s(2) & \varphi_a(3)\chi_s(3) \\ \varphi_b(1)\chi_s(1) & \varphi_b(2)\chi_s(2) & \varphi_b(3)\chi_s(3) \\ \varphi_c(1)\chi_r(1) & \varphi_c(2)\chi_r(2) & \varphi_c(3)\chi_r(3) \end{vmatrix} \tag{4.23}$$

all with $r = +$ and $s = -$, or the converse, corresponding to one spin up and two down, or the converse. Of all these determinants, the spin functions factor out of only the two quartet states of (4.20). Two more quartet states must therefore be intimately mixed in with the two sets of doublet states in the determinants of (4.21–23).

We see that the determinantal method does not necessarily generate eigenfunctions of S^2_{tot}, although by construction all determinants are eigenfunctions of S^z_{tot}.

The doublet states can be projected from the states of (4.21–23) by means of the obviously constructed projection operator,

$$\boldsymbol{D} \equiv \frac{1}{3}\left(\frac{15}{4} - \boldsymbol{S}^2_{\text{tot}}\right) = \frac{1}{2} - \frac{2}{3}(\boldsymbol{S}_1\cdot\boldsymbol{S}_2 + \boldsymbol{S}_2\cdot\boldsymbol{S}_3 + \boldsymbol{S}_3\cdot\boldsymbol{S}_1) \tag{4.24}$$

which has zero eigenvalue on the quartet states, and eigenvalue unity on the doublet states. These states are not single Slater determinants, but linear combinations of them.

Problem 4.2. Find the eigenfunctions of the projection operator of (4.24) among the set of determinantal functions of (4.20–23). Expand them in terms of the space functions $\psi_1, \psi_2, \ldots, \psi_6$ of the three-atom problem of Chap. 2.

Trial functions to calculate the low-lying eigenvalues of a Hamiltonian for three electrons could be made of linear combinations of the eight determinantal functions. Moreover, if $\mathscr{H}$ doesn't involve the spins explicitly, we can always decompose it into noninteracting blocks by using the projection operators $\boldsymbol{D}$ and 1-$\boldsymbol{D}$. The "quartet," i.e., fourfold degenerate states, $\boldsymbol{S}_{\text{tot}} = \frac{3}{2}$ belong to the totally antisymmetric space function. The two sets of doublet states, $\boldsymbol{S}_{\text{tot}} = \frac{1}{2}$, give rise to two distinct doubly degenerate doublet levels. (In the case of three hydrogen atoms which we analyzed previously the two doublets were accidentally degenerate, however, and corresponded to the four solutions of type $\boldsymbol{v}_{23}$. Another space function which we had constructed (by permutations of space coordinates alone) was the totally symmetric function, which is not admissible and does not even enter via the determinantal approach. The sixth function was the totally antisymmetric quartet state.)

Problem 4.3. Construct the projection operator $\boldsymbol{R}$ which has zero eigenvalue for triplet states and eigenvalue unity for singlet states of two electrons. Operate with $\boldsymbol{R}$ on the most general Slater determinant of two electrons, and show that the singlet state is always symmetric under the interchange of the spatial coordinates of the two particles, and the triplet antisymmetric.

4.4 Eigenfunctions of Total S^2 and S^z

If a function of space and spin coordinates is totally antisymmetric, we shall denote it a "Pauli function": the Slater determinants are a special case. If in addition it is an eigenfunction of S^2_{tot} and S^z_{tot}, then there are severe limitations on the type of spin and space functions which can be used to construct such a Pauli function.

Let us assume first that the Pauli function is an eigenfunction of S^z_{tot} with eigenvalue M, that is,

$$S^z_{\text{tot}}\,{}^M\Psi = M\,{}^M\Psi \tag{4.25}$$

indicating the quantum number explicitly as a superscript. Let us now expand (4.25) in a complete set of spin functions, the product functions

$$\chi_{r1}(\xi_1)\chi_{r2}(\xi_2)\ldots\chi_{rN}(\xi_N). \tag{4.26}$$

Normally 2^N such functions would be needed, but the restriction to a definite M value means that

$$M + \tfrac{1}{2}N \text{ have spin up } (+) \text{ and } -M + \tfrac{1}{2}N \text{ have spin down } (-). \tag{4.27}$$

We can take a typical such function, for example,

$$^{M}\chi(\xi_1, \xi_2, \ldots, \xi_{M+1/2N} | \xi_{M+1/2N+1}, \ldots, \xi_N) \equiv \chi_+(\xi_1) \cdots \chi_+(\xi_{M+1/2N}) \cdots \chi_-(\xi_N) \tag{4.28}$$

and generate all the others by permutations. The permutations of the spins *up* among themselves to not generate a new function, and neither do permutations of the spins *down* among themselves. We therefore call these the *trivial permutations* of the spin function $^{M}\chi(\xi_1, \ldots | \ldots, \xi_N)$. This function is totally symmetric under the trivial permutations, those of spin coordinates to the left of the vertical bar amongst themselves, or of the spin coordinates to the right of the bar amongst themselves.

There are, however, $[N!/(\frac{1}{2}N + M)!\,(\frac{1}{2}N - M)!] = n_M$ *nontrivial* permutations generating n_M orthogonal spin functions including the original function in (4.28), and these comprise the complete set for the expansion of $^{M}\psi$. Note that

$$\sum_{M=-1/2N}^{+1/2N} n_M = \sum_{M=-1/2N}^{+1/2N} \frac{N!}{\left(\frac{1}{2}N + M\right)!\left(\frac{1}{2}N - M\right)!} = 2^N \tag{4.29}$$

the dimensionality of all the M subspaces adding up to the total correct number 2^N of orthogonal spin functions.

When $^{M}\psi$ is expanded in these functions, the coefficient of $^{M}\chi$ in the expansion must be a space function of the type,

$$^{M}f(r_1, r_2, \ldots, r_{1/2N+M} | r_{1/2N+M+1}, \ldots, r_N) \tag{4.30}$$

which is totally antisymmetric under the trivial permutations of coordinates to the left of the bar amongst themselves, or of coordinates to the right of the bar amongst themselves. In general, however, it has complicated transformation properties under the n_M nontrivial permutations (see Problem 4.4).

Problem 4.4 In the case of three hydrogen atoms, there were two states for $M = \frac{1}{2}$, notably $\boldsymbol{v}_{23}$ and $\boldsymbol{v}'_{23}$. Express the *nontrivial* permutations of each of these two states in terms of the other, and of the partly symmetric functions $\boldsymbol{v}_{23\text{sy}}$ and $\boldsymbol{v}'_{23\text{sy}}$.

If in addition, the Pauli function is an eigenfunction of S^2_{tot}, with eigenvalue S,

$$S^2_{\text{tot}}\,{}^{M}\Psi^S = S(S+1)\,{}^{M}\Psi^S \qquad S^z_{\text{tot}}\,{}^{M}\Psi^S = M\,{}^{M}\Psi^S \tag{4.31}$$

then we can actually figure out the nontrivial symmetries of the space functions $^{M}f^S$ using only simple notions about angular momentum. We do this systematically, starting with $^{-S}f^S$, which is the coefficient of the spin function of (4.28) when $M = -S$. Recall that a fundamental property of angular momentum is that $|M| \leq S$, and therefore it should be impossible to further decrease M. In more formal terms,

$$S_{\text{tot}}^{-}({}^{-S}\Psi^{S}) \equiv 0. \tag{4.32}$$

This requires that there be some linear relationship among the f functions and their nontrivial permutations, and indeed, by use of (4.28, 30, 32), we find

$$\begin{aligned} &{}^{-S}f^{S}(r_1, \ldots, r_{1/2N-S} \,|\, r_{1/2N-S+1}, \ldots, r_N) \\ &\quad - \sum_{j=1/2N-S+1}^{N} \mathscr{P}_{ij}\, {}^{-S}f^{S}(r_1, \ldots, r_{1/2N-S} \,|\, r_{1/2N-S+1}, \ldots, r_N) = 0 \end{aligned} \tag{4.33}$$

with

$$i = 1, 2, \ldots, \frac{1}{2}N - S$$

and $\mathscr{P}_{ij}$ being the transposition operator for the coordinates of the particles i and j. But this is very reminiscent of (4.18) if we interpret the above as an attempt to increase the number of coordinates to the right of the vertical bar (in which f is totally antisymmetric) by the addition of one member. This attempt results in zero: the set to the right cannot be further increased, i.e., the bar cannot be moved to the left.

We can construct the f functions belonging to higher M values, using the ${}^{-S}f^{S}$ which properly obey the equation above. This is done by repeated applications of the total-spin raising operator,

$$S_{\text{tot}}^{+}\, {}^{-S}\Psi^{S} = {}^{-S+1}\Psi^{S} \qquad \text{unnormalized} \tag{4.34}$$

and leads directly to

$${}^{-S+1}f^{S}(r_1, \ldots, r_{1/2N-S+1} \,|\, r_{1/2N-S+2}, \ldots) = {}^{-S}f^{S} - \sum_{j=1}^{1/2N-S} \mathscr{P}_{j,1/2N-S+1}\, {}^{-S}f^{S} \tag{4.35}$$

with

$${}^{-S}f^{S} = {}^{-S}f^{S}(r_1, \ldots, r_{1/2N-S} \,|\, r_{1/2N-S+1}, \ldots) \tag{4.36}$$

a function which obeys (4.33).

Thus, the attempt to move the bar to the right does *not* fail. By further application of the spin-raising operator, or of the antisymmetrization rules of (4.18), we progressively generate ${}^{-S+2}f^{S}$, ..., up to ${}^{-S}f^{S}$, *all expressed as linear combinations of the original function* ${}^{-S}f^{S}$ *and its nontrivial permutations.* Finally, the angular-momentum rule

$$S_{\text{tot}}^{+}({}^{+S}\Psi^{S}) \equiv 0 \tag{4.37}$$

gives us an additional set of constraints on the f functions, similar to the first ones we found *supra*. But now, these constrains are that the bar cannot be moved farther to the *right*:

$$ {}^{S}f^{S}(r_1, \dots, r_{1/2N+S} | r_{1/2N+S+1}, \dots, r_N) - \sum_{j=1}^{1/2N+S} \mathscr{P}_{1,q}\,{}^{S}f^{S}(r_1, \dots, r_{1/2N+S} | r_{1/2N+S+1}, \dots, r_N) = 0 \tag{4.38} $$

with

$$ q = \frac{1}{2}N + S + 1, \dots, N. $$

Given an arbitrary space function, it is always a simple matter to generate a function of the type ${}^{M}f$, that is, the space partner of the spin eigenfunction of S^z_{tot}. We merely apply the rules of (4.18), and antisymmetrize in the first set of $p = \frac{1}{2}N + M$ coordinates, then antisymmetrize in the set of the remaining coordinates. Only those permutations which we have denoted as "trivial" are used in this process, the others being used to raise or lower M, as we have already seen. If for some reason f is unsuitable, the process of antisymmetrization will yield identically *zero* instead of the desired function. For example: if f is of the form ${}^{0}f^{0}$ and we wish to construct a function ${}^{M}f$ $M \neq 0$ with it, we get zero instead (i.e., like trying to antisymmetrize a symmetric function).

There is no correspondingly simple rule for constructing functions of the type ${}^{M}f^{S}$ from an arbitrary function. It is necessary to projct out the unwanted components belonging to values of the total spin other than S by repeated use of the spin raising and lowering operators $S^{\pm}_{\text{tot}}$, after first constructing ${}^{M}f$, with $M = \pm S$. But in some special cases this can be done by inspection, as in the following example:

Example: A product wavefunction describes noninteracting particles, and also has its uses in variational approximations, as we have seen. Let us use such a product space function as the raw material in an example where we construct the proper space functions corresponding to definite M and S eigenvalues.

Starting with such a product as

$$ \varphi_1(r_1)\varphi_2(r_1) \dots \varphi_N(r_N) \tag{4.39} $$

we can immediately write the ${}^{M}f$ function as a product of two determinants

$$ {}^{M}f = \det \begin{vmatrix} \varphi_1(r_1) & \varphi_1(r_2) & \cdots & \varphi_1(r_{1/2N+M}) \\ \varphi_2(r_1) & \cdot & & \cdot \\ \cdot & \cdot & & \cdot \\ \varphi_{1/2N+M}(r_1) & \cdot & \cdots & \varphi_{1/2N+M}(r_{1/2N+M}) \end{vmatrix} $$

$$\times \det \begin{vmatrix} \varphi_{1/2N+M+1}(r_{1/2N+M+1}) & \cdots & \varphi_{1/2N+M+1}(r_N) \\ \cdot & \cdot & \cdot \\ \cdot & \cdot & \cdot \\ \varphi_N(r_{1/2N+M+1}) & \cdots & \varphi_N(r_N) \end{vmatrix} \tag{4.40}$$

which is explicitly antisymmetric in the two sets of variables. Now assume $M \leq 0$. (The case $M > 0$ can be handled by simply interchanging the two determinants.) Thus the first determinant is smaller than the second. Now, if all the functions $\varphi_1, \varphi_2, \ldots, \varphi_{1/2N+M}$ which appear in the smaller determinant are also present in the bigger one, then according to the rules of (4.32, 33), the antisymmetrization procedure of (4.33) will obviously yield *zero*, and therefore the function of (4.40) belongs to the definite S value,

$$S = |M|. \tag{4.41}$$

Of course this requires that the various φ_j not be all distinct functions. When the φ_j are all distinct, the construction of a function of definite S is possible but more tedious, but we shall not further investigate this topic. The various φ_j in *each* determinant must be distinct or else ${}^{M}f$ will vanish identically. But these can appear in both determinants, which is merely another expression of Kramers' degeneracy; for the appearance of a φ_j in the first determinant means there is an electron spin "up" in orbital φ_j, whereas its appearance in the second determinant means there is an electron spin "down" in that orbital.

This completes the demonstration, of how the Pauli wavefunction could be totally antisymmetric while the space functions, presumably eigenfunctions of a space Hamiltonian, could have complex transformation properties under the permutation group. We have also shown that these transformation properties are a consequence of the rules concerning total spin angular momentum. We now proceed to specific applications.

4.5 Ground State of Two Electrons: A Theorem

In many problems of differential equations, the lowest eigenvalue can be shown to belong to the nodeless solution. Can something analogous be proved in the many electron problem, and if so, does it have any bearing on the problem of ferromagnetism?

The answer to both queries is affirmative. If we first examine the atoms and the periodic table, we find that in helium the ground state is nodeless, and we draw therefrom important consequences. For heavier atoms, conversely, the ground state has a certain number of nodes, so as to satisy Hund's rules. These nodes result in atomic spin and angular momentum in the ground state which, as we shall see, apparently violate a two-particle theorem. We shall return to the extension of the theorem to N particles and its apparent violation in three dimensions in subsequent sections, and then resolve the paradox.

Consider first the helium atom, slightly generalized to arbitrary potential

$$\mathcal{H} = \frac{1}{2m}(\boldsymbol{p}_1^2 + \boldsymbol{p}_2^2) + V(\boldsymbol{r}_1, \boldsymbol{r}_2). \tag{4.42}$$

The Hamiltonian is assumed and real, invariant under the interchange of the two (indistinguishable) particles. The eigenfunctions $f(\boldsymbol{r}_1, \boldsymbol{r}_2)$ of Schrödinger's equation,

$$\mathcal{H}f(\boldsymbol{r}_1, \boldsymbol{r}_2) = Ef(\boldsymbol{r}_1, \boldsymbol{r}_2) \tag{4.43}$$

are therefore real, and either even or odd under the interchange of the two particles. Because of the simplicity of the problem, it is not necessary to label f with M and S value: recall merely that the even solutions must be associated with odd spin function.

$$\frac{1}{\sqrt{2}}[\chi_+(\xi_1)\chi_-(\xi_2) - \chi_+(\xi_2)\chi_-(\xi_1)] \tag{4.44}$$

belonging to $S_{\text{tot}} = 0$, and the odd solutions with one of the even spin functions

$$[\chi_+(\xi_1)\chi_+(\xi_2)] \qquad \text{or} \qquad [\chi_-(\xi_1)\chi_-(\xi_2)] \qquad \text{or}$$
$$\frac{1}{\sqrt{2}}[\chi_+(\xi_1)\chi_-(\xi_2) + \chi_+(\xi_2)\chi_-(\xi_1)] \tag{4.45}$$

belonging to $\boldsymbol{S}_{\text{tot}} = 1$, $M = 1, -1$, or 0. Then the product is a proper Pauli wavefunction. We shall make use of the variational theorem, stated in the following terms: The ground-state energy always lies below the variational energy except when an exact ground-state eigenfunction is used as a variational trial wavefunction. Explicitly,

$$E_{\text{grd state}} < E_{\text{var}} \equiv \frac{\int g^*(\boldsymbol{r}_1, \boldsymbol{r}_2)\mathcal{H}g(\boldsymbol{r}_1, \boldsymbol{r}_2)\, d^3r_1\, d^3r_2}{\int |g|^2\, d^3r_1\, d^3r_2} \tag{4.46}$$

and the strict inequality holds *except* if the trial function $g(\boldsymbol{r}_1, \boldsymbol{r}_2)$ obeys the differential equation

$$(\mathcal{H} - E_{\text{grd state}})g(\boldsymbol{r}_1, \boldsymbol{r}_2) = 0 \tag{4.47}$$

for all values of $\boldsymbol{r}_1$ and $\boldsymbol{r}_2$. In that case only,

$$E_{\text{var}} = E_{\text{grd state}}. \tag{4.48}$$

We need consider only functions g which are real, and either even or odd under the interchange of $\boldsymbol{r}_1$ and $\boldsymbol{r}_2$, as in the case of the eigenfunctions f. We can now state the following:

Theorem. The ground-state eigenfunction is nodeless, ***ergo*** it belongs to $S_{\text{tot}} = 0$. Moreover, if $V(\boldsymbol{r}_1, \boldsymbol{r}_2)$ is invariant under rotation of the coordinate axes, i.e. commutes with the total angular momentum, then in the ground state also $L_{\text{tot}} = 0$. Thus helium in the ground state, for example, has precisely zero total angular momentum and spin and is totally nonmagnetic [4.5].

Proof. We shall first assume that the ground-state eigenfunction has nodes (that is, has at least one change of sign), and then show that this hypothesis leads to a contradiction. Let the ground-state eigenfunction with nodes be f_0, belonging to energy eigenvalue E_0. By hypothesis, $E_0 = E_{\text{grd state}}$.

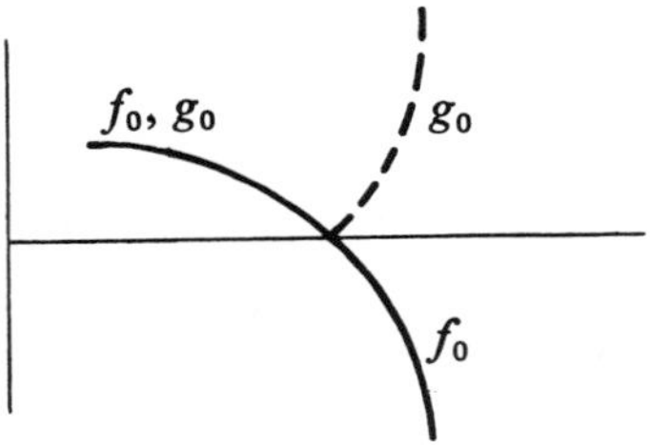

Fig. 4.1. The Wavefunction f_0 and its absolute value $g_0 = |f_0|$ plotted along a direction normal to a nodal surface of f_0

Consider the nodeless function $g_0 \equiv |f_0|$, the absolute value of f_0. Is it an admissible variational wavefunction? If f_0 is constrained to vanish at ∞, or is square integrable, then the same follows for g_0. If f_0 is odd, or if it is even, g_0 is even. Since f_0 is real, so is g_0. Therefore it indeed fulfills all the requirements for a variation trial function.

But g_0 is not itself an eigenfunction, for along the nodal surface $f_0(\boldsymbol{r}_1, \boldsymbol{r}_2) = 0$, it has discontinuous normal derivatives. See sketch in Fig. 4.1. Therefore the strict inequality of (4.46) is applicable, $E_{\text{var}} > E_{\text{grd state}}$.

The variational energy is calculated by either of two methods with identical results: 1) Observe that g_0 obeys Schrödinger's equation with eigenvalue E_0 everywhere except on the nodal surfaces. Because g_0 itself vanishes on these surfaces, the variational energy is precisely E_0. 2) By partial integration the variational integral in (4.46) can be brought into the canonical form, from which Schrödinger initially derived his wave equation

$$E_{\text{var}} = \frac{\int \left[\frac{\hbar^2}{2m}(|\nabla_1 g_0|^2 + |\nabla_2 g_0|^2) + V|g_0|^2\right] d^3r_1\, d^3r_2}{\int |g_0|^2\, d^3r_1\, d^3r_2}. \tag{4.49}$$

But $|g_0|^2 = |f_0|^2$ everywhere, and $|\dot{g}_0|^2 = |\dot{f}_0|^2$ everywhere, letting the dot over the letter stand for any derivative. [In deriving (4.49), we have neglected only surface terms at ∞, where both g_0 and f_0 vanish by the boundary condition]. By direct substitution, we now also obtain $E_{\text{var}} = E_0$. Hence $E_0 > E_{\text{grd state}}$.

This result contradicts the hypothesis that $E_0 = E_{\text{grd state}}$, if f_0 has nodes; or the hypothesis that f_0 has nodes, if $E_0 = E_{\text{grd state}}$. Alternatively, if it is assumed that the ground-state wavefunction is nodeless, then its absolute value is just $\pm$ the same function, and there results no such contradiction.

Consequently, *the ground state is nondegenerate.* For if there is more than one eigenfunction belonging to $E_{\text{grd state}}$, the previous arguments indicate that *each* must be nodeless. However, there is only one nodeless eigenfunction of $\mathscr{H}$, because the eigenfunctions of a Hamiltonian form a complete, orthogonal set of functions, and two nodeless functions *cannot be orthogonal.*

If the problem has rotational symmetry, then the eigenfunctions also have definite total angular momentum. The nodeless function must belong to zero eigenvalue, because it is not orthogonal to the constant function 1, itself an eigenfunction of total angular momentum, obviously with *zero* eigenvalue. Q.E.D.

Problem 4.5. Prove the following unconventional equation for the ground-state energy $E_{\text{grd state}}$ and ground-state wavefunction $f_{\text{grd state}}$ of the helium atom and molecule:

$$E_{\text{grd state}} = \frac{\int e^2(1/r_{12} - 2/r_{11} - 2/r_2) f_{\text{grd state}}\, d^3r_1\, d^3r_2}{\int f_{\text{grd state}}\, d^3r_1\, d^3r_2}.$$

Find an analogous expression for the ground state of the N-Boson problem [4.6]. For the ground state of three or more Fermions? Hydrogen atom and molecule?

The actual atomic structures of helium and also of the heavier atoms are just as well understood theoretically[1] as experimentally. The well-known calculations of Hartree and others in the late 1920s and 1930s, and modern high-speed electronic computers have reduced atomic calculations to what is almost an exact science The present theorem has merely served to show that the well-known ground state properties of helium are not the results of a particular calculation, but are shared by arbitrary two-Fermion systems subject to arbitrary spatial potentials.

4.6 Hund's Rules

The Hartree-Fock procedure[1] allows one to solve the many-electron Schrödinger equation approximately by supposing each electron to move in a self-consistent,

[1] Even in the old quantum theory, see, e.g. [4.7], or any modern text on atomic structure or quantum mechanics.

averaged potential.[2] Numerical solution of the effective one-body Schrödinger equations is still required, as the self-consistent potentials are not strictly Coullombic so that the eigenfunctions are not obtainable in terms of elementary functions. Still, the one-electron angular momentum is useful to organize the electrons into shells, and the periodic structure of the atomic table is understood to be the consequence of the filling up of successive shells.

We have already examined the first closed shell, that of helium and found that without net spin or angular momentum, it was magnetically inert. Let us add two more electrons, to make Be. It may be thought the outer shell obeys an effective two-electron Hamiltonian, the solutions of which must be orthogonalized to the core (helium) states. If the theorem of the preceding section holds for the two new electrons, then the total spin and angular momentum of Be must also vanish, which is the case.

Next, we add two more electrons to form C, and we find the first surprise. Carbon has $S_{tot} = 1$, and $L_{tot} = 1$, and the theorem fails completely for the third pair of electrons! Indeed, if we look further, of the first eighteen elements to have an even number of electrons (which takes us up to krypton, $Z = 36$), only eight have no spin or angular momentum in the ground state; and of these, half are inert gases. Why does the argument go awry? One might argue that the theorem breaks down for the following mathematical reason: that orthogonality to core states is equivalent to nonlocal potential forces, for which the theorem is inapplicable. But there are more convincing physical explanations. How can one predict the magnetic properties of an atom in the ground state?

The answers are given by Hund's rules. These laws, originally based on the abundant spectroscopic evidence, were of course confirmed by complete atomic calculations.[2] Fortunately, there exists a less exhaustive explanation of Hund's rules in terms of exchange, or the vector model, of which we shall give some examples.

First Rule: Electrons occupy the $2 \times (2l + 1)$ states of a shell in such a manner as to maximize the total spin. This may determine the configuration uniquely. When it does not, then one appeals to the second rule.

Second Rule: Any ambiguities in the first rule are resolved in favor of the highest value of L_{tot}. But the second rule is not always a unique prescription either, although these two rules are adequate to describe *grosso modo* the magnetic properties of the atom. Other rules determine whether the spin and angular momenta are parallel or not, but it is not necessary for us to become too involved with atomic structure. For with the exception of the rare earths, angular momentum is most always quenched in the process of chemical binding or in the solid state, and only the spin and Hund's first rule survive.

---

[2] This variational extension of Hartree's calculations by *Fock*, and *Slater*, has proved capable of very high accuracy in atomic-structure theory.

Qualitatively, both Hund's rules follow from degenerate-state perturbation theory. Electrons in the outer shells move in the spherical potential of the nucleus and of the filled inert-gas configurations. To a first approximation, all the configurations within the outer shell are degenerate. It is the Coulomb repulsion between these electrons which lifts the degeneracy. Now if any two electrons are in the same orbital state, the Coulomb repulsion is maximal. These configurations should be avoided, just as the ionized configurations were avoided in the Heitler-London theory.

Therefore, in a first crude attempt to "prove" Hund's rules, or rather to make them more plausible without too much work, only those configurations of the outer shell electrons need be considered in which every electron occupies a different orbital state. This would imply that whenever a shell contains $2l + 1$ electrons, all the degenerate orbital states in that shell are occupied: for each ml there is a $-ml$; and therefore presumably $L_{\text{tot}} = 0$.

The validity of this simple guess is striking: there appear to be no exceptions in the periodic table. For example, the elements with a half-filled p shell are N, P, As, Sb, Bi, and all have vanishing orbital angular momentum; the elements with half-filled d shell, Cr, Mn, Mo, etc., also obey this rule, etc.

For shells less than half-filled, one speaks of electrons, whereas for shells more than half-filled, one commonly speaks of "holes." There is indeed a great similarity between the ground terms of elements on either side of half-filled shells in the periodic table: with a general exception that the total J value (spin *plus* orbital angular momenta) equals $|S - L|$ for electrons and $S + L$ for holes. This difference is caused by spin-orbit coupling, a relatively small energy responsible for the phenomenon of "magnetic anisotropy" in the solid state.

To understand Hund's rules, the vector model, and particularly their physical origins, it is important to work out some examples. This is done in the following sections, but only for illustrative purposes, and the reader who wishes to become proficient in the true art of atomic calculation and quantum chemistry is referred to specialized treatises and review articles.

Hund's rules give only the ground state of the atom. The methods of atomic calculation which we shall illustrate give the (very important) low-lying excited states as well, together with their energy, spin, and angular-momentum quantum numbers. These results are central to the theory of magnetism, for it is impossible to understand magnetic solids without understanding the magnetic atoms and the magnetic molecules first.

Now we can return to our earlier question: why doesn't the two-particle theorem hold for the valence electrons of carbon? Are Hund's rules contradictory to our theorem? The answer, briefly, is this. The orbital states, in atoms heavier than helium, must be orthogonal to the (inert) core, and therefore cannot be nodeless. The theorem, particularly the method of proof we used, just does not apply to these "excited state" orbitals. On the contrary; if the functions are already required to suffer a number of nodes, and to pay the ensuing price in kinetic energy, it is then advantageous for the potential energy to be minimized *in exactly the manner predicted by Hund's rules,* as we shall now see by actual calculation.

4.7 p^3 Configuration

First, recall that four atomic quantum numbers specify each electron: the principal quantum number $n = 1, 2, \ldots$, the orbital angular momentum $l \leq n - 1$, the magnetic quantum number $m_l = -l, \ldots, +l$, and the spin eigenvalue $m_s = \pm\frac{1}{2}$. But in the description of an electron, instead of giving $l = 0, 1, 2, 3, \ldots$, it is conventional to use the letters $s(l = 0)$, $p(l = 1)$, $d(l = 2)$, $f(l = 3)$, $g(l = 4)$, and thence in alphabetical sequence. Therefore, $n = 2$ and $l = 0$ *is a* $2s$ *electron.* One does not indicate m_l or m_s, and the number of particles with given n and l is shown as a superscript. For example, two electrons with $n = 3$, $l = 2$ are denoted $3d^2$; three such electrons by $3d^3$. The m values not being specified, there are many possible wavefunctions for a given configuration. With the total angular momentum of the configuration indicated by a capital letter, for example, $S, P, D, F, \ldots$ for $L = 0, 1, 2, 3$, and the total spin S_{tot} *via* an exponent: $2S_{tot} + 1 =$ multiplicity, the various configurations are indicated in Table 4.3. The ground state has *maximum* multiplicity.

Consider nitrogen: outside the helium core, two electrons are in the $2s$ shell, and three in the $2p$ shell. The s electrons have no angular momentum or spin, and we ignore them in zeroth order. The important particles are the three $2p$ electrons (the $2p^3$ configuration, in atomic notation). If such is the case, the ground-state and low-lying terms of any atom or ion with a p^3 configuration

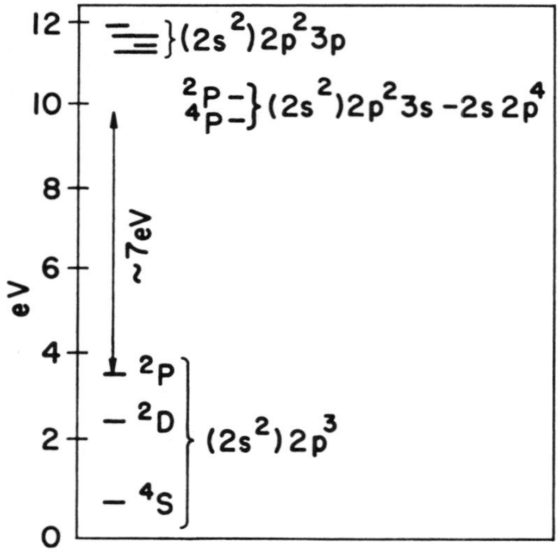

Fig. 4.2. Energy-level diagram of nitrogen (N). Vector model applies to lowest three levels only; but note that they are well below the other levels. The ground state belongs to maximum $S_{tot} = 3/2$

outside of closed shells must be similar. Figure 4.2 and Table 4.1 confirm this, and show the good agreement between theory and experiment.

We first tackle this problem by the same technique used in the study of three hydrogen atoms. The more conventional approach is harder, and is discussed in the next section. First we write out the spatial basis functions. Instead of atoms a, b, c, with which to associate the electrons, there are now the three p states specified by the quantum numbers $m_l = -1, 0, +1$. We allow each to be occupied just once: this reduces both the Coulomb repulsion, and the number of spatial configurations necessary to consider.

The $3! = 6$ spatial configurations are

$$\begin{aligned}
\psi_1 &= \varphi_1(r_1) \quad \varphi_0(r_2) \quad \varphi_{-1}(r_3)\\
\psi_2 &= \varphi_1(r_2) \quad \varphi_0(r_1) \quad \varphi_{-1}(r_3)\\
\psi_3 &= \varphi_1(r_3) \quad \varphi_0(r_2) \quad \varphi_{-1}(r_1)\\
\psi_4 &= \varphi_1(r_1) \quad \varphi_0(r_3) \quad \varphi_{-1}(r_2)\\
\psi_5 &= \varphi_1(r_2) \quad \varphi_0(r_3) \quad \varphi_{-1}(r_1)\\
\psi_6 &= \varphi_1(r_3) \quad \varphi_0(r_1) \quad \varphi_{-1}(r_2)\,.
\end{aligned} \tag{4.50}$$

The subscripts on the one-particle functions refer to m_i^l, the eigenvalue of the one-electron operators L_i^z. There are only two major differences between these functions and those in the three-atom case: the present functions are orthonormal, and they are not all three equivalent. The significance of this will become immediately clear.

The functions $\varphi_{m_l}(r)$ are by definition solutions of the best possible self-consistent one-particle Hamiltonian, $\mathscr{H}_0$. As $\mathscr{H}_0$ usually turns out to be spherically symmetric, we shall assume this to be the case here, and factor the functions as follows:

$$\varphi_{m_l}(r) = Y_{l,m}(\theta, \phi)R(r)\,. \tag{4.51}$$

Fortunately, everything is known about the angular part of the wavefunction, the spherical harmonics, $Y_{l,m}$, and as for the common radial function which is unknown, it will be eliminated from the problem.

The perturbation is the sum of 2-body interactions:

$$\mathscr{H}' = e^2\left(\frac{1}{r_{12}} + \frac{1}{r_{23}} + \frac{1}{r_{31}}\right). \tag{4.52}$$

We may write down an eigenvalue equation for the energies in terms of the matrix elements of $\mathscr{H}'$ between the six basis states, and the overlap matrix. The justification for this was given in the chapter on exchange, in terms of making the energies stationary, or diagonalizing $\mathscr{H}'$ within the restricted subspace. The overlap matrix is in the present case just the unit matrix, by orthonormality. It might be thought that the $\mathscr{H}'$ matrix could be obtained from our earlier result also by setting the overlap integral $= 0$, but that doesn't quite work as

$$\mathscr{H}'_{1,2} \neq \mathscr{H}'_{1,3} \tag{4.53}$$

because permuting functions with $|\Delta m_l| = 1$ is not quite the same as when $|\Delta m_l| = 2$. Otherwise, we proceed much as before. We define the diagonal term as

$$A \equiv \mathscr{H}'_{1,1} = \cdots = \mathscr{H}'_{6,6} > 0 \tag{4.54}$$

and nondiagonal elements $B_1 A$ by

$$\begin{aligned} B_1 A &\equiv \mathscr{H}'_{1,2} = \mathscr{H}'_{1,4} = \cdots = \mathscr{H}'_{4,6} \\ &= \int \frac{e^2}{r_{12}} [\varphi_1^*(r_1)\varphi_0(r_1)\varphi_0^*(r_2)\varphi_1(r_2)]\, d^3r_1\, d^3r_2 \,. \end{aligned} \tag{4.55}$$

Other nondiagonal elements $B_2 A$ are defined as,

$$\begin{aligned} B_2 A &\equiv \mathscr{H}'_{1,3} = \cdots = \mathscr{H}'_{6,2} \\ &= \int \frac{e^2}{r_{13}} [\varphi_1^*(r_1)\varphi_{-1}(r_1)\varphi_{-1}^*(r_3)\varphi_1(r_3)]\, d^3r_1\, d^3r_3 \,. \end{aligned} \tag{4.56}$$

All matrix elements are proportional to the same constant A, also all turn out to be real or can be made real. We have made use of the orthonormality of the one-electron functions to eliminate all the irrelevant terms in $\mathscr{H}'$ from these expressions, and have indicated the difference between the two types of exchange integrals by the subscript B_1 or B_2. The third type of matrix element, for example, $\mathscr{H}'_{1,5}$, automatically vanishes. One finds this out either by setting the overlap integral $= 0$ in the previous problem, or by noticing directly that the exchange between functions differing by more than a simple transposition must vanish in the case of two-body forces and orthonormal basis functions. Finally, we obtain the eigenvalue equation,

$$\mathscr{H}' \cdot v = A \begin{bmatrix} 1 & B_1 & B_2 & B_1 & 0 & 0 \\ B_1 & 1 & 0 & 0 & B_1 & B_2 \\ B_2 & 0 & 1 & 0 & B_1 & B_1 \\ B_1 & 0 & 0 & 1 & B_2 & B_1 \\ 0 & B_1 & B_1 & B_2 & 1 & 0 \\ 0 & B_2 & B_1 & B_1 & 0 & 1 \end{bmatrix} \cdot v = Ev. \tag{4.57}$$

Recall that there was a useless totally symmetric eigenvector, and a totally antisymmetric eigenvector which we identified as the space partner of the quartet state, $S_{tot} = \frac{3}{2}$. Then there were two doublet states; these are no longer degenerate, and we must find the proper linear combination to diagonalize $\mathcal{H}'$. Let us examine the two states v_{23} and v'_{23} which were antisymmetric in particles 2 and 3. (They cannot mix with the two functions which are symmetric in those particles, v_{23sy}, and v'_{23sy}). One finds by construction, the new eigenvectors

$$w_{23} = \frac{2}{3}v_{23} + v'_{23} \quad \text{and} \tag{4.58}$$

$$w'_{23} = -2v_{23} + v'_{23} \ . \tag{4.59}$$

Problem 4.6. Find the proper linear combinations of v_{23sy} and v'_{23sy} which are eigenvectors of $\mathcal{H}'$.

There is an additional operator of which we should like to know the eigenvalues, that is L^2_{tot}:

$$\begin{aligned} L^2_{tot} &= L^2_1 + L^2_2 + L^2_3 + 2(L^z_1L^z_2 + L^z_2L^z_3 + L^z_3L^z_1) \\ &\quad + (L^+_1L^-_2 + L^+_2L^-_3 + L^+_3L^-_1 + \text{H.c.}) \ . \end{aligned} \tag{4.60}$$

All that is required to calculate the matrix elements of this operator on our set of states is such information as [cf. (3.44)]

$$L^2_i = 1(1+1) = 2 \quad \text{and} \tag{4.61}$$

$$(m_i + 1|L^+_i|m_i) = \sqrt{2}(\delta_{m_i,-1} + \delta_{m_i,0}) \tag{4.62}$$

for angular momentum 1. Thus in the diagonal matrix element $(L^2_{tot})_{1,1}$ there enter three contributions of the form of (4.61), making 6, minus 2 from $2L^z_1L^z_3$, for a total of 4. The other matrix elements are just as easily found, and finally,

$$L^2_{tot} = \begin{bmatrix} 4 & 2 & 0 & 2 & 0 & 0 \\ 2 & 4 & 0 & 0 & 2 & 0 \\ 0 & 0 & 4 & 0 & 2 & 2 \\ 2 & 0 & 0 & 4 & 0 & 2 \\ 0 & 2 & 2 & 0 & 4 & 0 \\ 0 & 0 & 2 & 2 & 0 & 4 \end{bmatrix} . \tag{4.63}$$

It is now a simple exercise (for the reader) to write the eigenvalues E as functions of A, B_1, and B_2 and calculate the value of the angular momentum to which they belong, with the eigenvectors already given. The spin of these same states has already been discussed and given several times.

The following fact may be extricated from *Condon* and *Shortley* [4.10] or proved directly by an expansion of $(r_{12})^{-1}$ in spherical harmonics (see Problem 4.8):

$$B_2 = 2B_1 \; . \tag{4.64}$$

It simplifies the results even more for the actual problem. Finally one finds for the eigenvalues

$$E(^2P) = A(1 + B_1) \tag{4.65}$$

$$E(^2D) = \mathrm{A}(1 - B_1) \tag{4.66}$$

$$E(^4S) = A(1 - 4B_1) \; . \tag{4.67}$$

In the usual spectroscopic notation, the multiplicity ($2S_{\text{tot}} + 1$) has been written as a prefixed exponent to the angular momentum $S, P, D, F, G, \ldots$ ($L_{\text{tot}} = 0, 1, 2, 3, 4, \ldots$). From these equations, we find for the level spacings,

$$\frac{E(^2P) - E(^2D)}{E(^2D) - E(^4S)} = \frac{2}{3} \tag{4.68}$$

a ratio independent of A and B_1, or of the integrals involving the radial parts of the wavefunctions.

The agreement with experiment is pleasant. Not only are all p^3 configurations found to possess low-lying terms of the type discussed, and ordered in energy as we have found them, but the ratio of (4.68) is experimentally well obeyed within our self-imposed limits of accuracy. In Fig. 4.2 is reproduced the term level scheme of nitrogen, together with the electronic configurations giving rise to the various terms. In Table 4.1 is given the experimentally observed ratio of (4.68) for various substances.

Table 4.1. Configuration and ratio of term splittings in various atoms and ions, and theory

		Experimental			
	Theory	N	O^+	S^+	As
Configuration	np^3	$2p^3$	$2p^3$	$3p^3$	$4p^3$
Ratio[a]	0.667	0.500	0.509	0.651	0.751

[a] See (4.68)
Source: *Condon* and *Shortley* [4.10]

Because each one-electron state is occupied, there is no difficulty in principle to the construction of an equivalent Heisenberg, or *vector model*, Hamiltonian for spins one-half. This is illustrated in Fig. 4.3.

From previous experience, we may almost guess that in $\mathscr{H}_{\text{Heis}}$

$$\mathscr{H}_{\text{Heis}} = -J_1(\boldsymbol{S}_{-1}\cdot\boldsymbol{S}_0 + \boldsymbol{S}_0\cdot\boldsymbol{S}_{+1}) - J_2(\boldsymbol{S}_{-1}\cdot\boldsymbol{S}_{+1}) \tag{4.69}$$

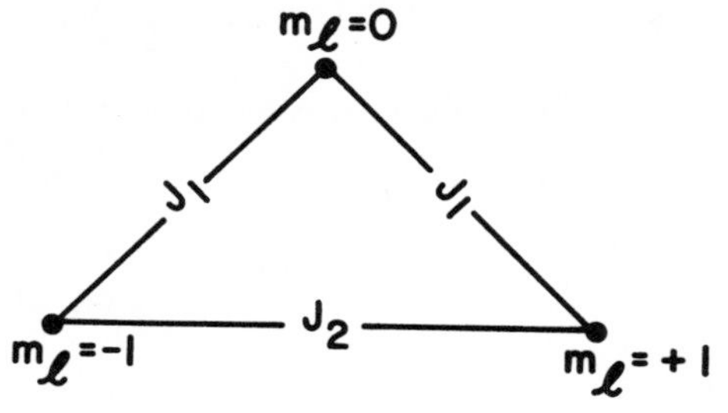

Fig. 4.3. Labels (m_l) and bonds in vector model of nitrogen atom

the proper choice of exchange constants is,

$$J_1 = 2B_1S \quad \text{and} \quad J_2 = 2B_2A \tag{4.70}$$

which are defined in (4.55, 56). The solutions are given as Problem 4.7.

Problem 4.7. Find the eigenvalues of the Heisenberg Hamiltonian for spins one-half, (4.69). Use the Pauli principle, and space inversion symmetry considerations (on the true Pauli wavefunction of space and spin) to classify the solutions of this exchange Hamiltonian according to L_{tot} as well as S_{tot}. Find the ratio $[E(^2P) - E(^2D)]/[E(^2D) - E(^4S)]$ and verify that it equals $\frac{2}{3}$ when $J_2 = 2J_1$. Hint: define $\boldsymbol{T} = \boldsymbol{S}_{-1} + \boldsymbol{S}_1$, express $\mathscr{H}_{\text{Heis}}$ in terms of $\boldsymbol{T}$ and $\boldsymbol{T} + \boldsymbol{S}_0$.

4.8 p^2 and p^4 Configurations

Carbon, silicon, germanium, tin all have two electrons in the p shell. Oxygen, sulphur, selenium, and others have two holes in the p shell. All have ^3P ground states. (The term level scheme for carbon, which is representative, is given in Fig. 4.4.) Can we show this theoretically, and find the low-lying terms as well?

This is an exercise in Hund's second rule, which has not come into play before. Also, here is a problem in which the Heisenberg Hamiltonian is of dubious validity since there are only *two* particles, with *three* orbital states in which to put them. This configuration is studied in lieu of the complex transition series atoms, or rare-earth series, which are of greatest interest in magnetism. For it is not intended to take the reader into involved computations, regardless of their relevancy, but only to provide a conceptual introduction to the theory of atomic structure.

It is essential to reduce the number of configurations that one has to consider to a manageable few. In the $L-S$ coupling scheme (so-called), states which can be obtained from each other by repeated applications of $S^{\pm}_{\text{tot}}$, or $L^{\pm}_{\text{tot}}$, are equivalent except for the splitting due to spin-orbit coupling, (which is usually treated separately). So, it is only necessary to consider the largest attainable value of M_S and M_L for each term, viz.,

$$M_S = S_{\text{tot}} \quad \text{and} \quad M_L = L_{\text{tot}}, \tag{4.71}$$

and in order to determine S_{tot} and L_{tot} and the associated energy eigenvalues, to remain in the subspace of $M_S = 0, 1$ and $M_L = 0, 1, 2$.

$M_L = 2$: The space function must be

$$\varphi_1(\boldsymbol{r}_1)\varphi_1(\boldsymbol{r}_2) \tag{4.72}$$

in which $M_L = m_{l1} + m_{l2} = 1 + 1 = 2$. As this is symmetric under interchange of the spatial coordinates, the Pauli principle requires an antisymmetric spin function, the singlet state

$$\frac{1}{\sqrt{2}}[\chi_+(\xi_1)\chi_-(\xi_2) - \chi_+(\xi_2)\chi_-(\xi_1)] . \tag{4.73}$$

Thus, $L_{\text{tot}} = 2$ and $S_{\text{tot}} = 0$, and this is the ^1D state. The energy will be computed subsequently, but now we continue the classification.

$M_L = 1$: The space function of interest must be

$$\frac{1}{\sqrt{2}}[\varphi_1(\boldsymbol{r}_1)\varphi_0(\boldsymbol{r}_2) - \varphi_1(\boldsymbol{r}_2)\varphi_0(\boldsymbol{r}_1)] \tag{4.74}$$

since the other possibility, the linear combination with + sign, is symmetric under the interchange of the spatial coordinates and is therefore the $M_L = 1$ projection of the $L_{\text{tot}} = 2$ state derived above. If the space part is antisymmetric, the spin part must be symmetric [the spin-triplet $\chi_+(\xi_1)\chi_+(\xi_2)$, or $\chi_-(\xi_1)\chi_-(\xi_2)$, or (4.73) with + replacing −]; thus $L_{\text{tot}} = 1$ and $S_{\text{tot}} = 1$, and this is the ^3P configuration.

$M_L = 0$: The three functions with $M_L = 0$ have $m_{l1} = m_{l2} = 0$, or $m_{l1} = -m_{l2} = \pm 1$. Two linear combinations of these must be the $M_L = 0$ projections of the two states found above, (4.72, 74). The linear combination orthogonal to both of these is the new function of interest,

$$\frac{1}{\sqrt{3}}[\varphi_0(\boldsymbol{r}_1)\varphi_0(\boldsymbol{r}_2) - \varphi_{-1}(\boldsymbol{r}_1)\varphi_1(\boldsymbol{r}_2) - \varphi_{-1}(\boldsymbol{r}_2)\varphi_1(\boldsymbol{r}_1)] . \tag{4.75}$$

$L^{\pm}_{\text{tot}}$ applied to this state yields identically zero, and it follows that this is the ^1S configuration, and the list is now complete (see Table 4.3).

The perturbation is

$$\mathscr{H}' = \frac{e^2}{r_{12}} \tag{4.76}$$

which we expand in Legendre polynomials,

$$\frac{1}{r_{12}} = \sum_{n=0}^{\infty} \frac{\boldsymbol{r}_<^n}{\boldsymbol{r}_>^{n+1}} P_n(\cos\omega) \qquad \text{where} \qquad \cos\omega = \frac{\boldsymbol{r}_1 \cdot \boldsymbol{r}_2}{r_1 r_2} \tag{4.77}$$

in which $\boldsymbol{r}_<$ is the lesser, and $\boldsymbol{r}_>$ is the greater of $\boldsymbol{r}_1$ and $\boldsymbol{r}_2$, and ω is the angle between the two vectors. Because one takes expectation values using p functions, only terms with $n = 0$ and $n = 2$ can survive the integration.

The atomic functions have normalized angular factors (cf. Table 3.1):

$$Y_{1,0} = (3/4\pi)^{1/2} \cos\theta \tag{4.78}$$

and

$$Y_{1,\pm 1} = \mp(3/8\pi)^{1/2} \sin\theta e^{\pm i\varphi} \tag{4.79}$$

as well as a common (unknown) radial factor $R(r)$.

The term energies are the expectation values of the perturbation in the various states listed previously. The leading term in the expansion, correspondint to $n = 0$, is

$$\frac{1}{\boldsymbol{r}_>} \tag{4.80}$$

which does not depend on angles and is therefore the same in all three states. This can be eliminated by a simple shift in the zero of energy.

The next term in the expansion, $n = 1$,

$$\frac{\boldsymbol{r}_<}{\boldsymbol{r}_>^2} P_1(\cos\omega) \tag{4.81}$$

vanishes by inversion symmetry, but finally $n = 2$,

$$\frac{\boldsymbol{r}_<^2}{\boldsymbol{r}_>^3} P_2(\cos\omega) \tag{4.82}$$

contributes differently in all three states. The radial factor is common to all three. It is readily seen or it can be proved by the triangle inequality (of p. 87) that all succeeding terms in the expansion in Legendre polynomials will not contribute. So that apart from an additive constant arising from (4.80), all the energies will be proportional to the same, albeit unknown, radial integral. And it remains a straightforward exercise in manipulating trigonometric identities to show that

$$\frac{E(^1S) - E(^1D)}{E(^1D) - E(^3P)} = \frac{3}{2}\,. \tag{4.83}$$

This relationship is of course independent of the additive energy shift from (4.80), or of the common radial integral. Its validity depends only on the validity of the Hartree-Fock procedure and on the validity of our neglect of states outside the p shell. In Table 4.2 we compare the theory with experiment.

Problem 4.8. Using the procedure indicated in the text, derive (4.83). Hint: no integrals need be evaluated to establish this result, but various integrals need to be compared. This can be done solely by use of the mathematical identity,

$$\int d\varphi_1 \int d\theta_1 \sin\theta_1 P_2(\cos\omega)\cos^2\theta_1 = -\int d\varphi_1 \int d\theta_1 \sin\theta_1 P_2(\cos\omega)\sin^2\theta_1 \,.$$

The agreement is *not* always as perfect as for the atoms in Table 4.2. For example, the failure of one or another of the hypotheses gives for La^+ the ratio 18.43, which differs from theory by an order of magnitude. In such instances more terms must be taken into account, and while such calculations yield agreement with experiment they may become very involved. But at present the ultimate validity of the manybody Schrödinger equation for the atoms is not in dispute, and even the remarkably simple truncation, such as the organization into atomic shells, gives reasonably accurate results with minimum calculation.

Table 4.2. Configurations and ratio of term splittings in various atoms vs theory

		Experimental				
	Theory[a]	C	Si	Ge	Sn	O
Configuration	$np^{3\pm1}$	$2p^2$	$3p^2$	$4p^2$	$5p^2$	$2p^4$
Ratio	1.50	1.13	1.48	1.50	1.39	1.14

[a] See (4.83)
Source: *Condon* and *Shortley* [4.10]

Thus, atomic magnetism which is the consequence of Hund's rule, is documented by experiment, explained by elementary theory, and confirmed by elaborate calculations.

In Fig. 4.4, we show the spectrum of carbon. It is to be noted that as in N, the terms *not* described by the simple theory lie several volts above the ^1S term. As a consequence, theories of neutral carbon in solids probably need take only the lowest three terms into account, which in the atom are adequately described by the sort of first-order perturbation theory just expounded.

Finally, in Table 4.3, there is given a list of possible terms (configurations) which can be constructed out of a specified number of equivalent electrons. The lowest, according to Hund's rule, is given in boldface.

Practically all known magnetic substances involve d- or f-shell electrons, the study of which would take us well beyond the scope of an introductory work. The interested reader will readily find specialized treatises on this subject, see, e.g. [4.12].

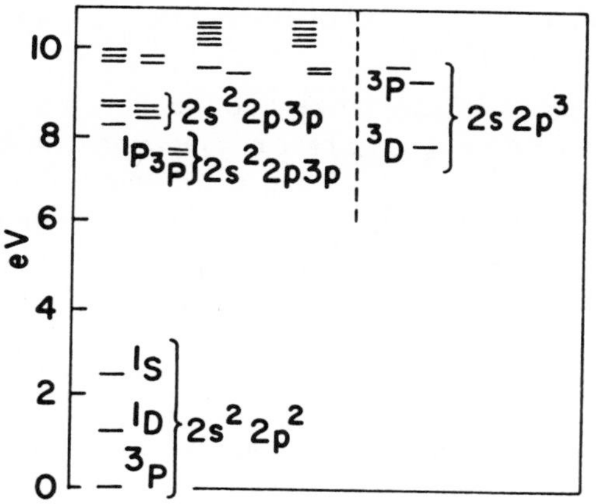

Fig. 4.4. Energy-level diagram of Carbon (C), as example of Hund's two rules

Table 4.3. Possible terms for a number of equivalent electrons

s		^2S						
s^2	^1S							
p or p^5		^{2}**P**						
p^2 or p^4	^1SD		^{3}**P**					
p^3		^2PD		^{4}**S**				
d or d^9		^{2}**D**						
d^2 or d^8	^1SDG		^3P**F**					
d^3 or d^7		^2PDFGH 2		^4P**F**				
d^4 or d^6	^1SDFGI 2 2 2		^3PDFGH 2 2		^{5}**D**			
d^5		$^2SPDFGHI$ 3 2 2		^4PDFG		^{6}**S**		
f or f^{13}		^{2}**F**						
f^2 or f^{12}	^1SDGI		^3PF**H**					
f^3 or f^{11}		$^2PDFGHIKL$ 2 2 2 2		^4SDFG**I**				
f^4 or f^{10}	$^1SDFGHIKLN$ 2 4 4 2 3 2		$^3PDFGHIKLM$ 3 2 4 3 4 2 2		^5SDFG**I**			
f^5 or f^9		$^2PDFGHIKLMNO$ 4 5 7 6 7 5 5 3 2		$^4SPDFGHIKLM$ 2 3 4 4 3 3 2		^6PF**H**		
f^6 or f^8	$^1SPDFGHIKLMNQ$ 4 6 4 8 4 7 3 4 2 2		$^3PDFGHIKLMNO$ 6 5 9 7 9 6 6 3 3		$^5SPDFGHIKL$ 3 2 3 2 2		^{7}**F**	
f^7		$^2SPDFGHIKLMNOQ$ 2 5 7 10 10 9 9 7 5 4 2		$^4SPDFGHIKLMN$ 2 2 6 5 7 5 5 3 3		^6PDFGHI		^{8}**S**

Note: A number under a term symbol indicates the number of different levels of this type. The lowest level is normally that with the highest L of the highest multiplicity (Hund's rule). It is indicated in bold type.

Source: [4.11].

4.9 Independent Electrons

It is possible to refine the study of atomic structure far beyond the short introduction provided in the preceding pages. Moreover, we have not yet approached the subject of magnetic elements. Usually magnetic substances are those composed of rare-earth or transition-series elements, with incomplete (magnetic) f shell or d shell as the case may be. Because magnetism is a phenomenon that occurs in the solid state, it is fitting however to interrupt the discussion of atomic properties, at the point where it begins to become complicated, and turn to the idealized model of electrons in the solid state: the independent particle model.

In this section we shall consider independent electrons in solids from a somewhat special point of view, ordering the ground state and elementary excited states according to their multiplicity, and proving some results which later will apply to a theorem about interacting electrons. What we shall now observe is that the degeneracies (which in the atom were lifted by interactions among the valence electrons, resulting in Hund's rules) will in solids be lifted in a different manner, without the need for any interactions; and that these *noninteracting* electrons will be *nonmagnetic,* except for second-order para- and diamagnetic effects which are discussed elsewhere.

Previously (see the section on overlap exchange) we had proved that interacting electrons are not ferromagnetic without overlap; here we shall show that overlapping electrons are not ferromagnetic if they do not correlate.

The Hamiltonian for N independent electrons (for simplicity assume $N = \text{even}$) is of the form

$$\mathscr{H} = \sum_{i=1}^{N} h(\boldsymbol{r}_i) \, . \tag{4.84}$$

It is not useful to specify further the one-particle Hamiltonian $h(\boldsymbol{r})$, which may include interactions with other electrons on the average, with the periodic array of atoms comprising the solid, with the surfaces and other external forces or impurities. In addition to these forces and the kinetic energy of the particle, we may allow any other potentials into $h(\boldsymbol{r})$, except those which involve the spin of the particle explicitly, and still arrive at the desired results. First, solve the eigenvalue equation,

$$h(\boldsymbol{r})f_n(\boldsymbol{r}) = e_n f_n(\boldsymbol{r}) \tag{4.85}$$

for the one-particle wavefunctions and eigenvalues f_n and e_n, arranging the eigenvalues in ascending order:

$$e_1 \leq e_2 \leq \cdots \tag{4.86}$$

etc. [It should be noted that the density of levels increases with the volume of the box; for a finite but arbitrarily large box, the levels can always be enumerated as in (4.86).] The many-electron wavefunctions are Slater determinants of space and spin functions. By the completeness of the set of f_n, the Slater determinants form a complete set of admissible Pauli wavefunctions for the many-electron system. However, we recall that they are not automatically eigenfunctions of S^2_{tot}, although they are always good eigenfunctions of S^z_{tot} with eigenvalue M. Recall also the results of (4.30) and those following: the spin function may be projected out of the Pauli function of definite M, leaving just the space partner

$$^{M}F(\boldsymbol{r}_1, \dots, \boldsymbol{r}_{1/2N+M} | \dots, \boldsymbol{r}_N) \tag{4.87}$$

which for independent particles is the product of two determinants. Because the one-electron functions may occur once in each determinant, the lowest energy belongs to a function of this type where the lowest possible e_n's are represented twice

$$^{0}F = \frac{1}{\left(\frac{1}{2}N\right)!} \det\|f_1(\boldsymbol{r}_1) \cdots f_{1/2N}(\boldsymbol{r}_{1/2N})\| \det\|f_1(\boldsymbol{r}_{1/2N+1}) \cdots f_{N1/1}(\boldsymbol{r}_N)\| \,. \tag{4.88}$$

This is automatically an eigenstate belonging to $M = 0$, with energy

$$E_0 = 2 \sum_{n=1}^{1/2N} e_n \,. \tag{4.89}$$

The factor of 2 comes ultimately from Kramers' degeneracy.

The function ^{0}F chosen in this manner has also the important property that it is an eigenfunction of S^2_{tot} with *zero* eigenvalue (compare p. 105). Therefore, it should be labeled $^{0}F^{0}$ according to the notation previously introduced. It is the absolute ground state, unless there is degeneracy, in which case it is among the ground states.

If next we inquire about the ground state in the $M = 1$ subspace, we must first eliminate $^{0}F^{0}$ which cannot be raised. But by reasoning similar to the above, the optimum second choice is found to be

$$\begin{aligned} ^{1}F = {} & \frac{1}{\sqrt{\left(\frac{1}{2}N + 1\right)!\left(\frac{1}{2}N - 1\right)!}} \det\|f_1(\boldsymbol{r}_1) \cdots f_{1/2N+1}(\boldsymbol{r}_{1/2N+1})\| \\ & \times \det\|f_1(\boldsymbol{r}_{1/2N+2}) \cdots f_{1/2N-1}(\boldsymbol{r}_N)\| \,. \end{aligned} \tag{4.90}$$

The energy of this state is given in (4.91). Now it is trivial to construct functions of the type ^{0}F and ^{-1}F starting with this new function; but the functions of type $^{\pm 2}F$, which we might attempt to construct with F^{1} and its permutations, vanish identically. This proves in fact that we have found ${}^{1}F^{1}$, an eigenfunction of S^2_{tot} with the indicated eigenvalue. The energy E_1,

$$E_1 = E_0 + e_{1/2N+1} - e_{1/2N} \geq E_0 \tag{4.91}$$

is the absolute ground-state energy in the subspace of eigenfunctions belonging to $S_{\text{tot}} \geq 1$, although it is greater than E_0.

One proceeds systematically through all the higher M subspaces, establishing always that the function of lowest energy of type ${}^{M}F$ is in fact ${}^{M}F^{M}$; and the concomitant result for the energy can be summarized most compactly as

$$E_0(S) \leq E_0(S+1) \tag{4.92}$$

defining $E_0(S)$ to be the lowest among all the energy eigenvalues $E(S)$ of the determinantal functions or linear combinations thereof, ${}^{M}F^{S}$.

We next examine the possible degeneracies. If $h(\boldsymbol{r})$ is real and one of the f_n is complex, then by complex conjugation of both sides of (4.85) we see that f_n^* is also an eigenfunction, with the same eigenvalue e_n. This is the only degeneracy possible in all generality. For consider an empty box, with dimensions L_x, L_y and L_z and a Hamiltonian

$$h(\boldsymbol{r}) = -\frac{\hbar^2}{2m}\nabla^2 \tag{4.93}$$

(kinetic energy only) which has eigenfunctions $e^{(i\boldsymbol{k}\cdot\boldsymbol{r})}$ and eigenvalues

$$e_{\boldsymbol{k}} = \frac{\hbar^2 \boldsymbol{k}^2}{2m} \qquad \boldsymbol{k} = 2\pi\left(\frac{n_1}{L_x}, \frac{n_2}{L_y}, \frac{n_3}{L_z}\right). \tag{4.94}$$

We see that if $L_x \neq L_y \neq L_z$ there is in fact no additional degeneracy; so that in the presence of potentials any added degeneracy would be "accidental" and could be eliminated by distorting the box a little bit, or by some other artifice.

For convenience N is now chosen to be a multiple of 4, which guarantees E_0 to be nondegenerate; for

$$e_{1/2N+1} \underset{\neq}{>} e_{1/2N} \qquad \text{and therefore} \qquad E_1 \underset{\neq}{>} E_0\,. \tag{4.95}$$

What is more, the eigenfunction ${}^{0}F^{0}$ of (4.88) is real; for it is both possible and necessary in the ground state that for every complex f_n the conjugate f_n^* also be present in each determinant. And even more compelling, is the argument that a complex ${}^{0}F^{0}$ would lead to a degenerate ground state, in violation of (4.95) above.

The total current operator is

$$\frac{e\hbar}{im}\sum_{j=1}^{N}\nabla_j = I_{op} \tag{4.96}$$

and its ground-state eigenvalue I is given by

$$I_{op}\,{}^0F^0 = I^0F^0\,. \tag{4.97}$$

Complex conjugation of both sides of this equation, and the reality of the eigenvalue I and of the eigenfunction ${}^0F^0$ establish the result $I^0 = 0$, regardless of the potential. Thus the current and the spin both vanish in the ground state, and the system is entirely nonmagnetic. This is the quantal generalization of the Bohr-Van Leeuwen theorem.

4.10 Electrons in One Dimension: A Theorem

> ... no one could move to the right or left to make way for passers-by, it followed that no Linelander could ever pass another. Once neighbors, always neighbors. Neighborhood with them was like marriage with us. Neighbors remained neighbors till death did them part [4.13].

Because electrons in one dimension can be ordered like beads on a string, the topology of their nodal surfaces turns out relatively simply. One can generalize the result found for two electrons, and prove the Lieb-Mattis theorem in one dimension, [4.5]: *In one dimension, not even interacting and overlapping electrons can be ferromagnetic.* A somewhat milder extension of this theorem, in which the type of interactions is restricted, can also be proved for N electrons in three dimensions (as we shall see in a subsequent section).

Before stating the theorem and proceeding to the proof, it is appropriate to comment on its relevance. This theorem occupies a special place as the guardian of our conscience when we speculate concerning the origins of ferromagnetism. Any fundamental explanation or theory, which conceivably would lead to ferromagnetism in one dimension, must *ipso facto* be false! And that is the reason we have discussed angular momentum, atomic theory, and the origins of Hund's rules, for these are intimately connected with the three-dimensionality of space, and, by virtue of the present theorem, are an essential part of the phenomena. Deferring further such considerations, we now go on with the

---

Theorem: For interacting electrons in one dimension, the ground state in any M subspace is nondegenerate, and belongs to $S_{tot} = |M|$. From this it follows that $E_0(S) < E_0(S+1)$, that the current vanishes in the ground state of any M subspace, and that electrons in one dimension are entirely nonferromagnetic.

Proof: What we have shown for noninteracting particles in the previous section will be adapted to the present case of interacting electrons, for which the Hamiltonian is

$$\mathscr{H} = -\frac{\hbar^2}{2m}\sum_{j=1}^{N}\frac{\partial^2}{\partial x_j^2} + V(x_1, x_2, \ldots, x_N)\,. \tag{4.98}$$

The potential V includes all external forces, periodic or other, and the interactions among electrons which may be taken to be arbitrary. The only exclusions are as follows: no velocity or spin-dependent forces or nonlocal potentials will be admitted, and V must be integrable, for reasons which are or will become clear. We shall adopt a standard boundary condition: If the size of Lineland is D, the boundary condition will be that the wavefunctions vanish whenever any x_n $\pm\frac{1}{2}D$. Admissible Pauli eigenfunctions are

$$^{M}\psi^{S}(1, 2, \ldots, N) \tag{4.99}$$

where, for compactness, $n = 1, 2, \ldots$ stands for the couple of space and spin coordinates (x_n, ξ_n). First, we write the eigenvalue equation,

$$\mathscr{H}\,^{M}\psi^{S} = E(S)\,^{M}\psi^{S} \tag{4.100}$$

noting that the energy cannot depend on M. (Because $\mathscr{H}$ commutes with the spin operators, we may apply spin raising and lowering operators to both sides of Schrodinger's equation, changing M without affecting the energy eigenvalue).

As usual, we now eliminate the spins, and consider the space functions $^{M}f^{S}$, which themselves obey Schrödinger's equation (4.100), with precisely the eigenvalue $E(S)$. All of them can be brought into the $M = 0$ subspace, by application of spin raising or lowering operators if necessary, and therefore we start by examining this particular subspace in which the ground-state eigenfunction is surely to be found. Consider an arbitrary eigenfunction, of unknown S_{tot},

$$^{0}f(x_1, \ldots, x_{1/2N} | x_{1/2N+1}, \ldots, x_N) \tag{4.101}$$

in the region R defined as follows:

$$R \equiv \begin{cases} -\frac{1}{2}D < x_1 < \cdots < \frac{1}{2}D \\ -\frac{1}{2}D < x_{1/2N+1} < \cdots < x_N < \frac{1}{2}D\,. \end{cases} \tag{4.102}$$

The natural boundaries of R occur where the inequalities are just barely violated, e.g.,

$$x_1 = -\frac{1}{2}D \qquad \text{or} \qquad x_n = x_{n\pm 1} \qquad \text{or} \qquad x_{1/2N} = \frac{1}{2}D\,, \tag{4.103}$$

etc., and it should be noted that ${}^0f$ vanishes identically on these boundaries. Although the total configuration space consists of the variables $x_1, \ldots, x_N$ in any ordering, *when ${}^0f$ is known in R it is known everywhere*, because the other regions can be found by trivial permutations $\mathscr{P}$ of the two sets of coordinates. (Recall the definition of *trivial permutations* given on p. 103) For example, if

$$(x_1, \ldots) \qquad \text{is a point in } R \tag{4.104}$$

and

$$(x_1', \ldots) \qquad \text{is a point in } R' \tag{4.105}$$

and $\mathscr{P}$ is a permutation which connects R to R',

$$\mathscr{P}R = R' \tag{4.106}$$

then we obtain ${}^0f$ in R' by the trivial permutation

$${}^0f(x_1', \ldots) \equiv (-1)^P \mathscr{P}[{}^0f(x_1, \ldots)] \,. \tag{4.107}$$

And in this self-evident manner, we can find the function everywhere. Indeed, with the sign convention used above, it is guaranteed that ${}^0f$ will have the proper antisymmetries under its trivial permutations $\mathscr{P}$. The nontrivial permutations, it is seen, are not needed to map R into the entire space; and thus they play only their usual role, which is raising and lowering M.

Now, we state and prove a *lemma* which is at the core of the theorem: The eigenfunction of $\mathscr{H}$ which belongs to the lowest energy eigenvalue and is subject only to the boundary condition that it vanish on the boundaries of R is *nodeless* in R.

The proof proceeds exactly as in the case of two electrons. If the ground-state function had nodes, then its absolute value would be a nodeless function, which would have identically the same variational energy but which would not itself be an eigenfunction. Since the variational energy of any but a ground-state eigenfunction must exceed the ground-state energy, there is a contradiction unless the absolute value of the ground-state eigenfunction is itself a ground-state eigenfunction; which is only possible if the ground state is nodeless. The nodeless state is, moreover, nondegenerate. For two nodeless functions cannot be orthogonal, but all the eigenfunctions of $\mathscr{H}$ can be made mutually orthogonal, and other than nodeless functions have been excluded by the preceding arguments. This nondegeneracy is stronger than we were able to prove for noninteracting particles (in the previous section) because we have restricted the arguments here to one dimension.

The nodeless property is independent of the magnitude of V, and so is the shape of R. We may therefore compare in R the ground state of a Hamiltonian $\mathscr{H}$ with arbitrary V, to that of the noninteracting Hamiltonian $\mathscr{H}_0$ with $V = 0$.

In both cases, the ground state is nodeless; the two corresponding functions cannot be orthogonal, therefore they share the same quantum numbers I and S_{tot}.

What these quantum numbers are for noninteracting electrons ($V = 0$) is something which was already discussed: viz., *zero*. And by the above it follows that both these quantum numbers vanish for arbitrary V as well, in the ground state.

Next, one introduces the fundamental region suitable for $M = 1$, which is

$$R \equiv \begin{cases} -\frac{1}{2}D < x_1 < \cdots < x_{1/2N+1} < \frac{1}{2}D \\ -\frac{1}{2}D < x_{1/2N+2} < \cdots < x_N < \frac{1}{2}D\,. \end{cases} \tag{4.108}$$

By analogy, one proves that the ground-state function in this region is nodeless and belongs to the same quantum numbers regardless of V. It follows by comparison with the results for $V = 0$ that this function has $S_{\text{tot}} = 1$; the proof for higher M proceeding in identically the same manner, until maximum S is attained and the theorem for the interacting particles is proved.

4.11 The Wronskian [4.14]

A somewhat different approach may serve to clarify the results obtained so far. Suppose we have a set of real space eigenfunctions of the Schrödinger equation

$$\mathscr{H} f_j(x_1, \dots) = E_j f_j(x_1, \dots) \tag{4.109}$$

with

$$\mathscr{H} = -\frac{\hbar^2}{2m}\sum \frac{\partial^2}{\partial x_n^2} + V(x_1, \dots)$$

just as in (4.98), and it is desired to order the eigenvalues E_j in a sequence of increasing energy.

The lowest energy belongs to the nodeless eigenfunction; since the permutation operators commute with the Hamiltonian, it can also be established that this function is totally symmetric under the interchange of any coordinates and therefore is not of any use for more than two Fermions. But there can only be one nodeless function (compare proof of analogous statements on p. 109), and the others can be examined as to the topology of their nodal surfaces. The nodal surface is defined as the locus of $f = 0$.

We shall not go into any detail beyond proving the simple but interesting *lemma*:

If there are two functions f_i and f_j such that a nodal surface S_i of f_i, encloses a region R_i which neither contains, nor is intersected by any nodal surface of f_j, *then* $E_i > E_j$. For example, consider any eigenfunction f_i together with the nodeless eigenfunction f_0; the lemma implies $E_i > E_0$, which is obviously true for all i.

To prove the lemma, it is necessary to multiply both sides of (4.109) on the left by f_i; to multiply the analogous equation for f_i on the left by f_j, substract the first from the second, and obtain

$$-\frac{\hbar^2}{2m}\nabla\cdot(f_j\nabla f_i - f_i\nabla f_j) = (E_i - E_j)f_if_j\,. \tag{4.110}$$

The potential has been eliminated from this expression, and the left-hand side involves only the kinetic energies or the gradient of the Wronskian, for which we have used the generalized Laplacian notation

$$\nabla\cdot\nabla = \nabla^2 \equiv \sum_{n=1}^{N}\frac{\partial^2}{\partial x_n^2}\,. \tag{4.111}$$

By assumption, neither function changes sign in R_i so we may, in considering this region, assume both functions are positive, and if not, make them so. Next, integrate over R_i, which by Green's theorem gives an integral of the Wronskian over S_i on the left-hand side, and a positive integral times the energy difference on the right.

$$-\frac{\hbar^2}{2m}\int_{S_i} dS\cdot(f_j\nabla f_i - f_i\nabla f_j) = (E_i - E_j)\int_{R_i} d\tau f_if_j \tag{4.112}$$

But $f_i = 0$ on the nodal surface S_i; and the normal derivative of f_i is negative on S_i because the function is positive inside the volume bounded by the surface and negative outside (see Fig. 4.1). So the left-hand side is positive, and the equation can only be satisfied if $E_i > E_j$, which proves the lemma.

Evidently, we have used nothing about dimensionality in this demonstration, therefore it is indeed equally valid for electrons in three dimensions [with $\nabla^2 \equiv \sum_{u=1}^{N}\nabla_n^2$, instead of (4.111)]. But the difficulty of finding nodal surfaces has so far frustrated attempts to apply these results to realistic problems for $N > 2$ in three dimensions.

4.12 Theorem in Three Dimensions

The great stumbling blocks to proving our theorem in three dimensions are the impossibility of ordering the particles in any sensible manner and the difficulty in locating the nodal surfaces along which the wavefunctions vanish. One may get around both these if he restricts the potential to the form

$$V(x_1, \ldots, x_N; y_1, \ldots, y_N; z_1, \ldots, z_N) \tag{4.113}$$

which is to say, "separately symmetric" under permutations of the x, y, or z coordinates alone. The periodic potential typical of the solid state may be included, as well as electron-electron interactions. For potentials of this class, we may prove the theorem in the same form as for noninteracting electrons

$$E_0(S) \leq E_0(S+1) \tag{4.114}$$

so that *for ferromagnetism to occur*, even in three dimensions, *it does not suffice for electrons to overlap and to interact, no matter how strong the potential.* The *form* of the potential shall play a peculiar role.

In the present section, we shall prove the above inequality for a special case of the potential V, viz.,

$$V = V(x_1, \dots, x_N) + V'(y_1, \dots, y_N) + V''(z_1, \dots, z_N) \tag{4.115}$$

the separable potential. Primes are used to distinguish the functions. Extension to the general class of potentials, (4.113), is straightforward.

When the kinetic energy operators are included, the Hamiltonian with a separable potential can be written as the sum of three Hamiltonians, one for each dimension. The space eigenfunctions are therefore product functions:

$$\psi = f(x_1, \dots)f'(y_1, \dots)f''(z_1, \dots) \tag{4.116}$$

and the energy eigenvalues are additive:

$$E = E + E' + E''\,. \tag{4.117}$$

We should now introduce the spin functions, minimize the sum of the energies E, E', and E'' subject to the Pauli principle, and finally try to discover the spin of the lowest allowable eigenfunction. The $f(x_1, \dots)$ function obeys a one-dimensional Schrödinger equation which is invariant under permutations of the x coordinates. Similar statements are true for f' and f''. Therefore we may choose f, f', and f'' to be simultaneous eigenfunctions of the permutation operators and write

$$f = f(x_1, \dots, x_p | x_{p+1}, \dots | x_{r+1}, \dots | \dots | \dots, x_N) \tag{4.118}$$

and similar expressions for f' and f''. The function is antisymmetric under the interchange of coordinates separated by commas; but has no particular symmetry properties under interchange of coordinates separated by one or more vertical bars.

Whereas the Pauli principle restricts the allowable functions in one dimensions to those with either zero or one vertical bar, in the case of electrons obeying the Pauli principle in two or more dimensions we must allow for *any number* of bars, and therein lies part of the difficulty.

Example: Consider a totally antisymmetric space function, F, which belongs to the totally symmetric (totally ferromagnetic) spin function; it can be of the form

$$F = f(x_1, \ldots, x_N) f'(y_1 | y_2 | \ldots | y_N) f''(z_1 | z_2 | \ldots | z_N) \tag{4.119}$$

where f' and f'' are both totally symmetric, and therefore have $N - 1$ vertical bars, and f is totally antisymmetric. Or all three functions could be totally antisymmetric; and F would be of the form

$$F = f(x_1, \ldots, x_N) f'(y_1, \ldots, y_N) f''(z_1, \ldots, z_N) \tag{4.120}$$

with no vertical bars at all. If we require F to have one bar, or if we allow linear combinations of products of f's, then the number of possibilities of bars for the f's becomes virtually endless.

Proof of Theorem in (4.114): Given an eigenfunction of $\mathscr{H}$ and S^2_{tot}, which is also a Pauli function, i.e., a totally antisymmetric function of the coordinates $\boldsymbol{r}_i = (x_i, y_i, z_i, m_i)$ (let $m_i = \uparrow$ or $\downarrow$ denote the spin eigenvalue), express it as a linear combination of product functions of the type

$$f(\boldsymbol{u}_1, \ldots, \boldsymbol{u}_p | \boldsymbol{u}_{p+1}, \ldots | \ldots | \ldots, \boldsymbol{u}_N) g(\boldsymbol{v}_1; \ldots; \boldsymbol{v}_p | \ldots | \ldots; \boldsymbol{v}_N) \tag{4.121}$$

where $\boldsymbol{u}_i = (x_i, m_i)$ and $\boldsymbol{v}_i = (y_i, z_i)$. In order that a totally antisymmetric Pauli function might be extractable from (4.121), it is required that g be symmetric in the variables separated by semicolons if f is antisymmetric in the variables separated by commas. The function above has the same energy as the original Pauli eigenfunction. This is because the latter is a linear combination of (4.121) and its permutations, and of course $\mathscr{H}$ is invariant under the permutations, and therefore the energy is unaffected.

A simplified proof of the theorem proceeds as follows: for all practical purposes, f is a Pauli function for p one-dimensional electrons $x_1, \ldots, x_p$, and separately also for $x_{p+1}, \ldots$, etc.; and if we take linear combinations of f-type functions so as to give these groups of variables definite spin, it is clear by virtue of the one-dimensional theorem that the lowest energy is achieved only if the total spin of the first p electrons is zero (or one-half, if the number is odd), and similarly for all the successive groups separated by vertical bars. Such linear combinations are of course once more degenerate with the original function, (4.121), as well as with the original Pauli function from which the latter was extracted. Notice that thc process by which the sets of electrons are given definite spin does not affect the antisymmetry, so that we may, without loss of generality, assume that in f the various sets of variables have definite spin.

Finally, a Pauli function is constructed from (4.121) in two steps: first, an eigenfunction of S^2_{tot} is constructed with Clebsch-Gordan coefficients; second, the function so obtained is totally antisymmetrized. The sets with zero spin do not contribute to the total spin; thus if all the sets contain an even number of particles, only $S_{tot} = 0$ is accessible. If N is even, as we usually assume, we may have none, two, four, six, ... sets with odd numbers of variables, for which the lowest energy is achieved with spin one-half. Two spins one-half may be combined to give spin zero, or spin one. Four may be combined to yield $S_{tot} = 0$, 1, 2, etc. So long as we stick to the functions with the lowest energy in each symmetry class, we see that if a function belonging to total spin S can be constructed, then so can another function which has the same energy but total spin $S - 1$, or less. Because the same cannot be said about $S + 1$, one immediately arrives at a statement of the theorem in the form of the inequality, (4.114).[3]

4.13 Ordering Theorem versus Hund's Rule

The ordering theorem, as we proved it in three dimensions, differs from its one-dimensional version by not being stated in terms of strict inequalities. And indeed the ground state for the interacting three-dimensional electrons could belong to some nonzero value of the total spin, say S, but then necessarily also to $S - 1$, $S - 2$, ..., 1, and 0. The maximum value of S in the ground state may be estimated, on the basis of noninteracting electrons in three dimensions, not to exceed the *order of magnitude* $S = \frac{1}{2}N^{1/2}$, regardless of how many electrons are present. Since ferromagnetism in the solid state requires that

$$\lim_{N\to\infty} \frac{S}{N} \neq 0 \tag{4.122}$$

we see that solids with separately symmetric potentials are never ferromagnetic. Such is not the case for a hypothetical atom with a separately symmetric potential. The atom is typically a collection of less than 100 electrons, and a total spin of the order of $S = 5$ is not negligible, and indeed is of the order of magnitude observed in atoms. Moreover, charges in nature interact by Coulomb potentials which are not "separately symmetric". The difference between the Coulomb interaction and the closest separately symmetric analog to it may be considered as a perturbation, capable of lifting the degeneracy amongst ground states belonging to S, $S - 1$, ..., etc. In the partly filled shells of atoms and molecules, the state satisfying Hund's rules does not merely lie among the ground states, but is found to be the lowest state of all. Simple calculations explaining this feature have been presented in earlier sections of this chapter.

Isolated atoms and molecules can therefore possess a small "permanent" magnetic moment, one which requires some additional energy to destroy. What happens to these individual moments when atoms are aggregated to form a solid, is a question to be addressed in the following chapters. Before pursuing

[3] For much more detailed and constructive proof including the extension to the general separately symmetric potentials, (4.113), see [4.5].

this line of inquiry, we shall find it advantageous to review some fairly general and useful techniques in the study of large numbers of identical particles.

4.14 Second Quantization

The essential information contained in the determinantal function (4.14), is that 3 particles satisfying the Pauli principle occopy (j, r) [using short-hand notation for the spatial state $f_j(r)$ and spin state $r = \pm\frac{1}{2}$], (k, s) and (n, t). The set of 3 pairs of quantum numbers summarizes this information, but does not explicitly reveal the antisymmetry of the wavefunction. If however, we introduced 3 operators c^*_{jr}, c^*_{ks} and c^*_{nt} with the following property

$$c^*_a c^*_b = -c^*_b c^*_a$$

where b, a are any of the three sets of quantum numbers above, then $c^*_{jr}\, c^*_{ks}\, c^*_{nt}$ *is* an operator carrying the requisite information and which changes sign upon interchange of any pair of particles. It is on this observation that the formalism of second quantization is based.

Second quantization allows one to deal with variable numbers of particles as conveniently as with a fixed number. The "particles" can be electrons as stated above, or any other variety of interest such as lattice vibrations (phonons), and spin excitations (magnons) with differing symmetry properties. We shall first analyze the case of electrons, then discuss examples not subject to the Pauli principle. It is convenient to deal with noninteracting particles at first, and then to express the interactions in the new language.

The unique state of zero particles is denoted the "vacuum" and written,

$$|0\rangle \qquad \text{(with} \quad \langle 0| \quad \text{the conjugate state).} \tag{4.123}$$

To introduce an electron of spin s at the spatial point $\boldsymbol{r}$ one uses the wave-operator $\Psi^*_s(\boldsymbol{r})$ as follows:

$$\Psi^*_s(\boldsymbol{r})|0\rangle, \text{ with } \langle 0|\Psi_s(\boldsymbol{r}) \text{ the conjugate state.} \tag{4.124}$$

The requirement that this state be normalized and orthogonal to that of a particle having a different spin index or at a different point of space is the following:

$$\langle 0|\Psi_{s'}(\boldsymbol{r}')\Psi^*_s(\boldsymbol{r})|0\rangle = \delta_{s,s'}\delta(\boldsymbol{r}-\boldsymbol{r}') \tag{4.125}$$

using the Kronecker delta for the discrete index and the Dirac delta for the continuous one. Equation (4.125) has two interpretations, the first as the inner product of a state (4.124) with a similar (but conjugate, such as column to row vectors) state; and the second as the inner product of the (conjugate) vacuum with the state

$$\Psi_{s'}(\mathbf{r}')\Psi_s^*(\mathbf{r})|0\rangle\,. \tag{4.126}$$

For $s = s'$ and $\mathbf{r} = \mathbf{r}'$ this must be proportional to the vacuum, thus (4.126) must not contain any particles. Hence the interpretation: successive applications of Ψ^* raise the number of particles in a state by 1, successive applications of Ψ lower it by 1. (For the conjugate states, the opposite is true.) As the vacuum has the lowest possible occupancy, we find

$$\Psi_s(\mathbf{r})|0\rangle \equiv 0 \qquad (\text{or} \quad \langle 0|\Psi_s^*(\mathbf{r}) = 0)\,. \tag{4.127}$$

The Pauli principle comes into play with 2 or more particles

$$\Psi_s^*(\mathbf{r})\Psi_{s'}^*(\mathbf{r}')|0\rangle\,. \tag{4.128}$$

It requires a change of sign under interchange of two particles, i.e.,

$$\Psi_{s'}^*(\mathbf{r}')\Psi_s^*(\mathbf{r})|0\rangle = -\Psi_s^*(\mathbf{r})\Psi_{s'}^*(\mathbf{r}')|0\rangle \tag{4.129}$$

and this must be true even if $|0\rangle$ is replaced by an arbitrary state. This requirement, and the normalization (4.125) can be conveniently combined into a compact set of *anti*commutation relations as follows:

$$\begin{aligned} \{\Psi_s(\mathbf{r}),\ \Psi_{s'}(\mathbf{r}')\} &= 0 = \{\Psi_s^*(\mathbf{r}),\ \Psi_{s'}^*(\mathbf{r}')\} \\ \{\Psi_s(\mathbf{r}),\ \Psi_{s'}^*(\mathbf{r}')\} &= \delta_{ss'}\delta(\mathbf{r}-\mathbf{r}') \end{aligned} \tag{4.130}$$

in which the anticommutator bracket $\{A, B\}$ is short-hand for $AB + BA$.

The linear combination of Ψ^* which corresponds to a given spatial wavefunction, e.g., to one of the states $f_j(\mathbf{r})$ used in the Slater determinantal wavefunction early in the chapter, is simply

$$c_{j,s}^* \equiv \int f_j(\mathbf{r})\Psi_s^*(\mathbf{r})\,d^3r \tag{4.131}$$

and thus the second-quantized equivalent of the three-particle wavefunction (4.14) is precisely

$$c_{jr}^* c_{ks}^* c_{nt}^*\ |0\rangle\,. \tag{4.132}$$

Problem 4.9. Prove that $\Psi_A^{\mathrm{Slater}}(\mathbf{r}_1, \mathbf{r}_2, \mathbf{r}_3)$ of (4.14) is given by:

$$\frac{1}{\sqrt{3!}}\sum_{s,s',s''}\langle 0|\Psi_s(\mathbf{r}_3)\Psi_{s'}(\mathbf{r}_2)\Psi_{s''}(\mathbf{r}_1)\,c_{jr}^* c_{ks}^* c_{nt}^*|0\rangle\,.$$

[Use (4.127, 130, 131)].

Problem 4.10. Prove the anticommutation relations

$$\{c_a, c_b\} = \{c_a^*, c_b^*\} = 0 \qquad \text{and} \qquad \{c_a, c_b^*\} = \delta_{a,b}$$

by use of (4.130, 131).

Given a state such as (4.128) or (4.132) with a certain number of particles, we can define the particle number operator $\mathfrak{n}$ as

$$\mathfrak{n} = \sum_{s=\pm 1/2} \int \Psi_s^*(\boldsymbol{r})\Psi_s(\boldsymbol{r})\, d^3r\,. \tag{4.133}$$

Given an Hamiltonian $\mathscr{H}$ for noninteracting particles with a complete set of one-particle eigenstates $f_j(\boldsymbol{r})$, we can always invert the type of relation (4.131) to obtain

$$\Psi_s(\boldsymbol{r}) = \sum_j f_j(\boldsymbol{r}) c_{js}\,. \tag{4.134}$$

Inserting this into (4.133) yields for the number operator

$$\mathfrak{n} = \sum_s \sum_j c_{js}^* c_{js} = \sum_s \sum_j n_{js}\,. \tag{4.135}$$

The occupation of the state (j, s) is given by the eigenvalue of the operator $c_{js}^* c_{js}$, often abbreviated n_{js}.

Like the number operator, the Hamiltonian of noninteracting particles is a quadratic form in field operators

$$\boldsymbol{H}_{\text{op}} \equiv \sum_s \int \Psi_s^*(\boldsymbol{r}) H(\boldsymbol{r}) \Psi_s(\boldsymbol{r})\, d^3r = \sum_s \sum_j \epsilon_j c_{js}^* c_{js}\,, \tag{4.136}$$

making use of (4.134) and $Hf_j(\boldsymbol{r}) = \epsilon_j f_j(\boldsymbol{r})$, the time-independent Schrödinger equation for the jth eigenfunction. The interpretation is fairly direct: the energy of the jth state is the product of the energy (ϵ_j) and the occupancy ($n_{j+} + n_{j-}$).

The simplicity ceases when two-body forces are introduced. Where in first-quantization they might be written

$$\mathscr{H}_2 = \sum_{i,j} V(\boldsymbol{r}_i, \boldsymbol{r}_j)\,, \tag{4.137}$$

they now appear as

$$\boldsymbol{H}_{2\text{op}} = \sum_s \sum_{s'} \iint \Psi_s^*(\boldsymbol{r}) \Psi_s(\boldsymbol{r}) V(\boldsymbol{r}, \boldsymbol{r}') \Psi_{s'}^*(\boldsymbol{r}') \Psi_{s'}(\boldsymbol{r}')\, d^3r\, d^3r'\,. \tag{4.138}$$

In terms of the raising and lowering operators of the unperturbed Hamiltonian (4.136), c_{js}^* and c_{js}, the interaction Hamiltonian takes the form

$$\boldsymbol{H}_{2\mathrm{op}} = \sum_{\substack{s,s' \\ j_1,j_2 \\ j_3,j_4}} K(j_1 j_2 j_3 j_4) c^*_{j_1 s} c_{j_2 s} c^*_{j_3 s'} c_{j_4 s'} \tag{4.139}$$

with

$$K(j_1, \ldots, j_4) = \int f^*_{j_1}(\boldsymbol{r}) f_{j_2}(\boldsymbol{r}) V(\boldsymbol{r}, \boldsymbol{r}') f^*_{j_3}(\boldsymbol{r}') f_{j_4}(\boldsymbol{r}') d^3r d^3r' \tag{4.140}$$

The first-quantized operators and wavefunctions refer to the explicit number of particles [e.g., the 3×3 determinant (4.14) for 3 particles], whereas the second-quantized operators are valid for all possible occupation numbers. The variational and perturbation-theoretic formalisms are equally valid in the new language as in the old; indeed, the pictorial Feyman diagrams are often used to describe the various terms in the perturbation-theoretic studies of $\boldsymbol{H}_{2\mathrm{op}}$. For example, (4.139) is represented pictorially as a vertex with four arrows.
For further details, we refer to any specialized text, see, e.g. [4.15].

A very important result, known as Wick's theorem [4.15], allows the computation of the matrix elements of $\boldsymbol{H}_{2\mathrm{op}}$ and higher powers of this operator, in terms of the simple contractions such as $\langle 0 | c_{j_1 s} c^*_{j_2 s} | 0 \rangle$, their products and sums, e.g., determinants. This theorem is available for fermion (anticommuting) operators of the type we have been discussing, for boson operators as we shall now discuss, *but not for spin operators* which, as previously studied in Chap. 3, have fairly complex commutation relations.

Particles subject to Bose-Einstein statistics include He_4 atoms (even numbers of fermions) as well as any system of particles for which the wavefunction remains symmetric under permutations of two or more of the particles. The main modification one makes to the relations developed above for fermions is the substitution of commutation relations for the anticommutators of (4.130). Thus,

$$\Phi(\boldsymbol{r})\Phi(\boldsymbol{r}') - \Phi(\boldsymbol{r}')\Phi(\boldsymbol{r}) \equiv [\Phi(\boldsymbol{r}), \Phi(\boldsymbol{r}')] = 0$$

and similarly,

$$[\Phi^*(\boldsymbol{r}), \Phi^*(\boldsymbol{r}')] = 0$$

together with

$$[\Phi(\boldsymbol{r}), \Phi^*(\boldsymbol{r}')] = \delta(\boldsymbol{r} - \boldsymbol{r}') \tag{4.141}$$

writing the wave-operator as Φ for the bosons, and omitting the spin index, if any. Among fields satisfying these commutation relations we can also count quantized lattice vibrations (phonons), photons, and quantized spin-waves (magnons). Concerning the latter we shall have much more to say in later sections of this text. Particle-conservation laws are far less stringent for bosons than for fermions, and it is not uncommon to have odd powers of Φ or Φ^* in the Hamiltonian in addition to the terms discussed above for the fermions.

The field operators, whether for fermions or bosons, are particularly useful in the construction of Green functions—functions of time, space and temperature which contain all the thermodynamic and transport-theoretic information relevant to a given Hamiltonian. We therefore return to this topic later in the book.

5. From Magnons to Solitons: Spin Dynamics

There exist striking similarities between the elementary excitations of a ferromagnet and those in an elastic solid. In the latter, we know that an atom displaced from equilibrium position will oscillate with the motion and frequencies of the *normal modes* of the crystal. The effect of quantum mechanics on this motion is to quantize the amplitudes of the individual normal modes, into units known as *phonons*. The analogous normal modes in the magnetic system are the spin waves; when the quantum mechanical nature of spins is taken into account, these also are quantized, with the basic unit being the *magnon*.

In the ferromagnet, for which the ground state has all spins parallel, the small amplitude oscillations are accurately described by harmonic-oscillator type dynamical variables. The interaction between neighboring oscillators is responsible for the dispersion, the dependence of frequency ω of a normal mode, on the wave vector $\boldsymbol{k}$ (the wavelength $\lambda = 2\pi/k$). Indeed, as the ground state of a ferromagnet is known exactly, we are able to extend the analysis to obtain the exact one- and two-magnon eigenstates. The latter number bound states and scattering states, the interactions being caused by an inherent nonlinearity in the spin dynamics, due to the coupling of the motion in various directions by the commutation relations elaborated in Chap. 3. The analysis of 3- or more-magnon states requires the solution of a full many-body problem.

In the *anti*ferromagnet (a solid in which neighboring spins tend to be antiparallel), the inherent nonlinearities even prevent us from obtaining the exact ground state or elementary excitations, with the exception of one dimension (and there, only for spins 1/2 or spins $s \to \infty$, the classical limit). Nevertheless, spin-wave theory is as useful for an approximate analysis, as are the phonons in the study of the *an*harmonic solid. The ground state and elementary excited states, and the ground state correlation functions, can all be computed with spin-wave theory. A check on the accuracy exists, in a variety of numerical experiments on representative systems.

There are instances where magnons yield only part of the answers, perhaps not even the qualitatively significant part. In two-dimensional magnets, the *vortex* has been found to be the most significant topological excitation, and in one-dimensional magnets, magnons and *solitons* are both required in the dynamics. Such relatively new developments are treated in this chapter which, with slight exceptions, is consecrated to the study of the Heisenberg Hamiltonian with nearest-neighbor interactions, i.e., to insulators.

5.1 Spin Waves as Harmonic Oscillators

The approximation of spin waves by harmonic-oscillator functions, as opposed to a description "based on creation and annihilation operators, is to a considerable extent only a semantic one, but nevertheless is probably of use to those readers to whom harmonic oscillators are more intuitive than the techniques of quantum-mechanical field theory."[1]

For concreteness, let us assume (for the present) a ferromagnetic Heisenberg Hamiltonian with nearest-neighbor interactions, in an external homogeneous magnetic field H

$$\mathscr{H} = -J \sum_{\substack{i>j= \\ \text{nearest neighb.}}}\sum \boldsymbol{S}_i \cdot \boldsymbol{S}_j - Hg\mu_{\mathrm{B}} \sum_i S_j^z \qquad J > 0 \tag{5.1}$$

where

$$\boldsymbol{S}_i \cdot \boldsymbol{S}_j \equiv S_i^x S_j^x + S_i^y S_j^y + S_i^z S_j^z = \frac{1}{2}(S_i^+ S_j^- + S_j^+ S_i^-) + S_i^z S_j^z \,.$$

Beside the isotropic interactions which are included therein, we may also consider anisotropic interactions of various origins. Those of lowest order are of "dipolar" structure,

$$\frac{1}{2}\sum_i \sum_j D_{ij} \frac{\boldsymbol{S}_i \cdot \boldsymbol{S}_j r_{ij}^2 - 3\boldsymbol{S}_i \cdot \boldsymbol{r}_{ij} \boldsymbol{S}_j \cdot \boldsymbol{r}_{ij}}{r_{ij}^2} \tag{5.2}$$

which represents the magnetic dipole-dipole interaction (the magnetostatic potential of each spin in the field of the others) with

$$D_{ij} = g^2 \mu_{\mathrm{B}}^2 r_{ij}^{-3} \,. \tag{5.3}$$

But there may be additional contributions to D_{ij}, principally those arising from the spin-orbit coupling mechanism [5.2]. These, like the exchange force itself, tend to be very short-ranged. For concreteness, let us limit the present discussion to cubic materials and the anisotropic interactions to (5.3) in lowest order.

In Chap. 3 on angular momentum, we have shown how rigorous representations of spins may be constructed with the aid of harmonic-oscillator operators; and in particular, the Holstein-Primakoff representation, and approximations to it, were introduced. Let us review the latter from a new point of view. First, the spins are defined by the three nonvanishing matrix elements:

$$\langle n_i | S_j^x | n_i + 1 \rangle = \langle n_i + 1 | S_i^x | n_i \rangle^* = \frac{1}{2}\sqrt{(n_i + 1)(2s - n_i)}$$

[1] The present section has been modeled on an excellent review of spin waves by *Van Kranendonk* and *Van Vleck* [5.1] from which this quotation is taken.

$$\langle n_i | S_i^y | n_i + 1 \rangle = \langle n_i + 1 | S_i^y | n_i \rangle^* = -\frac{1}{2} i \sqrt{(n_i + 1)(2s - n_i)} \tag{5.4}$$

$$\langle n_i | S_i^z | n_i \rangle = s - n_i, \qquad 0 \leqslant n_i \leqslant 2s, \qquad \hbar = 1 .$$

The quantum numbers of other spins (n_j) have been suppressed, because the components of $\boldsymbol{S}_i$ cannot change these; but all occupation numbers $n_1, n_2, \ldots,$ n_N must be specified in a definite state of the N spins, and therefore there are exactly $(2s + 1)^N$ such states if all spins have the same magnitude s.

Matrix elements of harmonic-oscillator operators have a structure very similar to the above. For example, if we symbolize the integrals

$$\int \psi_n^*(x_i)(\mathrm{op})\psi_m(x_i)\, dx_i$$

where $\psi_n(x)$ is the nth harmonic-oscillator function, by means of the Dirac bracket notation, $\langle n | (\mathrm{op}) | m \rangle$, then for the very important operators x_i coordinate and p_i momentum, we have

$$\begin{aligned} \langle n_i | x_i | n_i + 1 \rangle &= \langle n_i + 1 | x_i | n \rangle^* = \sqrt{\frac{\hbar}{2m\omega}(n_i + 1)} \\ \langle n_i | p_i | n_i + 1 \rangle &= \langle n_i + 1 | p_i | n_i \rangle^* = -\mathrm{i}\sqrt{\frac{\hbar m\omega}{2}(n_i + 1)} \end{aligned} \tag{5.5}$$

and the quantum number n_i may also be defined as the eigenvalue of an operator, now the energy operator

$$\frac{p_i^2}{2m} + \frac{1}{2}m\omega^2 x_i^2 - \frac{1}{2}\hbar\omega | n_i \rangle = n_i \hbar\omega | n_i \rangle . \tag{5.6}$$

The connection between harmonic oscillator and spin is now estabished, by use of the dimensionless canonical variables P, Q, defined by

$$Q_i = x_i \sqrt{\frac{m\omega}{\hbar}} \qquad P_i = p_i \sqrt{\frac{1}{\hbar m\omega}} \qquad [P_i, Q_j] = \delta_{ij}\frac{1}{\mathrm{i}} , \tag{5.7}$$

by comparison of (5.4) and (5.5).
In terms of the P, Q the matrix elements are

$$\begin{aligned} \langle n_i | Q_i \sqrt{s} | n_i + 1 \rangle &= \langle n_i + 1 | Q_i \sqrt{s} | n_i \rangle^* = \frac{1}{2}\sqrt{(n_i + 1)2s} \\ \langle n_i | P_i \sqrt{s} | n_i + 1 \rangle &= \langle n_i + 1 | P_i \sqrt{s} | n_i \rangle^* = -\frac{1}{2}\mathrm{i}\sqrt{(n_i + 1)2s} \\ \langle n_i | s - \frac{1}{2}(P_i^2 + Q_i^2 - 1) | n_i \rangle &= s - n_i , \end{aligned} \tag{5.8}$$

quite similar to (5.4) *for small values of the* n_i. Note that for values of $n_i \approx s$, the harmonic-oscillator approximation becomes quantitatively incorrect, although the matrix structure is still qualitatively similar to the correct one. The error however becomes catastrophic only when some n_i equals or exceeds the value $2s$, for although (5.8) continues to define matrix elements for the harmonic oscillators, there is no corresponding structure in angular-momentum space. So it must be understood that the whole theory which will be developed on the base of similarity between spins and harmonic oscillators will only be valid for low occupation numbers of every harmonic oscillator.

In the Hamiltonian of (5.1) we now make the substitutions,

$$S_i^x = Q_i\sqrt{s}\,, \qquad S_i^y = P_i\sqrt{s}\,, \qquad S_i^z = s - \frac{1}{2}(P_i^2 + Q_i^2 - 1) \tag{5.9}$$

and in accord with the above instructions, systematically discard cubic and quartic terms. One obtains the "linearized" Hamiltonian (i.e., the equations of motion are linearized, the Hamiltonian itself is of course quadratic)

$$\mathscr{H}_{\text{lin}} = E_0 + g\mu_B(H + H_0)\sum_i \frac{1}{2}(P_i^2 + Q_i^2 - 1) - Js \sum_{\substack{\text{nearest}\\ \text{neighb.}}} (P_iP_j + Q_iQ_j) \tag{5.10}$$

where

$$H_0 = \frac{Jsz}{g\mu_B} \tag{5.11}$$

(z = number of nearest neighbors of any given spin) is precisely the molecular field constant which is introduced in the molecular field approximation, in a later chapter. The constant energy term

$$E_0 = -NHg\mu_B s - \frac{1}{2}NzJs^2 \tag{5.12}$$

is the energy of the completely saturated state in which all spins are parallel to the applied field.

The linearized Hamiltonian bears an obvious resemblance to the Hamiltonian of lattice vibrations, which also involves interactions among neighboring harmonic oscillators. An important difference arises from the presence of "velocity-dependent forces" P_iP_j, to which may be attributed the difference between the spectra of spin waves and that of sound.

We now make the plane-wave transformation to the new set of generalized coordinates and momenta

$$Q_i = \frac{1}{\sqrt{N}}\sum_k e^{i\mathbf{k}\cdot\mathbf{R}_i} Q_k \qquad P_i = \frac{1}{\sqrt{N}}\sum_k e^{i\mathbf{k}\cdot\mathbf{R}_i} P_k$$

note

$$Q_k^* = Q_{-k} \text{ etc.,} \qquad \text{and} \tag{5.13}$$

$$[P_k^*, Q_{k'}] = \frac{\delta_{kk'}}{\mathrm{i}}$$

where the components k_x, k_y, k_z are integer multiples of $2\pi/L$. A discussion of this is given following (5.112). This transformation diagonalizes (5.10) for which we find

$$\mathscr{H}_{\text{lin}} = \sum_k \frac{1}{2}(P_k^* P_k + Q_k^* Q_k - 1)\hbar\omega(\boldsymbol{k}) + E_0 = \sum_k \mathfrak{n}_k \hbar\omega(\boldsymbol{k}) + E_0. \tag{5.14}$$

It may be verified that the operators, P_k, $Q_{k'}^*$, etc., which refer to different momenta, commute, and so we have N new oscillators, which are uncoupled in the above approximation. For each of these, the quadratic "number operator"

$$\mathfrak{n}_k \equiv \frac{1}{2}(P_k^* P_k + Q_k^* Q_k - 1) \tag{5.15}$$

has integer eigenvalues $= 0, 1, 2, \ldots$. Fortunately, it is not at all necessary that each of these eigenvalues be restricted to the range $\ll s$, and it is sufficient for the validity of the linearization procedure merely that $\sum n_k \ll Ns$.

As for the energies of the plane-wave "magnons,"

$$\begin{aligned} \hbar\omega(\boldsymbol{k}) &= Hg\mu_{\text{B}} + Js \sum_{\boldsymbol{\delta}} (1 - \cos \boldsymbol{k} \cdot \boldsymbol{\delta}) \\ &\simeq Hg\mu_{\text{B}} + Jsa^2\boldsymbol{k}^2 + O(\boldsymbol{k}^4) \end{aligned} \tag{5.16}$$

the sum runs over the vectors $\boldsymbol{\delta}$ which connect a typical spin to the z nearest neighbors with which it interacts; $a = |\boldsymbol{\delta}|$ for the simple-cubic (sc) lattice, and the wave vectors $\boldsymbol{k}$ are restricted to the range $|k_x|$, $|k_y|$, $|k_z| < \pi/a$.

Let us now include the dipolar interactions, to see what is their effect on the magnon energy spectrum. Defining the direction cosines, α, β, and γ of the unit vector joining the spins,

$$\frac{\boldsymbol{r}_{ij}}{r_{ij}} = (\alpha_{ij}, \beta_{ii}, \gamma_{ij}) \tag{5.17}$$

we find three contributions to the dipolar Hamiltonian of (5.2), after omitting cubic and higher terms outside the scope of the linearized theory

$$\mathscr{H}_{d,0} = \frac{1}{2}s^2 \sum_i \sum_j D_{ij}(1 - 3\gamma_{ij}^2) \tag{5.18}$$

$$\mathscr{H}_{d,1} = -3s^{3/2}\sum_i\sum_j D_{ij}(\alpha_{ij}\gamma_{ij}Q_j + \beta_{ij}\gamma_{ij}P_j) \tag{5.19}$$

$$\mathscr{H}_{d,2} = \frac{1}{2}s\sum_i\sum_j D_{ij}[(1-3\alpha_{ij}^2)Q_iQ_j - 3\alpha_{ij}\beta_{ij}(Q_iP_j + P_iQ_j) + (1-3\beta_{ij}^2)P_iP_j - (1-3\gamma_{ij}^2)(P_i^2 + Q_i^2 - 1)]. \tag{5.20}$$

For the present, it is convenient to imagine that our (cubic) ferromagnet has the shape of an ellipsoid with one of the principal axes in the z direction. The The zeroth order term above is then readily interpreted in terms of the classical demagnetizing factor N_z,

$$N_z = a^3\sum_j \frac{1}{r_{ij}^3}(1-3\gamma_i^2) + \frac{4\pi}{3} \tag{5.21}$$

with α and β replacing γ above in the similar definitions of N_x and N_y. The sum in (5.21) being independent of $\boldsymbol{r}_i$, provided this point is within the crystal, we may express the zeroth order dipolar contribution in terms of the demagnetizing factor, introducing

$$M_0 = \frac{g\mu_B s}{a^3} = \text{saturation magnetization per unit volume} \tag{5.22}$$

where a^3 = volume per spin, and we find

$$\mathscr{H}_{d,0} = -\frac{1}{2}VM_0\left(\frac{4\pi}{3}M_0 - N_zM_0\right). \tag{5.23}$$

The factor in brackets is the sum of the Lorentz field $(4\pi/3)\,M_0$ and the demagnetizing field (which is uniform in an ellipsoidally shaped material) $-\,N_zM_0$, and is therefore the "effective field" acting at a given lattice site; $V = Na^3$ is the total volume.

Aside from this constant energy term and the constant E_0 in the exchange Hamiltonian, (5.10), the dynamic terms are the nontrivial terms in (5.10) plus $\mathscr{H}_{d,1}$ and $\mathscr{H}_{d,2}$, which all together are of the form

$$\mathscr{H} = \sum_{i,j}(V_{ij}P_iP_j + W_{ij}Q_iQ_j + U_{ij}P_iQ_j) - 3s^{3/2}\sum_j[(\sum_i D_{ij}\alpha_{ij}\gamma_{ij})Q_j + (\sum_i D_{ij}\beta_{ij}\gamma_{ij})P_j] \tag{5.24}$$

allowing V_{ij}, W_{ij}, and U_{ij} to indicate schematically the coefficients in (5.10, 20). The linear term may be eliminated altogether by means of the canonical transformation,

$$P_i \to P_i + f_i \qquad \text{and} \qquad Q_i \to Q_i + g_i \tag{5.25}$$

where f_i, g_i are appropriately determined constants. As it happens, perturbation theory also converges for these linear terms, and the shift in ground-state energy which can be calculated by treating $\mathscr{H}_{d,1}$ by second-order perturbation theory agrees identically with the result of the exact canonical transformation above. Clearly the result is small, for it is second order in such quantities as

$$\sum_j D_{ij}\alpha_{ij}\gamma_{ij} \tag{5.26}$$

which vanish at any point $\boldsymbol{r}_i$ which is a point of symmetry. This would be at *every* point not on the surface of the cubic material if D_{ij} were not of such long range. Calculation of this term is left as an exercise for the reader, a tedious one.

The complete elimination of $\mathscr{H}_{d,1}$ still leaves the first line in (5.24), that is $\mathscr{H}_{\text{lin}} + \mathscr{H}_{d,2}$ unaffected; and this is now to be diagonalized by a transformation, first to running waves, (5.13),

$$\mathscr{H} \rightarrow \frac{1}{2}\sum_{\boldsymbol{k}}[\mathrm{A}(\boldsymbol{k})Q_k^*Q_k + \mathrm{B}(\boldsymbol{k})P_k^*P_k + 2C(\boldsymbol{k})Q_k^*P_k] + \text{const} \tag{5.27}$$

where, with $\hbar\omega(\boldsymbol{k})$ given in (5.16),

$$\begin{aligned} A(\boldsymbol{k}) &= \hbar\omega(\boldsymbol{k}) + A_{xx}(\boldsymbol{k}) - A_{zz}(\boldsymbol{0}) \\ B(\boldsymbol{k}) &= \hbar\omega(\boldsymbol{k}) + A_{yy}(\boldsymbol{k}) - A_{zz}(\boldsymbol{0}) \\ C(\boldsymbol{k}) &= A_{xy}(\boldsymbol{k}). \end{aligned} \tag{5.28}$$

The various A's are seen by comparison with the original equations, (5.10, 20), to be dipolar lattice sums,

$$A_{xx}(\boldsymbol{k}) = \frac{s}{N}\sum_{i,j} D_{ij}(1 - 3\alpha_{ij}^2)\mathrm{e}^{\mathrm{i}\boldsymbol{k}\cdot\boldsymbol{R}_{ij}} \tag{5.29}$$

and A_{yy}, A_{zz} are given similarly by replacing α_{ij} by β_{ij} or γ_{ij}. Also

$$A_{xy}(\boldsymbol{k}) = \frac{-3s}{N}\sum_{i,j} D_{ij}\alpha_{ij}\beta_{ij}e^{\mathrm{i}\boldsymbol{k}\cdot\boldsymbol{R}_{ij}} \tag{5.30}$$

and A_{xz}, etc., can also be obtained by obvious permutations. These dipolar sums occur in several other problems of interest in solid-state physics, and have been investigated numerically, quite thoroughly, by *Cohen* and *Keffer* [5.3].[2] As an example of the size effect, these authors give $A_{xx}(\boldsymbol{k})$ calculated for a spherical sample of radius R, and $\boldsymbol{r}_i$ near the center of the crystal

$$A_{xx}(\boldsymbol{k}) \sim \left(1 - 3\frac{k_x^2}{k^2}\right)\left[1 - \frac{3j_1(kR)}{kR}\right] \tag{5.31}$$

[2] Is the energy properly extensive ($\propto N$)? This is answered in the affirmative by *Griffiths* [5.4].

(omitting constant multiplicative factors). We see a not entirely unexpected phenomenon: as k is decreased to a value $\approx 10/R$, the dipolar sums in a finite crystal begin to differ significantly from their value in an infinite material, and the shape of the surface, and the position of the origin become of importance. However, precisely at $\boldsymbol{k} = \boldsymbol{0}$, this lattice sum approaches a well-defined limit related to the demagnetizing factor defined in (5.21), and we shall make use of this fact.

Assuming, then, that the coefficients $A(\boldsymbol{k})$, $B(\boldsymbol{k})$, and $C(\boldsymbol{k})$ are known, it is a straightforward matter to perform a canonical transformation to a new set of normal modes

$$P'_{\boldsymbol{k}} = a_{\boldsymbol{k}} P_{\boldsymbol{k}} + b_{\boldsymbol{k}} Q_{\boldsymbol{k}} \qquad Q'_{\boldsymbol{k}} = c_{\boldsymbol{k}} Q_{\boldsymbol{k}} + d_{\boldsymbol{k}} P_{\boldsymbol{k}} \tag{5.32}$$

choosing the numerical coefficients $a_{\boldsymbol{k}}, \dots, d_{\boldsymbol{k}}$ so as to diagonalize the Hamiltonian yet preserve the canonical commutation relations,

$$[P'^{*}_{\boldsymbol{k}_1}, Q'_{\boldsymbol{k}_2}] = \frac{\delta_{\boldsymbol{k}_1 \boldsymbol{k}_2}}{i} \tag{5.33}$$

all other commutators $= 0$. The final result is,

$$\mathscr{H} = \sum \frac{1}{2}(P'^{*}_{\boldsymbol{k}} P'_{\boldsymbol{k}} + Q'^{*}_{\boldsymbol{k}} Q'_{\boldsymbol{k}} - 1)\hbar\omega'(\boldsymbol{k}) + \text{const} \tag{5.34}$$

with

$$\hbar\omega'(\boldsymbol{k}) = \sqrt{A(\boldsymbol{k})B(\boldsymbol{k}) - C^2(\boldsymbol{k})} \tag{5.35}$$

an anisotropic function of the direction of $\boldsymbol{k}$, as is seen from the definition of $A(\boldsymbol{k})$, $B(\boldsymbol{k})$, and $C(\boldsymbol{k})$ given in (5.28), and as plotted in Fig. 5.1.

In *ferromagnetic resonance* an oscillating electromagnetic field is applied to the sample. Usually, the electromagnetic wavelengths are sufficiently long that only the $\boldsymbol{k} = \boldsymbol{0}$ mode need be considered, and we therefore obtain a resonance phenomenon at applied frequencies in the neighborhood of $\omega'(\boldsymbol{0})$. To calculate

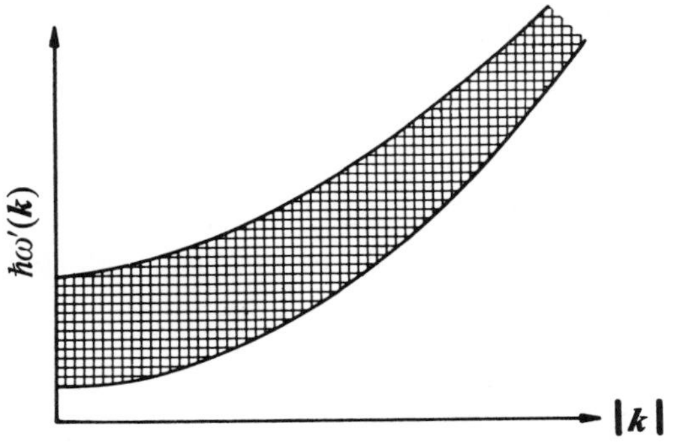

Fig. 5.1. Spectrum of energies $\hbar\omega'(\boldsymbol{k}) = \sqrt{A(\boldsymbol{k})B(\boldsymbol{k}) - C^2(\boldsymbol{k})}$. Spread, and gap at $\boldsymbol{k} = \boldsymbol{0}$, are caused by anisotropy

the magnitude of this frequency, it is reasonable to set $A_{xy}(\mathbf{0}) = 0$ [note the resemblance of this quantity, defined in (5.30), to the small term discussed in (5.26)]; whereas the other coefficients may be expressed in terms of the various demagnetizing factors to obtain

$$\hbar\omega'(\mathbf{0}) = g\mu_B\sqrt{[H + (N_x - N_z)M_0][H + (N_y - N_z)M_0]}\,. \tag{5.36}$$

This differs from a well-known empirical law

$$\hbar\omega(\mathbf{0}) = g\mu_B\sqrt{[H + (N_x - N_z)M][H + (N_y - N_z)M]} \tag{5.37}$$

only when the true magnetization M becomes far less than the saturation magnetization M_0, as happens near T_c

It should be noted that the effects of dipolar forces (and of pseudodipolar anisotropy which may be treated similarly) are limited to the long wavelengths. Thus the predictions of the theory based on the exchange Hamiltonian alone are sufficient for obtaining many of the gross properties of the ferromagnet, and also for calculating the spin-wave energy over most of the Brillouin zone excluding the neighborhood of $\boldsymbol{k} \cong \mathbf{0}$. See, for example, Fig. 5.1 for a plot of $\omega'(\boldsymbol{k})$, where a typical frequency spectrum is given. The actual, complicated anisotropic curves quickly become asymptotic to the idealized parabolic dispersion law of a pure exchange ferromagnet. This is a justification for the systematic neglect of dipolar and other anisotropic effects in many fundamental investigations of ferromagnetism and in the remainder of this chapter. For spherically shaped samples, $N_x = N_y = N_z = 4\pi/3$ and therefore only the *applied* field H (if any) appears in the expression for the frequency of the long wavelength magnons. In the remainder of this book, the geometry will always be assumed to be spherical because of this convenient simplification of the formulas, which more than compensates for the loss of generality.

5.2 One-Magnon Eigenstates in Ferromagnets

So far, we have not taken advantage of the unique fact that the ferromagnetic ground state of $\mathscr{H}$ is known, where, reversing the sign of the applied field,

$$\mathscr{H} = -J \sum_{\substack{i>j=\\ \text{nearest neighb}}}\!\!\sum \boldsymbol{S}_i \cdot \boldsymbol{S}_j + Hg\mu_B \sum S_i^z \tag{5.38}$$

and for this direction of applied field, the state in question consists of all spins "down" and will be hereinafter denoted

$$\Psi_0 = |0) \qquad \text{energy } E_0 = -NHg\mu_B s - \frac{1}{2}NzJs^2 \tag{5.39}$$

the "vacuum" state. There are N different orthogonal and normalized states containing one spin deviate each

$$\psi_i = (2s)^{-1/2} S_i^+ |0) \tag{5.40}$$

corresponding to all the choices of $\boldsymbol{R}_i$. Now, amazingly enough, one finds that when $\mathscr{H}$ is applied to the state ψ_i, it generates other states of the same type but no states with two or more spin deviations are introduced. This would not be the case if we had included the anisotropic interactions in the Hamiltonian, and it is a simplifying property of the Heisenberg ferromagnetic exchange Hamiltonian. Thus $\mathscr{H}$ may be diagonalized within the N-dimensional subspace of the functions of (5.40).

If, as we shall assume, there is translation invariance (and periodic boundary conditions), the energy is immediately diagonalized by the introduction of plane waves

$$\psi_{\boldsymbol{k}} \equiv \frac{1}{\sqrt{N}} \sum_i e^{i\boldsymbol{k}\cdot\boldsymbol{R}_i} \psi_i \quad \text{and} \quad \mathscr{H}\psi_{\boldsymbol{k}} = E(\boldsymbol{k})\psi_{\boldsymbol{k}}, \tag{5.41}$$

and the energy eigenvalues are

$$E(\boldsymbol{k}) = E_0 + \hbar\omega(\boldsymbol{k}) \tag{5.42}$$

with E_0 and $\hbar\omega(\boldsymbol{k})$ exactly as defined in (5.12, 16). This is precisely the result we would have obtained in the harmonic-oscillator approximation to the isotropic Hamiltonian by setting $n_{\boldsymbol{k}} = 1$ and all other $n_{\boldsymbol{k}'} = 0$ in (5.14).

5.3 Two-Magnon States and Eigenstates in Ferromagnets

We have found that the ferromagnetic ground state, and the one-magnon eigenstates have precisely the energies predicted in the linearized harmonic-oscillator approximation. This is very encouraging, but not entirely unexpected in view of the agreement with the earlier semiclassical analysis. One might even be tempted to predict that states of two or more plane wave magnons are approximate eigenstates with an error of no more than $O(1/N)$. But this prediction is only partly correct, as we shall see in the present study of two-magnon states. For although most of the two-plane-wave magnon states undergo negligible scattering and energy shifts (which may be ascribed to the nonlinear corrections to the harmonic-oscillator approximation), a number of *bound states* appear. These are totally unexpected on the basis of our previous considerations, although they are of considerable importance in the demise of the spin-wave picture at or below the Curie temperature. Fortunately, the two magnon problem can be solved *exactly* an any number of dimensions, and for any magnitude of the spins s, and so the relative importance of these bound states can be estimated under the various regimes [5.5–7].

Let us introduce a notation, in which to develop qualitative arguments for why nonlinearities should be insignificant at long wavelengths. This will be followed by an exact analysis. Let the normalized, two-spin-deviate basis states be

$$\psi_{ij} = C_{ij}\, S_i^+ S_j^+ |0) = \psi_{ji}\,. \tag{5.43}$$

For $s > \frac{1}{2}$, there are $N(N+1)/2$ states in this orthonormal set. For $s = 1/2$, the nonexistence of states of type ψ_{ii} reduces the number to $N(N-1)/2$. In any case, the energy and wavefunctions of N of these may be found without further calculation, using only previously derived results. To see this, consider the two-plane-wave states

$$\psi_{\boldsymbol{k}\boldsymbol{k}'} = C \sum_{i,j} \mathrm{e}^{\mathrm{i}(\boldsymbol{k}\cdot\boldsymbol{R}_i + \boldsymbol{k}'\cdot\boldsymbol{R}_j)} \psi_{ij} = \psi_{\boldsymbol{k}'\boldsymbol{k}} \tag{5.44}$$

where C is an appropriate normalization constant. When $\boldsymbol{k} = \boldsymbol{0}$, (5.44) is just a one-magnon eigenstate, to which the total spin raising operator $S^+ = \sum S_i^+$ has been applied

$$\psi_{\boldsymbol{0}\boldsymbol{k}'} = C' S^+ \psi_{\boldsymbol{k}'}\,. \tag{5.45}$$

Now while S^+ (as well as S^- and S^z) commutes with the exchange part of the Hamiltonian, it is a raising operator for S^z. (The total spin is, of course, a good angular momentum).

$$[S^z, S^+] = S^+ \qquad (\text{for } \hbar = 1) \tag{5.46}$$

that is,

$$S^z(S^+\psi) = S^+(S^z + 1)\psi.$$

Thus, for the Hamiltonian $\mathscr{H}$ of (5.38),

$$\mathscr{H}\psi_{\boldsymbol{0}\boldsymbol{k}'} = C' S^+(\mathscr{H} + g\mu_{\mathrm{B}} H)\psi_{\boldsymbol{k}'} = [\mathrm{E}(\boldsymbol{k}') + g\mu_{\mathrm{B}} H]\psi_{\boldsymbol{0}\boldsymbol{k}'} \tag{5.47}$$

and therefore,

$$E(\boldsymbol{0}\boldsymbol{k}') = E_0 + \hbar\omega(\boldsymbol{k}') + g\mu_{\mathrm{B}} H = E_0 + \hbar\omega(\boldsymbol{k}') + \hbar\omega(\boldsymbol{0}). \tag{5.48}$$

In the absence of an external field H, $\omega(\boldsymbol{0}) = 0$, and this particular two-magnon state is degenerate with the one-magnon state to which it is related by a spatial rotation. Since it is in any case an eigenstate, there is no "scattering," and there is no energy shift in a magnetic field beyond the simple change in Zeeman energy. By continuity, we may expect that if $\boldsymbol{k}$ and $\boldsymbol{k}'$ are both reasonably small,

$$E(\boldsymbol{kk'}) = E_0 + \hbar\omega(\boldsymbol{k}) + \hbar\omega(\boldsymbol{k'}) + \frac{1}{N}\delta E(\boldsymbol{kk'}) \tag{5.49}$$

with $\delta E(\boldsymbol{kk'}) \to 0$ when either $\boldsymbol{k}$ or $\boldsymbol{k'} \to 0$; and this is indeed the computed result, if the states, (5.44), are taken as a set of variational states, and the energy is computed by the variational formula

$$E(\boldsymbol{kk'}) = (\psi^*_{\boldsymbol{kk'}}|\mathscr{H}|\psi_{\boldsymbol{kk'}})/(\psi^*_{\boldsymbol{kk'}}|\psi_{\boldsymbol{kk'}}) \,. \tag{5.50}$$

But such a computation would not be exact—nor even adequate, for several important reasons, as follows:

1) The functions $\psi_{\boldsymbol{kk'}}$ are not orthogonal. Their failure to be orthogonal, although it is only to $O(1/N)$, leads to what Dyson has denoted[3] the "kinematical interactions" which must be taken into account in solving for the energy eigenvalues. Note that for $s = \frac{1}{2}$, the set of two-magnon states is *overcomplete*: there are $\frac{1}{2}N(N+1)$ functions to describe $\frac{1}{2}N(N-1)$ physical states, which shows that the functions in that case cannot all be orthogonalized, *even in principle.*

2) These functions do not diagonalize the Hamiltonian, except when $\boldsymbol{k}$ or $\boldsymbol{k'}$ vanish; to the extent that both wave vectors are finite, there is scattering, i.e., "dynamical interactions" in Dyson's language, which may even lead to bound states. To see this, it is necessary to go beyond the approximate treatment and diagonalize $\mathscr{H}$ exactly within the proper subspace of states ψ_{ij}. Fortunately, the Hamiltonian has no matrix elements connecting any of these states with a state outside the subspace, so that the problem is quite well defined, as it was in the case of the one-magnon eigenstates.

3) If the number of magnons $\sum n_{\boldsymbol{k}}$ is not 1 or 2 but is $O(Ns)$, the magnon–magnon interactions add up to a finite contribution to the energy of each magnon, and must be taken into account in a correct statistical mechanics *even at low T*. Equation (5.49) can be generalized to this purpose, by writing

$$E_{\text{Total}} = E_0 + \sum_{\boldsymbol{k}} \hbar\omega(\boldsymbol{k})n_{\boldsymbol{k}} + \frac{1}{2N}\sum_{\boldsymbol{k},\boldsymbol{k'}}\sum \delta E^{\text{R}}(\boldsymbol{kk'})n_{\boldsymbol{k}}n_{\boldsymbol{k'}} \,, \tag{5.51}$$

an expression that must be valid at finite, but low, density of excitations, with δE^{R} the real part of the binary interaction energy [calculated in (5.78)]. The imaginary part of this term also has a direct interpretation in terms of scattering lifetimes. We define the inverse scattering lifetime of a given magnon ($\boldsymbol{k}$) in the presence of a finite, but low, density of other excitations as

$$1/\tau(\boldsymbol{k}) = \frac{2}{N}\sum_{\boldsymbol{k'}} \delta E^{\text{I}}(\boldsymbol{kk'})n_{\boldsymbol{k'}} \,. \tag{5.52}$$

Thus, all the important properties of the low-density many-body magnon

---

[3] The interaction of magnons was first studied correctly by *Dyson* [5.8]; this work was greatly expanded by *Boyd* and *Callaway* and placed on a rigorous footing by *Hepp* [5.9].

fluid may be expressed in terms of the *complex* binary interaction function $\delta E(\boldsymbol{kk}')$ that we shall calculate in this section.

The exact 2-magnon eigenstate is a linear combination of all possible configurations of 2 spin "flips", i.e., is of the form

$$\psi = \sum_{i>j} f_{ij} S_i^+ S_j^+ |0) \qquad f_{ij} \equiv f_{ij} \,. \tag{5.53}$$

Let us solve the eigenvalue equation,

$$\mathscr{H}\psi = E\psi \tag{5.54}$$

for the amplitudes f_{ij}. There is no need to use the normalized configurations as the normalization constants may be considered absorbed in the f_{ij}. Also, eigenstates belonging to different energies are automatically orthogonal. One takes the inner product of both sides of the Schrödinger equation, (5.54), with every basis state

$$(0|S_i^- S_j^- \,,$$

to obtain the equations obeyed by the amplitudes f_{ij}, which are,

$$\begin{aligned}(E - E_0 - 2g\mu_B H - 2sJz)f_{ij} &+ s\sum_n (J_{nj}f_{in} + J_{in}f_{nj}) \\ &= \frac{1}{2}J_{ij}(f_{ii} + f_{jj} - f_{ij} - f_{ji})\end{aligned} \tag{5.55}$$

where each of the bonds $J_{ij}, J_{nj}, J_{in} = J$, when their respective scripts (ij) (nj) and (in) are nearest neighbor pairs, and vanish otherwise. This equation is correct for all values of s. Even when $s = \frac{1}{2}$, the unphysical amplitudes f_{ii}, etc., cancel between both sides of (5.55). In that case, nevertheless, it is convenient to define fictitious f_{ii} in order to simplify the calculation. There are at least two manners of doing this: The way chosen by Bethe in his solution of the linear chain problem (see section on his one-dimensional solution, ahead) was to set $f_{ii} + f_{jj} = f_{ij} + f_{ji}$ when j, i are nearest neighbors, a sort of "boundary condition" ensuring that the homogeneous equation is obeyed at every pair of sites. But in general, it is more satisfactory to treat the special case $s = \frac{1}{2}$ on the same footing as $s > \frac{1}{2}$, and allow (5.55) with $i = j$ to *define* the unphysical amplitude f_{ii}. Note that in any event, J_{ij} vanishes unless i, j are nearest neighbors, so that with this exception, the right-hand side of (5.55) vanishes.

Periodic boundary conditions require

$$f_{ij} = f_{iJ} \qquad \text{where } \boldsymbol{R}_j - \boldsymbol{R}_J \equiv (L, 0, 0) \qquad \text{or} (0, L, 0), \tag{5.56}$$

etc., for a volume L^3 in three dimensions, a square of area L^2 in two dimensions, or a length L in one dimension. And if the right-hand side of (5.55) did in fact vanish everywhere (instead of *nearly* everywhere), then the above boundary conditions could be met by the plane-waves

$$f_{ij} = e^{i(\boldsymbol{k}\cdot\boldsymbol{R}_i+\boldsymbol{k}'\cdot\boldsymbol{R}_j)} + e^{i(\boldsymbol{k}'\cdot\boldsymbol{R}_i+\boldsymbol{k}\cdot\boldsymbol{R}_j)} ,$$

with the Cartesian components of the wavevectors

$$k_x, \dots \text{ and } k'_x, \dots = \frac{2\pi}{L}\cdot\text{integer}. \tag{5.57}$$

These *are* the eigenstates when either $\boldsymbol{k}$ or $\boldsymbol{k}' = \boldsymbol{0}$. For nonzero $\boldsymbol{k}, \boldsymbol{k}'$, it is necessary, because of the interaction, to take linear combinations of such plane waves to obtain the eigenstates. It is convenient to separate out the center of mass motion by introducing the center of mass wave vector, and coordinate,

$$\boldsymbol{K} = \boldsymbol{k} + \boldsymbol{k}' \qquad \boldsymbol{R} = \frac{\boldsymbol{R}_i + \boldsymbol{R}_j}{2} \tag{5.58}$$

and a relative wave vector, and coordinate,

$$\boldsymbol{q} = \frac{\boldsymbol{k} - \boldsymbol{k}'}{2} \qquad \boldsymbol{r} = \boldsymbol{R}_i - \boldsymbol{R}_j \tag{5.59}$$

and in terms of these, to express f_{ij} in the form

$$f_{ij} = A e^{i\boldsymbol{K}\cdot\boldsymbol{R}} \{\sum_{\boldsymbol{q}} e^{i\boldsymbol{q}\cdot\boldsymbol{r}} f(\boldsymbol{q})\} \equiv A e^{i\boldsymbol{K}\cdot\boldsymbol{R}} F(\boldsymbol{r}) \tag{5.60}$$

where A is a normalization constant, required for $\sum |f_{ij}|^2 = 1$.

Because spin operators on different sites commute,

$$F(\boldsymbol{r}) = F(-\boldsymbol{r}) \text{ or equivalently, } f(-\boldsymbol{q}) = f(\boldsymbol{q}). \tag{5.61}$$

In writing f_{ij} in the simple form above, one takes advantage of the fact that the interactions J_{ij} depend on the coordinates $\boldsymbol{R}_i$ and $\boldsymbol{R}_j$ only through the relative coordinate $\boldsymbol{r}$ and therefore the total momentum (wave vector $\boldsymbol{K}$) is a constant of the motion. We shall solve for $F(\boldsymbol{r})$, or rather, $f(\boldsymbol{q})$, at fixed $\boldsymbol{K}$. Once the solutions are obtained, they can be expressed in the original variables $\boldsymbol{k}, \boldsymbol{k}'$ by means of (5.58, 59).

Inserting f_{ij} in the form above into the set of difference equations (5.55), we have

$$(E - E_0 - 2g\mu_B H - 2sJz)F(\boldsymbol{r}) + 2sJ\sum_{\boldsymbol{\delta}}\left(\cos\frac{\boldsymbol{K}}{2}\cdot\boldsymbol{\delta}\right)F(\boldsymbol{r}+\boldsymbol{\delta})$$
$$= J(\boldsymbol{r})\left[\left(\cos\frac{\boldsymbol{K}}{2}\cdot\boldsymbol{r}\right)F(\boldsymbol{0}) - \frac{F(\boldsymbol{r}) + F(-\boldsymbol{r})}{2}\right] \tag{5.62}$$

in which the vectors $\boldsymbol{\delta}$ connect a spin to its z nearest neighbors for which the interaction $J(\boldsymbol{\delta}) = J \neq 0$. At all other distances $J(\boldsymbol{r})$ vanishes. Next, we multiply the equation for $F(\boldsymbol{r})$ by $\exp(-i\boldsymbol{q}\cdot\boldsymbol{r})$, and sum on all $\boldsymbol{r}$ to obtain the equation for $f(\boldsymbol{q})$

$$\left[E - E_0 - 2g\mu_B H - 2sJ\sum_{\delta}\left(1 - \cos\frac{\boldsymbol{K}}{2}\cdot\boldsymbol{\delta}\cos\boldsymbol{q}\cdot\boldsymbol{\delta}\right)\right]f(\boldsymbol{q})$$
$$= \frac{J}{N}\sum_{\delta}\cos\boldsymbol{q}\cdot\boldsymbol{\delta}\sum_{k}\left(\cos\frac{\boldsymbol{K}}{2}\cdot\boldsymbol{\delta} - \cos\boldsymbol{k}\cdot\boldsymbol{\delta}\right)f(\boldsymbol{k}). \tag{5.63}$$

We have made use of the usual identity,

$$\frac{1}{N}\sum_{r} e^{i(\boldsymbol{k}-\boldsymbol{k}')\cdot\boldsymbol{r}} = \delta_{\boldsymbol{k},\boldsymbol{k}'}. \tag{5.64}$$

For brevity, define a symbol which will be used repeatedly:

$$\gamma_{\boldsymbol{k}}(\boldsymbol{q}) \equiv E_0 + 2g\mu_B H + 2sJ\sum_{\delta}\left(1 - \cos\frac{\boldsymbol{K}}{2}\cdot\boldsymbol{\delta}\cos\boldsymbol{q}\cdot\boldsymbol{\delta}\right) \tag{5.65}$$

which is, basically, the "incoming" 2-magnon energy $E_0 + \hbar\omega(\boldsymbol{k}) + \hbar\omega(\boldsymbol{k}')$ of (5.49) *Proof:* assume a simple-cubic (sc) lattice in 3D, a square (sq) one in 2D, and for either of these (as well as for the linear chain in 1D) we have the obvious symmetries $\gamma_{\boldsymbol{K}}(\boldsymbol{q}) = \gamma_{-\boldsymbol{K}}(\boldsymbol{q}) = \gamma_{\boldsymbol{K}}(-\boldsymbol{q})$. In terms of the original $\boldsymbol{k},\boldsymbol{k}'$ we can write

$$\gamma_{\boldsymbol{K}}(\boldsymbol{q}) = \gamma(\boldsymbol{k}\boldsymbol{k}') = E_0 + 2g\mu_B H + sJ\sum_{\delta}(2 - \cos\boldsymbol{k}\cdot\boldsymbol{\delta} - \cos\boldsymbol{k}'\cdot\boldsymbol{\delta})$$
$$= E_0 + \hbar\omega(\boldsymbol{k}) + \hbar\omega(\boldsymbol{k}'). \qquad \text{QED.} \tag{5.66}$$

Bound states, if they exist, must be outside the continuum of states lying between the maximum and minimum vaue of $\gamma_{\boldsymbol{K}}(\boldsymbol{q})$ and so at fixed $\boldsymbol{K}$ must lie either lower than

$$E_0 + 2g\mu_B H + 2sJ\sum_{\delta}\left(1 - \left|\cos\frac{\boldsymbol{K}}{2}\cdot\boldsymbol{\delta}\right|\right)$$

or higher than

$$E_0 + 2g\mu_B H + 2sJ\sum_{\delta}\left(1 + \left|\cos\frac{\boldsymbol{K}}{2}\cdot\boldsymbol{\delta}\right|\right). \tag{5.67}$$

Let us start by an analysis of the scattering solutions in the continuum. It is conventional to distinguish between "incoming" and "scattered" waves. This is easily done in (5.63) by singling out

$$f(\boldsymbol{q}_0) = f(-\boldsymbol{q}_0) = \frac{1}{2}N \tag{5.68}$$

to be the amplitude of the "incoming" wave and solving for the remaining amplitudes $f(\boldsymbol{k})$ [$= O(1)$] as a smooth function of the continuous variable $\boldsymbol{k}$ in the large N limit. Equation (5.63) can be put in a most compact form by defining D-dimensional vectors as follows:

$$\boldsymbol{C}(\boldsymbol{q}) = (\cos q_x\delta_x, \dots) \tag{5.69}$$

and

$$\boldsymbol{V} = (V_x, \dots)$$

where

$$V_x = \frac{1}{N}\sum_{\boldsymbol{k}}\left(\cos\frac{K_x}{2}\delta_x - \cos k_x \cdot \delta_x\right) f(\boldsymbol{k}) \qquad \boldsymbol{k} \neq \pm \boldsymbol{q}_0 \,. \tag{5.70}$$

This yields for (5.63):

$$f(\boldsymbol{q}) = \frac{2J}{E - \gamma_{\boldsymbol{K}}(\boldsymbol{q})} \boldsymbol{C}(\boldsymbol{q}) \cdot \left[\boldsymbol{C}\left(\frac{1}{2}\boldsymbol{K}\right) - \boldsymbol{C}(\boldsymbol{q}_0) + \boldsymbol{V}\right] \tag{5.71}$$

for $\boldsymbol{q} \neq \pm\boldsymbol{q}_0$ and

$$E = \gamma_{\boldsymbol{K}}(\boldsymbol{q}_0) + 4(J/N)\, \boldsymbol{C}(\boldsymbol{q}_0) \cdot \left[\boldsymbol{C}\left(\frac{1}{2}\boldsymbol{K}\right) - \boldsymbol{C}(\boldsymbol{q}_0) + \boldsymbol{V}\right] \tag{5.72}$$

at $\boldsymbol{q} = \pm \boldsymbol{q}_0$. The procedure is now to solve the integral equation (5.71) for f($\boldsymbol{q}$) and $\boldsymbol{V}$, then to obtain the $1/N$ correction to the energy, (5.72). In so doing, we run into expression such as

$$(2J/N) \sum_{\boldsymbol{q} \neq \pm \boldsymbol{q}_0} \frac{(\cos 1/2K_\alpha\delta_\alpha - \cos q_\alpha\delta_\alpha)\cos q_\beta\delta_\beta}{E - \gamma_{\boldsymbol{K}}(\boldsymbol{q})} \equiv M_{\alpha,\beta}(E) \tag{5.73}$$

with $\alpha, \beta = x, y$, or z. In such a sum the value of E accurate to $O(1/N)$ given in (5.72) can be used, but the denominators can vanish at values of $\boldsymbol{q} \neq \pm \boldsymbol{q}_0$ (in fact, everywhere "on the energy shell") therefore it is convenient to write $E = \gamma_{\boldsymbol{K}}(\boldsymbol{q}_0) - \mathrm{i}\varepsilon$, with ε an infinitesimal positive quantity that we may allow to vanish at the conclusion of the calculation. It is now permissible to replace the sum in (5.73) by an integration.

The integrals are complex

$$M_{\alpha,\beta}(\boldsymbol{q}_0) = 2J\left(\frac{a}{2\pi}\right)^D \int d^D q (\cos 1/2K_\alpha\delta_\alpha - \cos q_\alpha\delta_\alpha)\cos q_\beta\delta_\beta$$
$$\times \left[\frac{\text{P.P.}}{\gamma_{\boldsymbol{K}}(\boldsymbol{q}_0) - \gamma_{\boldsymbol{K}}(\boldsymbol{q})} + \mathrm{i}\pi\delta[\gamma_{\boldsymbol{K}}(\boldsymbol{q}_0) - \gamma_{\boldsymbol{K}}(\boldsymbol{q})]\right] \tag{5.74}$$

where we use the identity,

$$\lim \varepsilon \to 0(1/x\text{-}\mathrm{i}\varepsilon) = \text{P.P.}(1/x) + \mathrm{i}\pi\delta(x)$$

with $\delta(x)$ the D-dimensional Dirac delta function and P.P. $(1/x) \equiv x/(x^2 + \varepsilon^2)$. Thus the contribution to the integral on the energy shell is entirely imaginary, the real part coming from the entire Brillouin Zone.

We combine (5.70,71), and with $\mathbb{M}$ the $D \times D$ matrix of which $M_{\alpha,\beta}$ is the matrix element, and $\boldsymbol{C}$ and $\boldsymbol{V}$ column vectors, we have

$$\boldsymbol{V} = \mathbb{M}\cdot[\boldsymbol{C}(1/2\boldsymbol{K}) - \boldsymbol{C}(\boldsymbol{q}_0) + \boldsymbol{V}] \,. \tag{5.75}$$

This is formally solved by multiplication by $[\mathbb{1} - \mathbb{M}]^{-1}$, with $\mathbb{1}$ the D-dimensional unit matrix. Inserting the result into (5.71, 72) yields

$$f(\boldsymbol{q}) = \frac{2J}{\gamma_{\boldsymbol{K}}(\boldsymbol{q}_0) - \gamma_{\boldsymbol{K}}(\boldsymbol{q})} \boldsymbol{C}(\boldsymbol{q})\cdot[\mathbb{1} - \mathbb{M}]^{-1}\cdot[\boldsymbol{C}(1/2\boldsymbol{K}) - \boldsymbol{C}(\boldsymbol{q}_0)] \qquad (\boldsymbol{q} \neq \pm\boldsymbol{q}_0) \tag{5.76}$$

and

$$E = \gamma_{\boldsymbol{K}}(\boldsymbol{q}_0) + 4\,(J/N)\,\boldsymbol{C}(\boldsymbol{q}_0)\cdot[\mathbb{1} - \mathbb{M}^{-1}]\cdot[\boldsymbol{C}(1/2\boldsymbol{K}) - \boldsymbol{C}(\boldsymbol{q}_0)] \,. \tag{5.77}$$

We observe that this $f(\boldsymbol{q})$ and the $1/N$ correction to the energy both vanish at $\boldsymbol{q}_0 = \pm\frac{1}{2}\boldsymbol{K}$; these are the special cases $\boldsymbol{k}$ or $\boldsymbol{k}' = \boldsymbol{0}$, for which we already anticipated the absence of scattering.

The integrals (5.74) making up the 3×3 matrix $\mathbb{M}$ are difficult to express in terms of elementary functions (although they simplify in the long wavelength limit ka, $k'a \to 0$, precisely the limit studied by *Dyson* [5.8] because of its applicability to low-temperature and low-energy behavior). In leading semiclassical approximation, we just ignore $\mathbb{M}$ which is $O(1/s)$, and evaluate

$$\begin{aligned}\delta E(\boldsymbol{k}\boldsymbol{k}') &\approx 4J\boldsymbol{C}(\boldsymbol{q}_0)\cdot[\boldsymbol{C}(1/2\boldsymbol{K}) - \boldsymbol{C}(\boldsymbol{q}_0)] \\ &= -(1/s)[\hbar\omega(\boldsymbol{k}) + \hbar\omega(\boldsymbol{k}') - \hbar\omega(\boldsymbol{k} - \boldsymbol{k}') - \hbar\omega(\boldsymbol{0})]\end{aligned} \tag{5.78}$$

with $\hbar\omega(\boldsymbol{k})$ given in (5.16), as usual.

In $1D$ $\mathbb{M}$ is just a number.

Equation (5.74) yields

$$\begin{aligned}M &= \frac{2J}{2\pi}\int_{-\pi}^{+\pi} dq \frac{(\cos 1/2K - \cos q)\cos q}{4sJ\cos 1/2K(\cos q - \cos q_0) + \mathrm{i}\varepsilon} \\ &= \frac{1}{2s\cos 1/2K}\left(\cos\frac{1}{2}K - \cos q_0\right)(1 - \mathrm{i}\cot q_0)\end{aligned} \tag{5.79}$$

or, in terms of k and k'

$$M = \frac{-1}{s\cos 1/2(k + k')}\sin\frac{1}{2}k\,\sin\frac{1}{2}k'\left[1 - \mathrm{i}\cot\frac{1}{2}(k - k')\right]. \tag{5.80}$$

Insertion into (5.76) yields the scattering amplitudes characteristic of the continuum of scattering states, expressed in terms of the unperturbed k values of (5.57). Later [see (5.150)] we shall re-examine this point; energy and momentum can be conserved with just two values of $\boldsymbol{k}$, provided the k values are shifted somewhat. This is part of the so-called *Bethe's ansatz*; its applications to date have been limited to $1D$, its relation to the above, somewhat obscure.

We now turn our attention to the two-magnon *bound* states, and look for E lying outside the range of values given in (5.67). As there is no corresponding (real) q_0 there is no special amplitude (5.68). The eigenvalue equation (5.63) is now an homogeneous equation, and
$f(\boldsymbol{q})$ is given by

$$f(\boldsymbol{q}) = \frac{2JC(\boldsymbol{q})\cdot \boldsymbol{V}}{E - \delta_{\boldsymbol{K}}(\boldsymbol{q})} \tag{5.81}$$

with

$$\boldsymbol{V} = \mathbf{M}\cdot \boldsymbol{V}, \text{ the matrix } M_{\alpha\beta}(E) \text{ given in (5.73)} \tag{5.82}$$

The bound state energy E is adjusted until $\boldsymbol{V}$ is an eigenvector of $\mathbf{M}$. This determines $\boldsymbol{V}$ to within a multiplicative constant, required for normalization.

5.4 Bound States in One Dimension

The easiest application of this theory is to the linear chain. The eigenvalue equation reduces to

$$1 = (J/\pi)\int_{-\pi}^{+\pi} dq \frac{(\cos 1/2K - \cos q)\cos q}{E - E_0 - 2g\mu_B H - 4sJ + \cos 1/2K \cos q} \tag{5.83}$$

For greater clarity, one introduces dimensionless parameters. Let,

$$A = (E - E_0 - 2g\mu_B H - 4sJ)/4sJ \cos \frac{1}{2}K\,.$$

and therefore,

$$\begin{aligned} 1 &= \frac{1}{2s\pi}\int_0^{\pi} dx \frac{1}{A + \cos x}\left(\cos x - \frac{\cos^2 x}{\cos 1/2K}\right) \\ &= \frac{1}{2s}\left(1 - \frac{|A|}{\sqrt{A^2 - 1}}\right)\left(1 + \frac{A}{\cos 1/2K}\right). \end{aligned} \tag{5.84}$$

There are solutions only for $A < -1$, proving that the bound state exists only below the continuum, and that there is no bound state above the continuum (which would correspond to positive values of A). This is not surprising in view of the fact that the effective interaction is attractive: two spin deviations have lower energy when they are nearest neighbors than when they are farther apart.

On the other hand, one might have thought of the effective interaction when two spin deviations occur on the same site, as a repulsive (or "hard core") potential which could lead to bound states above the continuum. The actual calculation shows this latter point of view to be false.

The equation is easiest to solve analytically when $s = \frac{1}{2}$, in which case we find

$$A \cos \frac{1}{2}K = -\frac{1}{2}\left(1 + \cos^2 \frac{1}{2}K\right) = -1 + \frac{1}{2} \sin^2 \frac{1}{2}K$$

and therefore,

$$E = E_0 + 2g\mu_B H + J \sin^2 \frac{1}{2}K \qquad \text{bound state.} \tag{5.85}$$

A similar formula will be obtained when we study solitons, (5.249). Equation (5.85) may be compared to the lowest energy of a scattering state, i.e., the bottom of the continuum, which occurs at $\boldsymbol{q} = \boldsymbol{0}$ [cf. (5.67)]

$$E_{\min}(\boldsymbol{q} = \boldsymbol{0}) = E_0 + 2g\mu_B H + 4J \sin^2 \frac{1}{4}K\,. \tag{5.86}$$

The bound state lies lower than this by a "binding energy" δE in the amount of

$$\delta E = 4J \sin^2 \frac{1}{4}K - J \sin^2 \frac{1}{2}K\,, \tag{5.87}$$

which ranges from

$$\frac{J}{4}\left(\frac{K}{2}\right)^4 \quad \text{for } K \ll 1 \qquad \text{to } J \quad \text{for } K = \pi\,.$$

A schematic plot is given in Fig. 5.2 for arbitrary value of the spin s.

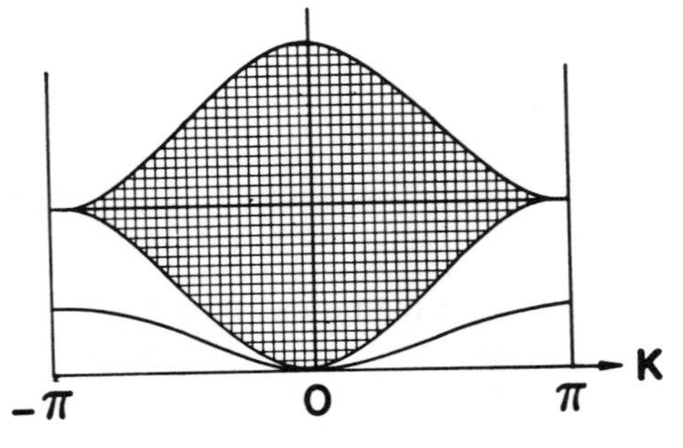

Fig. 5.2. Energy of 2-magnon states in linear chain. Shading indicates continuum, single line represents bound state

5.5 Bound States in Two and Three Dimensions

The results obtained by *Wortis* [5.5–7] show that in two dimensions there is at least one bound state below the continuum *for all* $\boldsymbol{K}$, and that for some $\boldsymbol{K}$ there may be *two* bound states. In three dimensions, he found that there are no bound states whatever for small $\boldsymbol{K}$, and it is only for sufficiently large values of $\boldsymbol{K}$ that one, two, or up to a maximum of *three* bound states make their appearance. This is a most important result, for the absence of any bound-state solution over a significant range of $\boldsymbol{K}$ confers validity to the linear spin-wave theory in three dimensions, even though the linearization be invalid in one or two dimensions.

Before considering the region of small $\boldsymbol{K}$, which because of the low spin-wave and bound-state energies will be of the greatest importance in thermodynamics, let us first examine the special case when all components of $\boldsymbol{K}$ are equal to π, the maximum possible value, in which case

$$\cos\frac{K_x}{2} = \cos\frac{K_y}{2} = \cos\frac{K_z}{2} = \cos\frac{\pi}{2} = 0$$

and all the integrals in the eigenvalue equation become trivial. Because the denominators are now all constant, the nondiagonal matrix elements $M_{x,y}$ defined in (5.73) all vanish, while the diagonal ones are all equal. The bound state will therefore be D-fold degenerate. Note that the continuum has collapsed to a single point at

$$\gamma_\pi \equiv \gamma_\pi(\boldsymbol{q}) \equiv E_0 + 2g\mu_B H + 2sJz \tag{5.88}$$

and therefore the integral equation reduces to an algebraic one. The bound-state solutions of (5.82) are easily found to have energy

$$E = \gamma_\pi - J\,, \tag{5.89}$$

by solving this algebraic equation. The bound state is identified as the complex formed by keeping the two spin deviates nearest neighbors; and the degeneracy can be understood as the D distinct directions along which two spins can be nearest neighbors.

The existence of at least one bound state for every value of $\boldsymbol{K}$ in *two dimension* is guaranteed by the behavior of the various integrals near $\boldsymbol{q} = \boldsymbol{0}$. Let us denote the energy of the lowest state in the continuum by $E_{\min}$, then the integrands in the vicinity of $\boldsymbol{q} = \boldsymbol{0}$ all contribute on the order of

$$\approx \int \frac{\text{numerator}}{E - E_{\min} - D'q^2}\, q\,dq$$

for suitable positive constant D', and appropriate nonvanishing numerator. This contribution, as a function of E, ranges from zero to $-\infty$ as E varies from $-\infty$

to $E_{\min}$, and in the limit, $E_{\min} - E = \delta E \to 0$, varies as $\approx \log \delta E$. Thus one expects, even without the benefit of explicit calculation, that the "binding energy" δE will depend exponentially on the various parameters. Indeed, for $K_x = K_y \approx 0$, Wortis found

$$\delta E \sim e^{+2\pi s C/(1-C)}$$

where $C = \cos(K_x/2)$; for more general $\boldsymbol{K}$ he obtained the binding energy in terms of various elliptic integrals.

Turning finally to the more realistic case of a three-dimensional structure, for which the integrals cannot be evaluated analytically, one may find it advantageous to represent them by Laplace transform methods. That is, the substitution

$$\frac{1}{D} = \int_0^\infty dt\, e^{-Dt}$$

combined with the definition of the Bessel functions of imaginary argument

$$I_n(z) = \frac{1}{\pi} \int_0^\pi e^{z \cos x} \cos nx\, dx$$

permits the replacement of three-dimensional integrals by a rapidly converging one-dimensional integral. But these substitutions are not required to determine the threshold for the existence of a bound-state solution.

For this, we may presume $K_x = K_y = K_z$, which yields the lowest edge of the continuum and is more likely to produce the bound state than some less isotropic direction. In this case the three diagonal matrix elements $M_{i,i}$ are all equal, and the off-diagonal elements are also equal to each other. This results in $V_x = V_y = V_z$ and in an eigenvalue equation

$$\begin{aligned} 1 &= M_{x,x} + 2M_{x,y} \\ &= \frac{2J}{N} \sum_q \frac{\cos q_x[3\cos(K_x/2) - \cos q_x - 2\cos q_y]}{E - E_0 - 2g\mu_B H - 12sJ + 4sJ(\cos q_x + \cos q_y + \cos q_z)\cos(K_x/2)} \\ &= \frac{E - E_0 - 2g\mu_B H - 12sJ\sin^2(K_x/2)}{24s^2 J\cos^2(K_x/2)} \left[1 - \frac{E - E_0 - 2g\mu_B H - 12sJ}{N} \sum_q \frac{1}{E - \gamma_{\boldsymbol{K}}(\boldsymbol{q})}\right]. \end{aligned} \tag{5.90}$$

Use has been made only of the cubic symmetry, and of the equality of the three components of $\boldsymbol{K}$ to arrive at the last, simplified expression. The bound-state threshold occurs for that $\boldsymbol{K}$ at which E just drops below the bottom of the continuum at $\gamma_{\boldsymbol{K}}(\mathbf{0})$. This occurs for K_{x0}

$$E = E_0 + 2g\mu_B H + 24sJ\sin^2\frac{K_{x0}}{4}. \tag{5.91}$$

Replacing the sum in (5.90) by an integral in the usual manner, we use the first line to determine the threshold K_{x0}

$$1 = \frac{-1 + \cos 1/2K_{x0}}{2s \cos 1/2K_{x0}}(1 - W), \text{ i.e., } \sin^2\frac{1}{4}K_{x0} = \frac{s}{2s + W - 1}, \tag{5.92}$$

where W is Watson's integral (see Note at end of section)

$$W = \frac{1}{(2\pi)^3}\iiint_{-\pi}^{\pi} \frac{dq_x\, dq_y\, dq_z}{1 - 1/3(\cos q_x + \cos q_y + \cos q_z)} = 1.516386\,. \tag{5.93}$$

Thus, the bound state exists in the range

$$0 \leq \cos\frac{K_x}{2} \leq \frac{0.516386}{2s + 0.516386} \tag{5.94}$$

provided $K_x = K_y = K_z$. This results in $140° \leq K_x \leq 180°$ for $s = \frac{1}{2}$, $157° \leq K_x \leq 180°$ for $s = 1$, $167° \leq K_x \leq 180°$ for $s = 2$, etc. As K is increased beyond the minimum, the other solutions make their appearance. For example, we have already found that there are three solutions at $K_x = K_y = K_z = 180°$, which is the largest total momentum wavevector in the positive direction. This small range of momentum space over which bound-state solutions are available, even in the most favorable case of spins one-half, is in sharp contrast with the results of the one- and two-dimensional calculations; and *in the correspondence limit $s \to \infty$, the bound states simply disappear.*

In Fig. 5.3, a sketch is given of the regions in which one, two, and three distinct bound states exist. The energy of the bound state increases with $\boldsymbol{K}$ (as does the binding energy). Thus (5.91) yields the minimum energy of a bound state in 3D, with (5.92) giving the value of K_{x0}. At temperatures, low compared with this

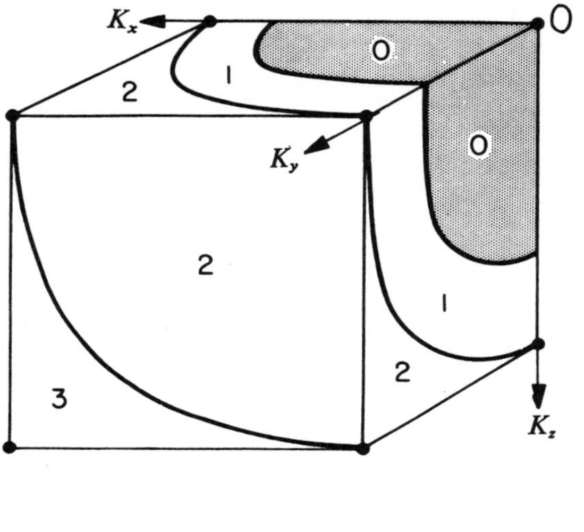

Fig. 5.3. Octant of Brillouin zone showing regions of 0, 1, 2, and 3 bound-state solutions for two spin waves, $s = \frac{1}{2}$. For $s > \frac{1}{2}$, the region of 0 solutions grows, and for $s \to \infty$ occupies the entire Brillouin zone. (Sketch after [5.5])

minimum energy $\div k_B$, spin-wave theory with scattering corrections is essentially exact.

Note: Concerning the magnetic origins of G.N. Watson's integral:

... it is strange that any one of his own free will would have investigated such integrals. Actually van Peijpe encountered the integral in constructing a theory of ferromagnetic anisotropy based on spin wave theory. Unable to evaluate it in closed form, he resorted to graphical integration. H. A. Kramers, van Peijpe's thesis supervisor, then put the problem ... to R. H. Fowler who communicated it to G. H. Hardy, whereupon, to quote Watson, "the problem then became common knowledge in Cambridge and subsequently in Oxford whence it made the journey to Birmingham without difficulty" [5.10].

5.6 One-Magnon Eigenstates in Heitler-London Solid

It is not to be supposed that the Heisenberg Hamiltonian has exactly the same spectrum of solutions as the Heitler-London Hamiltonian, due to the nonorthogonality of the electronic wavefunctions used in the latter. As this is discussed in the chapter on exchange, we refer to that chapter for further elaboration, and proceed forthwith to calculate the one-magnon eigenstates in the H-L theory. This is simple to do *in principle*, for by translational symmetry the eigenfunctions are plane waves. We then obtain the energies as functions of the interaction and normalization parameters.

Suppose, then, there is a single unpaired electron at each of N sites labeled $\boldsymbol{R}_j$, and that we use as the basis product functions

$$\psi(m_1, \dots, m_N) = \prod_{j=1}^{N} \varphi(\boldsymbol{r}_j - \boldsymbol{R}_j)\chi_{m_j}(\xi_j) \tag{5.95}$$

where $\boldsymbol{r}_j$ and ξ_j are the space and spin coordinates of the electron, respectively, and the 2^N sets of m_j's ($\pm\frac{1}{2}$) provide the same number of degrees of freedom as in a similar Heisenberg model. In evaluating energy, normalization, etc., the totally antisymmetrized wavefunctions

$$\Psi = \frac{1}{\sqrt{N!}} \sum_{\mathscr{P}} (-1)^{P} \mathscr{P} \psi(m_1, \dots, m_N) \tag{5.96}$$

must be used. Let us denote the totally ferromagnetic product function by ψ_0,

$$\psi_0 \equiv \psi\left(-\frac{1}{2}, \dots, -\frac{1}{2}\right) \tag{5.97}$$

and one spin-deviate states by ψ_i,

$$\psi_i = S_i^+ \psi_0 \qquad i = 1, \dots, N \tag{5.98}$$

and use the same indices on the corresponding totally antisymmetrized (determinantal) wavefunctions, Ψ_0 and Ψ_i, constructed from the above by means of the prescription (5.96). Because of the overlap of the one-electron wavefunctions, the Ψ_i do not form an orthonormal set. Therefore, instead of the usual Schrödinger equation, the energy variation principle yields an equation involving an overlap matrix Ω

$$\mathscr{H}_{ii} f_i + \sum_{j \neq i} \mathscr{H}_{ij} f_j = E(\Omega_{ii} f_i + \sum_{j \neq i} \Omega_{ij} f_j) \tag{5.99}$$

for an eigenfunction assumed of the form

$$\Phi = \sum_i f_i \Psi_i \, . \tag{5.100}$$

The Hamiltonian matrix elements are

$$\mathscr{H}_{ij} = \int d\tau \Psi_i^* \mathscr{H} \Psi_j \tag{5.101}$$

and the overlap matrix elements are simply

$$\Omega_{ij} = \int d\tau \Psi_i^* \Psi_j \tag{5.102}$$

including the spin variables in both integrations.

$\mathscr{H}$ presumably depends on the spatial coordinates only, and is translationally invariant. Therefore both sets of matrix elements depend on i and j only through the relative distance $\boldsymbol{R}_{ij}$, and the equations can be solved by setting

$$f_i = \mathrm{e}^{i\boldsymbol{k} \cdot \boldsymbol{R}_i} \, . \tag{5.103}$$

Consequently, the one-magnon eigenvalues $E(k)$ are simply ratios,

$$E(\boldsymbol{k}) = \frac{\sum \mathscr{H}_{ij} \mathrm{e}^{i\boldsymbol{k} \cdot \boldsymbol{R}_{ij}}}{\sum \Omega_{ij} \mathrm{e}^{i\boldsymbol{k} \cdot \boldsymbol{R}_{ij}}} = \frac{\mathscr{H}(\boldsymbol{k})}{\Omega(\boldsymbol{k})} \, . \tag{5.104}$$

By measuring these energies relative to the ferromagnetic energy ($\boldsymbol{k} = \boldsymbol{0}$), we find the spectrum of excitations

$$E(\boldsymbol{k}) - E(\boldsymbol{0}) = \frac{\Omega(\boldsymbol{0})[\mathscr{H}(\boldsymbol{k}) - \mathscr{H}(\boldsymbol{0})] - \mathscr{H}(\boldsymbol{0})[\Omega(\boldsymbol{k}) - \Omega(\boldsymbol{0})]}{\Omega(\boldsymbol{0})\{\Omega(\boldsymbol{0}) + [\Omega(\boldsymbol{k}) - \Omega(\boldsymbol{0})]\}} \approx \frac{D\boldsymbol{k}^2 + O(\boldsymbol{k})^4}{1 + O(\boldsymbol{k}\,)^2} \, . \tag{5.105}$$

The various parameters $\mathscr{H}_{ij}$ and Ω_{ij} may be evaluated by the theory of determinants (cf. section on "the method of Löwdin and Carr" in the nonorthogonality problem; however the present integrals are somewhat more difficult to evaluate than the ones solved in that section). They have reasonable magnitudes (i.e., they are not off by factors of N); and in fact (5.105) would be in exactly the form of the Heisenberg theory result given in (5.42, 16), were it not for the $\boldsymbol{k}$-dependence of the denominator. At long wavelengths this affects $\boldsymbol{k}^4$ (and higher-order) terms in the expansion of $E(\boldsymbol{k}) - E(\mathbf{0})$, and may eventually lead to some experimental determination of the importance of effects of nonorthogonality.

It may be presumed that a linearized version of H-L theory can be constructed which for many magnons in three dimensions would reproduce some of the results of "harmonic-oscillator spin-wave" theory in the Heisenberg ferromagnet. But this has not yet been attempted. Nor have the two-magnon problem and the existence of two-magnon bound states been studied in the H-L theory, although this would involve only slight generalizations of the analysis of the preceding section.

5.7 Nonlinear Spin-Wave Theory

If it is desired to go beyond the harmonic oscillator approximation and calculate nonlinear corrections to the Heisenberg Hamiltonian, without the rigors of many-body scattering theory, a fairly direct manipulation of the spin operators themselves lends itself to this purpose. We start with the Holstein-Primakoff representation (cf. Chap. 3),

$$S_i^+ = a_i^* \sqrt{2s}\sqrt{\left(1 - \frac{\mathfrak{n}_i}{2s}\right)} \quad S_i^- = \sqrt{2s}\sqrt{\left(1 - \frac{\mathfrak{n}_i}{2s}\right)}a_i \quad S_i^z = \mathfrak{n}_i - s. \tag{5.106}$$

The a's are boson operators, satisfying ordinary commutation relations

$$a_i a_j^* - a_j^* a_i \equiv [a_i, a_j^*] = \delta_{i,j} \qquad [a_i, a_j] = [a_i^*, a_j^*] = 0 \tag{5.107}$$

with $\mathfrak{n}_i = a_i^* a_i$ the Boson number operator measuring the deviation from spin "down". The Heisenberg Hamiltonian takes on the form

$$\mathscr{H} = -\frac{1}{2}\sum_{i,j} J_{ij}\left[\frac{1}{2}a_i^* \sqrt{\left(1 - \frac{\mathfrak{n}_i}{2s}\right)\left(1 - \frac{\mathfrak{n}_j}{2s}\right)}a_j + \text{H.c.} + (\mathfrak{n}_i - s)(\mathfrak{n}_j - s)\right]$$
$$+ g\mu_B H \sum_i (\mathfrak{n}_i - s)\,. \tag{5.108}$$

It is convenient to rationalize the square roots (cf. Maleev transformation, Chap. 3), by the similarity transformation

$$a_i \to \left(1 - \frac{\mathfrak{n}_i}{2s}\right)^{-1} a_i \quad \text{and} \quad a_i^* \to a_i^*\left(1 - \frac{\mathfrak{n}_i}{2s}\right). \tag{5.109}$$

This transformation preserves the number operator as well as the commutation relations (5.107) within the physically relevant Hilbert space.

Equation (5.108) now becomes

$$\mathscr{H} = E_0 - s\sum_{i,j} J_{ij}(a_i^* - a_j^*)\left(1 - \frac{\mathfrak{n}_i}{2s}\right)a_j + g\mu_B H \sum \mathfrak{n}_i\,, \tag{5.110}$$

the constant E_0 being the usual ground-state energy first defined in (5.12). This non-Hermitean operator has all the eigenvalues of the Heisenberg Hamiltonian, plus an infinite number of unphysical ones. Whether an eigenstate is physically admissible or not can be tested by seeing whether it is an eigenfunction of the positive semidefinite operator

$$\left[\frac{1}{N}\sum_{i=1}^{N}(a_i^*)^{2s+1}(a_i)^{2s+1}\right] \tag{5.111}$$

with eigenvalue 0, or not.

For the purposes of spin-wave theory, one now transforms the above $\mathscr{H}$ to running waves. We now give a discussion of this and a definition of the Brillouin zone which will be useful in the next chapter as well. Let

$$a_i = \frac{1}{\sqrt{N}}\sum_k e^{i\boldsymbol{k}\cdot\boldsymbol{R}_i}a_k \qquad a_i^* = \frac{1}{\sqrt{N}}\sum_k e^{-i\boldsymbol{k}\cdot\boldsymbol{R}_i}a_k^*$$

$$\boldsymbol{k} = \frac{2\pi}{L}(n, m, l) \tag{5.112}$$

where $n, m, l =$ integers. It may be verified that this transformation is canonical, by showing that the a_k's obey a set of commutation relations identical to (5.107). For if one inverts the transformation, he obtains

$$a_k = \frac{1}{\sqrt{N}}\sum_i e^{-i\boldsymbol{k}\cdot\boldsymbol{R}_i}a_i \qquad a_k^* = \frac{1}{\sqrt{N}}\sum_i e^{i\boldsymbol{k}\cdot\boldsymbol{R}_i}a_i^* \tag{5.113}$$

and verifies they also satisfy the commutation relations (5.107). It is sufficient in (5.112) to use the set of smallest integers: $n = 0, \pm 1, \pm 2, \dots, \pm \frac{1}{2}N_x$, $m = 0, \pm 1, \dots, \pm\frac{1}{2}N_y$, etc., for a solid $N_x \times N_y \times N_z$. This defines the *first Brillouin Zone*; an arbitrary $\boldsymbol{k}$ can always be reduced to the first BZ by subtracting an integer multiple of $(2\pi/a, 0, 0)$ or $(0, 2\pi/a, 0)$ or $(0, 0, 2\pi/a)$, where $a=$ lattice spacing. Denoting by $\boldsymbol{K}_n$ any integer multiple of these, we note

$$e^{i\boldsymbol{K}_n\cdot\boldsymbol{R}_i} = 1$$

hence $a_{k+K_n} = a_k$ [by (5.113)], and all relevant operators lie in the first BZ. The Hamiltonian in terms of a_k's is

$$\mathscr{H} = E_0 - s\sum_{i,j} J_{ij} \sum_{k_1 k_2} \frac{a^*_{k_1}}{\sqrt{N}} (\mathrm{e}^{\mathrm{i}k_1\cdot R_i} - \mathrm{e}^{\mathrm{i}k_i\cdot R_j})$$
$$\times \left(1 - \frac{1}{2s}\sum_{k_3 k_4} \frac{a^*_{k_3}}{\sqrt{N}} \mathrm{e}^{\mathrm{i}k_3 R_i} \frac{a_{k_4}}{\sqrt{N}} \mathrm{e}^{-\mathrm{i}k_4\cdot R_i}\right) \frac{a_{k_2}}{\sqrt{N}} \mathrm{e}^{-k_2\cdot R_j}$$
$$+ g\mu_B H \sum_i \sum_{k,k'} \frac{a^*_k}{\sqrt{N}} \mathrm{e}^{\mathrm{i}k\cdot R_i} \frac{a_{k'}}{\sqrt{N}} \mathrm{e}^{-\mathrm{i}k'\cdot R_i} . \quad (5.114)$$

Let us isolate, in this large sum of terms, the contribution of operators diagonal in the occupation number representation (that is, of operators such as constants and functions of $\mathfrak{n}_k = a^*_k a_k$) from the nondiagonal terms. Terms bilinear in the a's, such as the last term in the magnetic field, are automatically diagonal by virtue of momentum conservation. Quartic terms are not, although a subset of them, for which

$$(\boldsymbol{k}_1, \boldsymbol{k}_3) = (\boldsymbol{k}_2, \boldsymbol{k}_4) \qquad \text{or} \qquad (\boldsymbol{k}_4, \boldsymbol{k}_2)$$

(see Fig. 5.4) involve only $\mathfrak{n}_k$'s and so will be included in the important diagonal part, which we shall call $\mathscr{H}_D$. As the Fourier transform of J_{ij} consistently occurs, it is convenient to denote it $J(\boldsymbol{k})$, defined as

$$J(\boldsymbol{k}) = \frac{1}{2N}\sum_{i,j} \mathrm{e}^{\mathrm{i}\boldsymbol{k}\cdot(\boldsymbol{R}_i - \boldsymbol{R}_j)} J_{ij} . \quad (5.115)$$

In the cubic structure, $J^*(\boldsymbol{k}) = J(-\boldsymbol{k}) = J(\boldsymbol{k})$. For example, in the *simple cubic structure nearest-neighbor interaction*

$$J(\boldsymbol{k}) = 2J(\cos k_x a + \cos k_y a + \cos k_z a) . \quad (5.116)$$

And in this manner one obtains the compact expression, which reproduces exactly the results of (5.51, 78) based on binary scattering

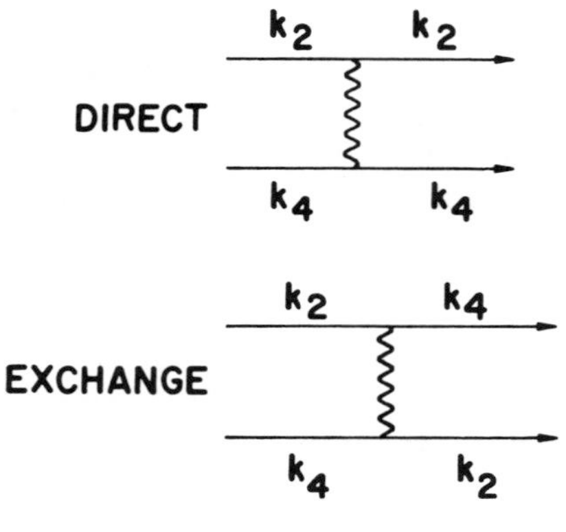

Fig. 5.4. Direct and exchange contributions to nonlinear magnon Hamiltonian

$$\mathscr{H}_D = E_0 + \sum_{\boldsymbol{k}} [2sJ(\mathbf{0}) - 2sJ(\boldsymbol{k}) + g\mu_B H]\mathfrak{n}_{\boldsymbol{k}}$$

$$- \frac{1}{N} \sum_{\boldsymbol{k},\boldsymbol{k}'} [J(\mathbf{0}) + J(\boldsymbol{k} - \boldsymbol{k}') - J(\boldsymbol{k}) - J(\boldsymbol{k}')]\mathfrak{n}_{\boldsymbol{k}}\mathfrak{n}_{\boldsymbol{k}'} . \tag{5.117}$$

No bound states are found but, as already noted (in Fig. 5.3), none are expected at large values of s.

The study of $\mathscr{H}_D$ involves only minor generalizations of the theory of the linear Hamiltonian. The only important question is the following. Given a state occupied by n magnons of wave vector $\boldsymbol{k}_1$, n_2 of wave vector $\boldsymbol{k}_2$, etc., how much energy is required to add one magnon of wave vector $\boldsymbol{k}$ to this state? Let $W(n_{\boldsymbol{k}})$ and $W(n_{\boldsymbol{k}} + 1)$ indicate the eigenvalue of $\mathscr{H}_D$ in the two states $|n_{\boldsymbol{k}})$ and $a_{\boldsymbol{k}}^* |n_{\boldsymbol{k}})$; then

$$\begin{aligned}
\mathscr{H}_D a_{\boldsymbol{k}}^* |n_{\boldsymbol{k}}) &= a_{\boldsymbol{k}}^* \mathscr{H}_D |n_{\boldsymbol{k}}) + [\mathscr{H}_D, a_{\boldsymbol{k}}^*] |n_{\boldsymbol{k}}) \\
&= W(n_{\boldsymbol{k}}) a_{\boldsymbol{k}}^* |n_{\boldsymbol{k}}) + \{2sJ(\mathbf{0}) - 2sJ(\boldsymbol{k}) + g\mu_B H \\
&\quad - \frac{2}{N} \sum_{\boldsymbol{k}'} [J(\mathbf{0}) + J(\boldsymbol{k} - \boldsymbol{k}') - J(\boldsymbol{k}) - J(\boldsymbol{k}')] \cdot \mathfrak{n}_{\boldsymbol{k}'}\} a_{\boldsymbol{k}}^* |n_{\boldsymbol{k}}) \\
&= W(n_{\boldsymbol{k}} + 1) a_{\boldsymbol{k}}^* |n_{\boldsymbol{k}})
\end{aligned} \tag{5.118}$$

and therefore

$$W(n_{\boldsymbol{k}} + 1) = W(n_{\boldsymbol{k}}) + [2sJ(\mathbf{0}) - 2sJ(\boldsymbol{k}) + g\mu_B H]$$

$$- \frac{2}{N} \sum_{\boldsymbol{k}'} [J(\mathbf{0}) + J(\boldsymbol{k} - \boldsymbol{k}') - J(\boldsymbol{k}) - J(\boldsymbol{k}')]\mathfrak{n}_{\boldsymbol{k}'} . \tag{5.119}$$

In terms of the magnon energy $\hbar\omega(\boldsymbol{k})$ as previously defined, for example, (5.14),

$$\begin{aligned}
W(n_{\boldsymbol{k}} + 1) - W(n_{\boldsymbol{k}}) &= \hbar\omega(\boldsymbol{k}) - \frac{1}{Ns} \sum_{\boldsymbol{k}'} [\hbar\omega(\boldsymbol{k}) + \hbar\omega(\boldsymbol{k}') - \hbar\omega(\boldsymbol{k} - \boldsymbol{k}') - \hbar\omega(\mathbf{0})]\mathfrak{n}_{\boldsymbol{k}'} \\
&\equiv \varepsilon(\boldsymbol{k})
\end{aligned} \tag{5.120}$$

with the sum representing the corrections to linear spin-wave theory. It is important to note that $\boldsymbol{k}' = \mathbf{0}$ spin waves do not contribute to this correction, for this confirms the rotationally invariant nature of the approximations. The nonlinear magnon energies are denoted $\varepsilon(\boldsymbol{k})$, and, for nearest-neighbor interactions,

$$\overline{\varepsilon(\boldsymbol{k})} \leq \overline{\hbar\omega(\boldsymbol{k})}$$

(indicating average over the directions of $\boldsymbol{k}$ by a bar). So, by increasing the occupation numbers $n_{\boldsymbol{k}}$, all magnon energies can be reduced, as we shall see elsewhere in a discussion of thermal effects. However the above inequality does not hold for arbitrary long-range interactions, and is *reversed* in nearest-neighbor *ferrimagnets*, so that it appears to be a special property of simple ferromagnets.

There is also an approximate superposition principle: the energy required to add *two* magnons of wave vectors $\boldsymbol{k}$ and $\boldsymbol{k}'$ to a given state equals the sum of the energies required for each, $\varepsilon(\boldsymbol{k}) + \varepsilon(\boldsymbol{k}')$, to within a correction term $O(1/N)$ arising from the interaction. The proof is given as Problem 5.1.

Problem 5.1. $W(n_k + 1, n_{k'} + 1) \equiv W(n_k + 1, n_{k'}) + W(n_k, n_{k'}, + 1) - W(n_k, n_{k'}) + w$. Find the correction term w; verify that it is $O(1/N)$.

Concluding this section, we note that the leading terms in $1/s$ are exactly the same as those obtained in the binary collision (2-magnon) analysis given earlier. Correction terms $O(1/s^2)$ compared with the leading term are calculable by perturbation theory.

5.8 Perturbation-Theoretic Correction

So far we have not focused any attention on the nondiagonal terms $\mathscr{H}' = \mathscr{H} - \mathscr{H}_D$. At his juncture we shall try to estimate the effect of such terms by perturbation theory. In principle, if carried out to all orders, this is a systematic means of diagonalizing the Hamiltonian. Once this is done, the proper and improper eigenstates can be separated out by the test that has been established in (5.111). In practice, the perturbation expansion might not converge, and we shall only carry it out to second order by the following means.

Consider the similarity transformation generated by T,

$$\text{op} \rightarrow e^{-T}\text{op}\, e^{+T} .$$

This is a unitary transformation if and only if $T^* = \neg T$, that is, if $T = \mathrm{i} \times$ Hermitean operator. Such a unitary transformation could not diagonalize the Hamiltonian, which in its present form is non-Hermitean and which when diagonalized will be explicitly real, diagonal, and therefore Hermitean. The connection between the two is possible only through an unrestricted T. To lowest order in the interaction, it is determined by expanding,

$$e^T = 1 + T + \frac{T^2}{2} + \cdots$$

and choosing T so as to eliminate $\mathscr{H}'$ from the Hamiltonian. Of what remains, we shall keep only the lowest-order diagonal terms. That is,

$$\begin{aligned}\mathscr{H} &\rightarrow \left(1 - T + \frac{T^2}{2}\right)\mathscr{H}_D\left(1 + T + \frac{T^2}{2}\right) + (1 - T)\mathscr{H}'(1 + T) \\ &\approx \mathscr{H}_D + ([\mathscr{H}_D, T] + \mathscr{H}') - T\mathscr{H}_D T + \frac{1}{2}(T^2\mathscr{H}_D + \mathscr{H}_D T^2) \\ &\quad - T\mathscr{H}' + \mathscr{H}'T + O(\mathscr{H}'^3) \end{aligned} \tag{5.121}$$

together with the relation

$$[\mathscr{H}_D, T] + \mathscr{H}' = 0 \tag{5.122}$$

specify the transformed Hamiltonian to the desired approximation. [Compare (5.122) with (5.126), where this equation is solved for the matrix elements of T]. The quantum numbers of the eigenstates of $\mathscr{H}_D$ are the sets of occupation numbers, and let $a, b, \ldots$ stand for such sets. One finds for the new eigenvalues of (5.121) after solving (5.122)

$$\mathscr{H}_{a,a} = (\mathscr{H}_D)_{a,a} - \sum_b \frac{\mathscr{H}'_{a,b}\mathscr{H}'_{b,a}}{(\mathscr{H}_D)_{b,b} - (\mathscr{H}_D)_{a,a}} \tag{5.123}$$

a familiar expression of the second-order perturbation correction to the energy. Substituting $\mathscr{H}'$ into this expression, one obtains the leading correction to $\mathscr{H}_D$

$$\begin{aligned}\mathscr{H} = \mathscr{H}_D - \frac{1}{4N^2} \sum_{\boldsymbol{k}_1\boldsymbol{k}_2\boldsymbol{k}_3\boldsymbol{k}_4} & \\ \times \frac{\Delta(\boldsymbol{k}_1 + \boldsymbol{k}_3 - \boldsymbol{k}_2 - \boldsymbol{k}_4)[J(\boldsymbol{k}_2) - J(\boldsymbol{k}_2 - \boldsymbol{k}_1)][J(\boldsymbol{k}_3) - J(\boldsymbol{k}_3 - \boldsymbol{k}_4)]}{W(n_{\boldsymbol{k}_1} + 1, n_{\boldsymbol{k}_2} - 1, n_{\boldsymbol{k}_3} + 1, n_{\boldsymbol{k}_4} - 1) - W(n_{\boldsymbol{k}_1}, n_{\boldsymbol{k}_2}, n_{\boldsymbol{k}_3}, n_{\boldsymbol{k}_4})} & \\ \times [(1 + n_{\boldsymbol{k}_1})(1 + n_{\boldsymbol{k}_3})n_{\boldsymbol{k}_2}n_{\boldsymbol{k}_4} - (1 + n_{\boldsymbol{k}_2})(1 + n_{\boldsymbol{k}_4})n_{\boldsymbol{k}_1}n_{\boldsymbol{k}_3})]. &\end{aligned} \tag{5.124}$$

This expression may be verified by use of the following properties of the interaction: $J(\boldsymbol{k}) = J(-\boldsymbol{k})$, and $J(\boldsymbol{k} + \boldsymbol{K}_n) = J(\boldsymbol{k})$, where $\boldsymbol{K}_n$ is one of the reciprocal lattice vectors, e.g., as defined following (5.113). The symbol $\Delta(\boldsymbol{k})$, which is often used to generalize the Kronecker delta to crystal physics, is defined to vanish except when its argument $\boldsymbol{k} = \boldsymbol{0}$ or $\boldsymbol{K}_n$, in which case it equals unity. The contribution of terms for $\boldsymbol{K}_n$ is called *umklapp*, after the German meaning *to be brought back into* (the Brillouin zone). The denominator equals

$$\varepsilon(\boldsymbol{k}_1) + \varepsilon(\boldsymbol{k}_3) - \varepsilon(\boldsymbol{k}_2) - \varepsilon(\boldsymbol{k}_4) + O\left(\frac{1}{N}\right).$$

Whereas the linear spin-wave theory gave magnon energies proportional to s; and whereas the nonlinear corrections in $\mathscr{H}_D$, (5.119), were independent of s; we now find that the perturbation-theoretic correction just obtained vanishes as $\sim s^{-1}$ and therefore will be negligible compared to $\mathscr{H}_D$ in the correspondence limit $s \to \infty$. This is due to the s-dependence of the denominator. Higher-order corrections vanish even more strongly as the number of energy denominators is increased.

Thus there appears to be a well-defined semiclassical limit $s \to \infty$, which in practice means $s \gg \frac{1}{2}$, where $\mathscr{H}_D$ is a sufficiently accurate approximation to the correct Hamiltonian that it may be used with confidence, provided

$$\sum n_{\boldsymbol{k}} \ll Ns\,.$$

$\mathscr{H}_D$ can therefore be considered as the effective Hamiltonian of nonlinear, semi-classical spin-wave theory, with perturbation theory and projection operators required to obtain further corrections.

Note: Even though the elimination of $\mathscr{H}'$ is not very important to the low-lying energies when s is large, it does contribute substantially toward the satisfaction of the sum rule,

$$\left[\frac{1}{N}\sum_i (a_i^*)^{2s+1}(a_i)^{2s+1}\right] = 0 \tag{5.111}$$

for physical states. That the sum rule is not obeyeed in the original spin-wave representation, can be seen by expanding the operators in plane waves. Retaining only the diagonal operators $\mathbf{n}_k$, we find there are $(2s + 1)!$ ways of pairing the creation and annihilation operators and thus obtain for the above,

$$(2s+1)!\left(\frac{1}{N}\sum n_k\right)^{2s+1} \tag{5.125}$$

which may not be large when $\sum n_k \ll N$, but is nevertheless nonzero.

After the similarity transformation, with

$$T = -\frac{1}{2N}\sum_{\substack{k_1k_2k_3k_4\\ k_4\neq k_1,k_3}} \frac{\Delta(\boldsymbol{k}_1+\boldsymbol{k}_3-\boldsymbol{k}_2-\boldsymbol{k}_4)[J(\boldsymbol{k}_2)-J(\boldsymbol{k}_2-\boldsymbol{k}_1)]a_{k_1}^*a_{k_3}^*a_{k_4}a_{k_2}}{W(n_{k_1}+1, n_{k_2}-1, n_{k_3}+1, n_{k_4}-1) - W(n_{k_1}, n_{k_2}, n_{k_3}, n_{k_4})} \tag{5.126}$$

where the denominator equals $\varepsilon(\boldsymbol{k}_1) + \varepsilon(\boldsymbol{k}_3) - \varepsilon(\boldsymbol{k}_2) - \varepsilon(\boldsymbol{k}_4) + O(1/N)$, we obtain for the diagonal part of the sum-rule operator, (5.111),

$$\begin{aligned}(1 - T + \cdots)&\left[\frac{1}{N^{2s+1}}\sum_{k_1\ldots k_{4s+2}} \Delta(\boldsymbol{k}_1 + \cdots + \boldsymbol{k}_{2s+1} - \boldsymbol{k}_{2s+2} - \cdots - \boldsymbol{k}_{4s+2})\right.\\
&\left.\times\, a_{k_1}^* \cdots a_{k_{2s+1}}^* a_{k_{2s+2}} \cdots a_{k_{4s+2}}\right](1 + T + \cdots)\\
&= (2s+1)!\left(\frac{1}{N}\sum n_k\right)^{2s+1}\left\{1 - \frac{(2s+1)(2s)}{2N^3[(1/N)\sum n_k]^2}\right.\\
&\left.\times \sum \frac{\Delta(\boldsymbol{k}_1+\boldsymbol{k}_3-\boldsymbol{k}_2-\boldsymbol{k}_4)[J(\boldsymbol{k}_2)-J(\boldsymbol{k}_2-\boldsymbol{k}_1)]}{\varepsilon_1+\varepsilon_3-\varepsilon_2-\varepsilon_4}F + \cdots\right\}\end{aligned} \tag{5.127}$$

with $F = (1 + n_{k_2})(1 + n_{k_4})n_{k_1}n_{k_3} - (1 + n_{k_1})(1 + n_{k_3})n_{k_2}n_{k_4}$, the ellipsis indicating nondiagonal and higher-order corrections. The improvement is schematically indicated in Fig. 5.5

While the above analysis gives one confidence in the nonlinear spinwave theory developed earlier, it certainly does not apply to 1D or 2D where the existence of bound states at arbitrarily low energies *ensures* the breakdown of ordinary perturbation theory. Indeed, careful study will show the situation to be quite different in the low-dimensional systems from what we have just concluded in three dimensions. We return to this point after a diversion into the study of antiferromagnets.

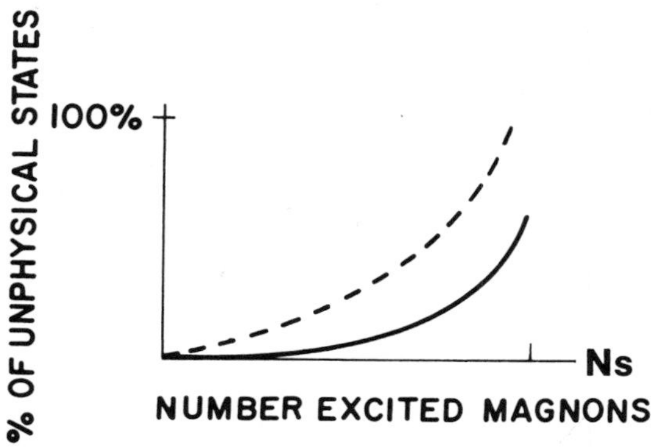

Fig. 5.5. Schematic effect of transformation T on elimination of unphysical states in nonlinear spin-wave theory. Upper curve is before, and lower curve after the transformation

5.9 Antiferromagnetic Magnons: The One-Dimensional XY Model

The following theorems can be proved about the Heisenberg *antiferromagnet*

$$\mathscr{H} = + \sum_{i,j=1}^{N} J_{ij} \boldsymbol{S}_i \cdot \boldsymbol{S}_j \tag{5.128}$$

with $J_{ij} > 0$ for (ij) = nearest neighbors and vanishing otherwise:

1) With some minor restrictions on the J_{ij}, and on the lattice geometry, the ground state is a nondegenerate singlet ($S_{\text{tot}} = 0$). What is more, the lowest energy of total spin S is lower than that of any state belonging to $S + 1$.

2) For spins one-half there is no energy gap against excited states, in the limit $N \to \infty$. With few exceptions the excited states in a many-body system ($N \to \infty$) form a continuum down to the ground state. The content of theorem 2 is that the continuum of $S = 0$ states starts at a lower energy than the continuum of $S = 1$ states, which in turn starts lower than $S = 2$, etc. The proofs of these theorems are given elsewhere,[4] they are related to the existence of long wavelength, low energy disturbances—spin waves. For when a spin deviation

[4] See [5.11–13]. The reader interested in a general introduction to the various aspects of the theory of antiferromagnetism should read the extensive review article [5.14].

propagates at very long wavelengths, the relative orientation of nearest-neighbor spins is maintained to a very good approximation, and consequently the cost in energy must be low.

In 1931 *Hans Bethe*, followed by Hulthen and others, succeeded in solving for the eigenstates of the Heisenberg linear chain model with nearest-neighbor interactions. We shall discuss this in the following section. Unfortunately, because of its difficulty, the Bethe solution has not given any insight into the question of long-range correlations, excited states, etc., and so it must be considered physically incomplete. But by truncating the full Hamiltonian somewhat, in fact by eliminating the z components of the interactions entirely, one arrives at a model which is completely soluble *in one dimension* for spins one-half and nearest-neighbor interactions. It has many of the features of the Bethe solution, but it is quasilinear and fully transparent. We call it the XY model.[5]

$$\mathscr{H}_{XY} = \frac{1}{2}\sum_{i=1}^{N}(S_i^+ S_{i+1}^- + \text{H.c.}), \qquad s = \frac{1}{2} \qquad \text{and} \qquad \boldsymbol{S}_{N+1} \equiv \boldsymbol{S}_1 \tag{5.129}$$

The vacuum (all spins down) is an eigenstate, with energy 0. By translational invariance, one may guess that the "one-particle" states ($N-1$ spins down, one spin up) are plane waves

$$\psi_{\boldsymbol{k}} = \frac{1}{\sqrt{N}}\sum_i \mathrm{e}^{i\boldsymbol{k}\cdot\boldsymbol{R}_i} S_i^+ |0) \tag{5.130}$$

with easily computed energy eigenvalues,

$$E_{\boldsymbol{k}} = \cos ka \qquad \text{with } |ka| < \pi\,. \tag{5.131}$$

Periodic boundary condition $\exp(ikNa) = 1$ results in the discrete set,

$$k = \frac{2\pi}{Na}\times\text{integer} = \frac{\pi(2p)}{Na} \qquad p = 0, \pm 1, \ldots \tag{5.132}$$

Next, a product of two plane waves does not vanish in the configurations $S_i^+ S_i^+$, as it should, but a determinant does. However, a determinant is antisymmetric under the interchange of the coordinates, whereas spins on different sites commute and therefore have a wavefunction symmetric under interchange. The following choice thus imposes itself:

$$\psi_{\boldsymbol{k},\boldsymbol{k}'} = \begin{cases} \dfrac{1}{\sqrt{N(N-1)}}\displaystyle\sum_{i,j>i}[\mathrm{e}^{\mathrm{i}(\boldsymbol{k}\cdot\boldsymbol{R}_i+\boldsymbol{k}'\cdot\boldsymbol{R}_j)} - \mathrm{e}^{\mathrm{i}(\boldsymbol{k}\cdot\boldsymbol{R}_j+\boldsymbol{k}'\cdot\boldsymbol{R}_i)}]S_i^+ S_j^+ |0) \\[2ex] \dfrac{-1}{\sqrt{N(N-1)}}\displaystyle\sum_{i,j<i}[\mathrm{e}^{\mathrm{i}(\boldsymbol{k}\cdot\boldsymbol{R}_i+\boldsymbol{k}'\cdot\boldsymbol{R}_j)} - \mathrm{e}^{\mathrm{i}(\boldsymbol{k}\cdot\boldsymbol{R}_j+\boldsymbol{k}'\cdot\boldsymbol{R}_i)}]S_i^+ S_j^+ |0)\,. \end{cases} \tag{5.133}$$

[5] See [5.12]. *Katsura* [5.15] has calculated the magnetic susceptibility in the XY model, and *Niemeijer* [5.16] the time-dependent spin correlations.

When j is increased to $N - 1$, then to N, and finally to $N + 1$, the second spin becomes the first and the wavefunction changes discontinuously from the upper form to the lower unless the proper boundary condition is imposed. Because the position of the origin of the numbering system is completely arbitrary for the cyclic problem we are solving, such discontinuities at a particular site are inadmissible. The resolution of this difficulty is to take, instead of (5.132),

$$e^{ikNa} = -1, \text{ or } \quad k = \frac{\pi(2p+1)}{Na} \tag{5.134}$$

(and similarly for k'), an *antiperiodic* boundary condition. The generalization to any number of spins up is straightforward. Let

$$\psi_{k_1k_2,\dots} = C \sum_{i_1,i_2,\dots} F^{i_1,i_2,\dots}_{k_1,k_2,\dots} S^+_{j_1} S^+_{j_2} \dots |0) \tag{5.135}$$

where C is the normalization constant, and F is the determinant

$$F^{i_1,\dots}_{k_1,\dots} = \varepsilon_P \begin{vmatrix} e^{ik_1R_{i_1}} & e^{ik_2R_{i_1}} & \dots \\ e^{ik_1R_{i_2}} & e^{ik_2R_{i_2}} & \dots \\ \vdots & \vdots & \end{vmatrix} \tag{5.135a}$$

$\varepsilon_P = +1$ when the spins are in a natural order, $i_1 < i_2 < i_3 < \dots$, or an even permutation of this order, and $\varepsilon_P = -1$ when the spins are arranged in an odd permutation of the natural order. If i_n is the farthest spin, then the translation of i_n from N to $N + 1$ involves a reordering equivalent to an odd permutation for $n + 1$ odd, and an even permutation for $n + 1$ even. Thus

$$n + 1 = \text{odd} \rightarrow k = \frac{\pi}{Na}(2p + 1)$$

and

$$n + 1 = \text{even} \rightarrow k = \frac{\pi}{Na}(2p) \,. \tag{5.136}$$

This is a very interesting situation. For although the many-spin wavefunctions at first appear to be essentially independent particle wavefunctions, yet when one "particle" is added (that is, when one more spin is turned up) *all* the other plane-wave states are modified. This is in the nature of a cooperative effect, due to the effective hard core repulsion [from $(S_i^+)^2 = 0$] of two nearby spin deviations. Whether the k's are members of the even or of the odd set, the energy corresponding to the wavefunction above is

$$E_{k_1,k_2,\dots} = \sum_{i=1}^{n} \cos k_i a = \sum_i \cos \frac{\pi}{N} \times \begin{cases} 2p_i \\ 2p_i + 1 \end{cases} \,. \tag{5.137}$$

Note that all k's must be distinct, or else $F \equiv 0$. The ground state energy is achieved by allowing all the states of negative energy to be occupied, that is, all k's in the range

$$\frac{\pi}{2} \leq ka \leq \frac{3\pi}{2} \text{ i.e., } \frac{N}{4} \leq p \leq \frac{3N}{4}. \tag{5.138}$$

However, we must consider *two* ground states: the ground state for the even k's and the ground state for the odd ones. Denote these by E_0' and E_0'', respectively, with values

$$E_0' = \sum_p \cos \frac{\pi}{N}(2p + 1) \qquad \text{and} \qquad E_0'' = \sum_p \cos \frac{\pi}{N}(2p)$$

$$\text{where} \lim_{N\to\infty} E_0' = E_0'' = -N/\pi \tag{5.139}$$

with the sums in either set restricted to the range, (5.138). The separation of energies E_0' and E_0'' depend on whether N is divisible by 4, or merely by 2, or whether it is odd. In any event, it is of the order of magnitude of the smallest elementary excitations.

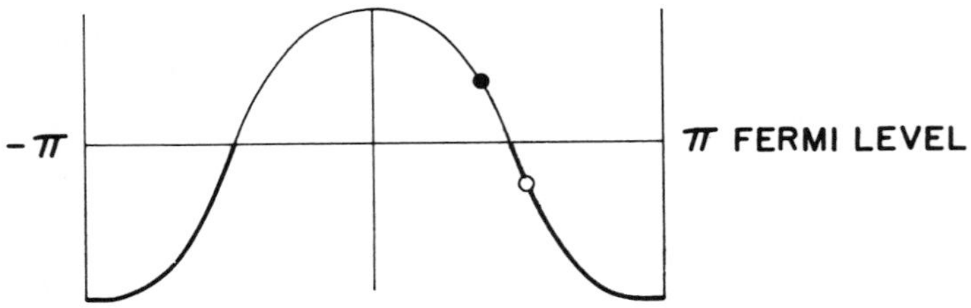

Fig. 5.6. Energy as a function of ka in XY model, showing occupied and unoccupied states. Bubble and dot indicate "quasihole" and "quasiparticle" created in elementary excitation

Either ground state may be represented by an energy-level diagram shown in Fig. 5.6. The Fermi level intersects the cosine curve at $ka = \pi/2$ and $3\pi/2$, and all allowed states below this are filled and all those above it are empty in either ground state. Excited states correspond to occupying states above the Fermi level, or emptying states below it. If k is a wave vector of either category in (5.136), as measured from $\pm\, \pi/2a$, the energy of an elementary excitation, $\varepsilon(q)$ is described by a doubly degenerate spectrum

$$\varepsilon(q) = |\sin q| \qquad \text{setting } q \equiv ka \tag{5.140}$$

in addition to any correction due to changes in parity of the state. Each elementary excitation represents an antiferromagnetic magnon, and it should be noted that *unlike the ferromagnetic magnons, their energy is linear in* q *for small* q. *This appears to be a model-independent property of all antiferromagnetics.*

These results are sufficient for energetic considerations, such as computing the statistical mechanical properties of the model at finite temperatures. They are *not* adequate for the dynamics. For example, in processes used experimentally to probe the excitation spectrum of a solid (optical absorption, neutron scattering, etc.) we must also know the relevant matrix elements. Generally, the response of the system is given in terms of the square of the matrix element, divided by the energy difference between the excited state and the ground state. The low-lying spectrum may in fact not be observable if the matrix element connecting it to the ground state vanishes in the particular experiment designed to probe the spectrum.

To write dynamic operators in the representation in which the Hamiltonian (5.129) is diagonal it is necessary to use the fermion representation given in (3.99) of Chap. 3, the operators c_i. Briefly,

$$S_i^+ S_{i+1}^- = c_i^* c_{i+1} \tag{5.141}$$

and similarly for the Hermitean conjugate expression. With some care taken at the periodic end ($S_N^+ S_1^- +$ H.c.) the Hamiltonian becomes a quadratic form diagonalized by a transformation to running waves [cf. (5.112)].

Problem 5.2. Carry out the indicated diagonalization, paying special attention to the operator Q_N [Chap. 3, (3.100a)] at the end of the periodic chain, for the two distinct cases of total occupancy $\sum c_i^* c_i =$ even or odd.

A typical dynamic matrix element involves operators such as

$$T_k \equiv N^{-1/2} \sum_{n=1}^{N} S_{x,n}\, \mathrm{e}^{ikna}\,. \tag{5.142}$$

In the fermion language,

$$T_k = N^{-1/2} \sum_{n=1}^{N} \frac{1}{2}[c_n^* \exp(\mathrm{i}\pi \sum_{j<n} c_j^* c_j) + \mathrm{H.c.}]\mathrm{e}^{ikna}\,. \tag{5.143}$$

The c_n's are then expressed in the running-wave operators in which the Hamiltonian is diagonal [analogous to (5.112, 113)]. The exponent involves expressions which are very complicated (but not impossible) to evaluate.

In this connection, we quote some preliminary results obtained by *Taylor* and his associates [5.17, 18] in their study of Praeseodymium chloride ($PrCl_3$).

The crystal structure of this compound (C_{6h}^2 symmetry) is such that the nearest neighbor Pr ions lie on the c axis, leading to predominantly one-dimensional interactions between nearest-neighbor Pr ions on each c-axis chain. These interactions involve the Kramers' doublet ground state of each ion and are characterized by a Hamiltonian

$$\mathscr{H} = \frac{1}{2} J \sum (S_i^+ S_{i+1}^- + \text{H.c.}) + \sum g\mu_B H_z S_{z,i} + \sum \gamma(\mathscr{E}_i^+ S_i^- + \text{H.c.}) \quad (5.144)$$

where H_z and $\mathscr{E}_i$ are externally applied magnetic and electric fields. The absorption of photons by the magnons occurs, if momentum and energy are conserved, via the coupling terms. The momentum of infrared photons is very small. We need to evaluate the matrix elements of T for very small k between the ground state and the low-lying excitations, viz., $(T_k)_{\alpha,0}$ and in principle obtain the absorption line shape. The experimental line is shown in Fig. 5.7,

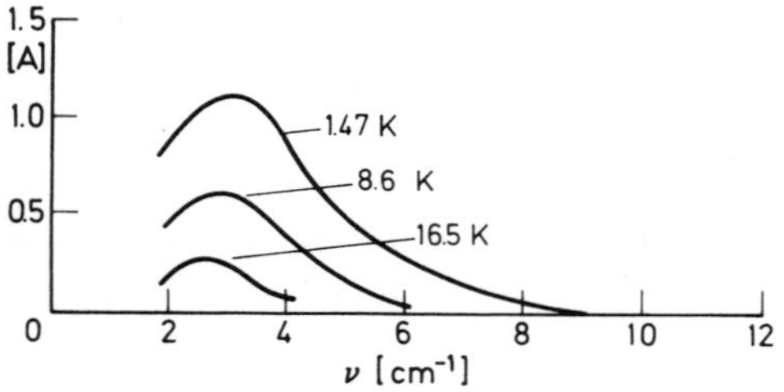

Fig. 5.7 Absorption coefficient A for $PrCl_3$ at various temperatures. [5.17,18]

and is seen to peak at 3 cm^{-1}. Instead of computing this line shape, one can try to calculate this peak (which corresponds to the average absorption energy as well). The result is

$$\bar{E}_k = J \frac{(0| -S_{y,m-1}S_{y,m} + S_{z,m-1}S_{z,m} \cos ka|0)}{1/4 + 2\sum_n \cos nka(0|S_{x,m-n}S_{x,m}|0)} \quad (5.145)$$

where m is any fixed site. The numerator is evaluated from the fermion representation, and yields $1/2\pi - \cos(ka/\pi^2)$. The denominator was evaluated numerically [5.17, 18] to yield 0.0386, at $k = 0$. The result is $\bar{E}_0 = 1.50J$. With $J = 2.0$ cm^{-1}, this leads to $\bar{E}_0 = 3$ cm^{-1}, in excellent agreement with the location of the experimental absorption peak. Of course, the evaluation of the exact matrix element and fitting of the absorption line shape, Fig. 5.7, would be even more compelling evidence for this theory, but the onerous calculations have not yet been performed.

5.10 Bethe's Solution of One-Dimensional Heisenberg Antiferromagnet

Progress in solving the Heisenberg antiferromagnet, even in one dimension, has been slow. Although *Bethe* [5.19] gave the wavefunctions in 1931, the ground-state energy was not obtained until 1938, by *Hulthen* [5.20]. It was subsequently generalized by *Orbach* [5.21], who discussed an anisotropic Hamiltonian, and *Walker* [5.22] who studied the analytic properties of the Orbach solutions. Finally, fully 32 years after the initial progress, *Des Cloizeaux* and *Pearson* [5.23] calculated the one-magnon spectrum and *Griffiths* the magnetic susceptibility [5.24].

It is desired to calculate the eigenstates of

$$\mathscr{H} = \sum_{i=1}^{N}\left[\frac{1}{2}(S_i^+ S_{i+1}^- + \text{H.c.}) + g S_i^z S_{i+1}^z\right] \qquad \text{for } s = \frac{1}{2},\ g = 1. \tag{5.146}$$

Other limits are:
$g = \infty$ (Ising model), $g = 0$ (XY model), $g = -1$ (transformable into ferromagnet by trivial rotation of every second spin). For *all* $g > -1$ the ground state belongs to $M = 0$, *i.e.* contains Ns "particles".

We start with 2 "particles", using a procedure valid only for spins one-half and assign to the unphysical amplitudes f_{ii} the value determined by

$$f_{i,i} + f_{i+1,i+1} = f_{i,i+1} + f_{i+1,i} = 2 f_{i,i+1} \tag{5.147}$$

where the physical amplitudes $f_{ij} (= f_{ji}$ for $j \neq i)$ obey the equations

$$(E - E_{\mathrm{f}} + 2) f_{ij} - \frac{1}{2}(f_{ij+1'} + f_{ij-1} + f_{i+1j} + f_{i-1j}) = 0 \tag{5.148}$$

This is the left-hand side of (5.55), with $s = \frac{1}{2}$, $J = -1$, $H = 0$, and $z = 2$. (5.147), playing the role of a boundary condition, ensures that the right-hand side of that equation always vanishes. The homogeneous equation is solved by plane waves, *via Bethe's ansatz*:

$$f_{ij} = \mathrm{e}^{\mathrm{i}(ki + k'j + 1/2\psi)} + \mathrm{e}^{\mathrm{i}(kj + k'i - 1/2\psi)} \qquad (j \geq i) \tag{5.149}$$

with f_{ij} in the range $j < i$ given by symmetry, $f_{ij} = f_{ji}$. The phase factor ψ ranges over the interval $-\pi, +\pi$. (In the XY model, (5.133), the effective ψ is a constant $\pm\pi$). We make no attempt to normalize solutions here, as it is not required in the calculation of energy eigenvalues and represents some fairly involved analysis.

Inserting the above form of f_{ij} into the boundary condition and performing some algebra, we obtain

$$2\cot\frac{1}{2}\psi = \cot\frac{1}{2}k - \cot\frac{1}{2}k' \tag{5.150}$$

with a second, periodic, boundary condition,

$$f_{iN} = f_{0i} \tag{5.151}$$

determining the modified k, k'

$$k = \frac{\pi(2p) + \psi}{N} \quad \text{and} \quad k' = \frac{\pi(2p') - \psi}{N} \tag{5.152}$$

with p,p' = integers. Thus, a two-magnon state is described by two wave vectors and a phase shift; there is refraction but no scattering. The sum $k + k'$ is independent of ψ and is the center of mass wave vector $\boldsymbol{K}$ which parametrized the solution in an earlier section of this chapter. The difference k-k' corresponds to the variable $\boldsymbol{q}$. The bound state that appeared below the continuum threshold at $\boldsymbol{q} = 0$ is now *above* the continuum, due to the change in sign of the interaction. Now, as k approaches k', (5.150) shows that ψ approaches $\pm\pi$, the upper sign applying if $k < k'$. From (5.152), if $p = p' - 1$, $k = k'$ and $\psi = \pi$ so that $f_{ij} = \exp ik\,(i + j) \times 2\cos\frac{1}{2}\pi \equiv 0$. Therefore, $|p - p'| \geq 2$.

The energy, measured from the ferromagnetic level $E_{\mathrm{f}} = \frac{1}{4}N$, is given by the above eigenvalue equation as

$$E - E_{\mathrm{f}} = -(1 - \cos k) - (1 - \cos k') \tag{5.153}$$

just the energy of two noninteracting magnons; the interactions are included implicitly, *via* the phase shifts in k, (5.150, 152)

Bethe's ansatz consists partly in the statement that in the many-particle state, the amplitudes are subject to phase shifts that are simply given, as the sum of the two-particle phase shifts. Thus, for a given ordering $i \leq j \ldots \leq n$,

$$f_{ij\cdots mn} = \left\{\exp\left[i\left(k_1 i + k_2 j + \ldots + k_n n + \frac{1}{2}\sum_{r<t}^{n}\sum^{n}\psi_{k_r k_t}\right)\right] + \begin{pmatrix}\text{all } n! - 1\\ \text{remaining permutations}\\ \text{of the } k\text{'s}\end{pmatrix}\right\}. \tag{5.154}$$

If we need $f_{i,j,\cdots}$ for any other ordering of the particles, we obtain it from the symmetry under permutations: $f_{i,j,\cdots,m,\cdots} = f_{i,m,\cdots,j,\cdots}$ so we need to know it only in the given interval. In this region, the amplitude f consists of a product over plane-wave factors, summed over all permutations of the wave vectors with the ψ's antisymmetric in their subscripts and satisfying the equations

$$2\cot\frac{1}{2}\psi_{kk'} = \cot\frac{1}{2}k - \cot\frac{1}{2}k' \tag{5.155}$$

$$k = \frac{\pi(2p) + \sum_{1}^{n}\psi_{kk'}}{N} \qquad p = 0, \pm 1, \ldots . \tag{5.156}$$

The sum in this equation, as well as in the next, is over the set of k's present in (5.154) The energy, measured relative to the ferromagnetic reference energy $E_{\mathrm{f}} = N/4$, is

$$E - E_{\mathrm{f}} = -\sum_{t=1}^{n}(1 - \cos k_t) . \tag{5.157}$$

Note that if we translate the entire chain by one site, that is, let $i, j, \ldots \to i + 1, j + 1, \ldots$, the wavefunction, (5.154), is multiplied by a phase factor

$$\exp\left(\mathrm{i}\sum_{\mathrm{i}}^{n} k\right) = \exp\left(\pi\mathrm{i}\sum_{1}^{n} 2p/N\right) . \tag{5.158}$$

The exponent is identified as the total momentum, modulo 2π, which is again independent of the phase shifts $\psi_{kk'}$. This is important, because the phase shifts themselves are now rather large. Each $\psi_{kk'}$ is $O(1)$; there are a number of them contributing to each k, and the total shift in k is $O(n/N)$ or a substantial fraction of k.

In the two-particle problem, there was no solution of (5.150, 152) for $\boldsymbol{k} - \boldsymbol{k}' \approx 0$. Similarly, in the many-particle state, one must choose the interval between k's such that no $p = 0$, and such that for all p and p',

$$|p - p'| \geq 2 \tag{5.159}$$

for a non-vanishing solution to exist. The ground state, for $N/2$ particles subject to this restriction, has the set of $\{p\}$:

$$\{p\} = 1, 3, \ldots, N - 1 . \tag{5.160}$$

Notice an interesting effect of the $S_i^z S_{i+1}^z$ "interaction," which is to spread the set of $\frac{1}{2}N$ integers p (restricted over the range $N/4 < p < 3N/4$ in the ground state of the XY model) to cover the entire range of phase space at present.

The regular spacing permits us to replace sums by integrals in the limit $N \to \infty$, and thus, with $x = p/N$ and $\Delta p = \frac{1}{2}Ndx$

$$E - E_f = -\frac{N}{2}\int_0^1 [1 - \cos k(x)]\, dx = -N\int_0^1 \sin^2 \frac{k(x)}{2} dx \quad \text{where} \tag{5.161}$$

$$2 \cot \frac{\psi(x, y)}{2} = \cot \frac{k(x)}{2} - \cot \frac{k(y)}{2} \qquad \text{and} \tag{5.162}$$

$$k(x) = 2\pi x + \frac{1}{2}\int_0^1 \psi(x, y) dy \, . \tag{5.163}$$

as $|\psi|$ cannot exceed π, (5.162) indicates that ψ has a jump discontinuity of -2π at $x = y$. More precisely, the derivative is

$$\frac{\partial \psi(x, y)}{\partial x} = -2\pi\delta(x - y) + \frac{1}{4}(\xi^2 + 1)\frac{dk(x)/dx}{1 + 1/4(\xi - \eta)^2} \tag{5.164}$$

where we use the short-hand notation

$$\xi \equiv \cot \frac{1}{2} k(x) \qquad \text{and} \qquad \eta \equiv \cot \frac{1}{2} k(y) \, . \tag{5.165}$$

We also find it convenient to introduce the functions

$$f(\xi) = -(d\xi/dx)^{-1} \qquad \text{and} \qquad f(\eta) = -(d\eta/dy)^{-1} \tag{5.166}$$

and to differentiate (5.163) w.r. to x to obtain

$$dk(x)/dx = \pi + \frac{1}{2}\int_{-\infty}^{+\infty} \frac{f(\eta)/f(\xi)}{1 + 1/4(\xi - \eta)^2} d\eta \, . \tag{5.167}$$

With the identity $f(\xi)\, dk/dx = 2(1 + \xi^2)^{-1}$ this becomes

$$\frac{2}{1 + \xi^2} = \pi f(\xi) + 2\int_{-\infty}^{+\infty} \frac{f(\eta)}{4 + (\xi - \eta)^2} d\eta \, . \tag{5.168}$$

This type of equation is soluble because the kernel is a function only of the difference $(\xi - \eta)$. One may take Fourier transforms,

$$F_k \equiv \int_{-\infty}^{\infty} d\theta f(\theta) e^{ik\theta} \qquad \text{and} \qquad f(\theta) = \int_{-\infty}^{\infty} \frac{dk}{2\pi} F_k e^{-ik\theta} \tag{5.169}$$

to obtain a simple solution of (5.168)

$$F_k = (2 \cosh k)^{-1} \, . \tag{5.170}$$

The ground state energy (5.161) has the form

$$E - E_f = -N \int_{-\infty}^{\infty} d\xi \frac{f(\xi)}{1+\xi^2} = -N \int_{-\infty}^{\infty} \frac{dkF_k}{2\pi} \int_{-\infty}^{\infty} d\xi \frac{e^{-ik\xi}}{1+\xi^2}$$

$$= -2N \int_0^{\infty} dkF_k e^{-|k|} = -2N \int_0^{\infty} \frac{dk e^{-2|k|}}{1+e^{-2|k|}} = -N \ln 2 \, . \qquad (5.171)$$

This is a very famous result, being one of the first exact solutions ever obtained of a nontrivial many-body problem in quantum mechanics.

In the usual terms of energy *per* spin, the Bethe-Hulthén result for E/N represents $-\ln 2 + \frac{1}{4} = -0.443$. We can compare this to the ground state energy *per* spin in the XY model [(5.139) $\times$ $1/N$] which is $-\pi^{-1} = -0.318$. While this is a variational upper bound to the Heisenberg model ground state energy, it is evidently not a very close one. On the other hand, $\frac{3}{2}$ $\times$ ground state energy of the XY model is a lower variational bound on the Heisenberg model (really, an "XYZ" model) and this yields $-3\pi/2 = -0.477$ reasonably close to the exact answer. By way of contrast, we note that the Ising model ($S_i^z S_{i+1}^z$) has energy -0.250 *per* spin in these same units, which is a variational upper bound (although a poor one) to both XY and Heisenberg models. Moreover, $2 \times (-0.25) = -0.500$ is a poor lower bound to the XY model and $3 \times (-0.25) = -0.750$ is an even worse lower bound to the Heisenberg energy *per* spin. (The reasons for these larger discrepancies may be attributed to the lack of spinwaves in the Ising model).

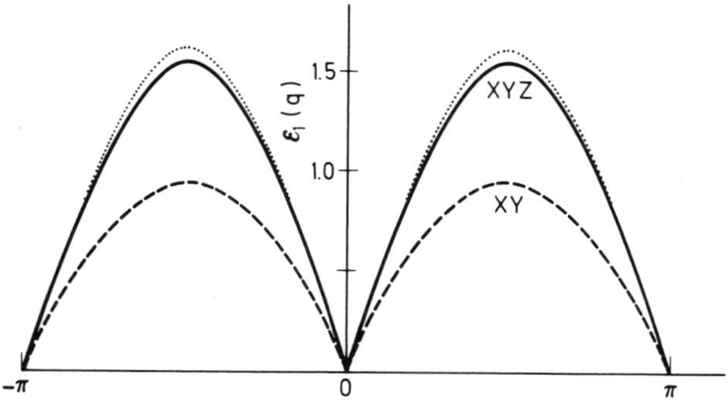

Fig. 5.8. Quasiparticle spectrum $\varepsilon_1(ka)$ obtained by Des Cloizeaux and Pearson for the infinite Heisenberg antiferromagnetic chain, $\pi/2 \sin ka$ (*upper curve*), compared with their numerical results for chain of 48 spins. Also, spectrum for infinite XY antiferromagnetic chain (*lower, dashed curve*). Anderson's linerarized antiferromagnons coincide precisely with the XY model

Des Cloizeaux and *Pearson* [5.23], Fig. 5.8, have extended Bethe's analysis to obtain the energy of some of the lowest excited states, which are known to be triplet states [5.11]. $N/2 - 1$ values of k are chosen by them, to obtain the eigenstates of lowest energy belonging to a finite total momentum $\sum k$. This procedure is very delicate, however, and somewhat too complicated to reproduce or to justify here. There is an unanswered question of whether bound states participate among the low-lying excitations. If they do, then Bethe's *ansatz* formalism is incapable of describing them since the validity of the various equations above seems predicated on having real k's. Nevertheless, it is entirely possible that further mathematical study of (5.152) and those following may allow the extension of the known results to complex k plane and that a complete classification scheme will result. For example, following Orbach one might consider the $S_i^z S_{i+1}^z$ coupling terms with variable parameter g, and study the energy levels as g is adiabatically increased. At $g = 0$, we recover the XY model (of which the solutions are completely determined) and then increase it to $g = 1$ which is the present Heisenberg model. As g is further increased, the Ising model is approached. In fact, $g = \pm 1$ are two "critical points": for $g > 1$ the properties approach those of an Ising antiferromagnet (e.g., the ground state energy approaches $-\frac{1}{4}g$ *per* spin at large g, the excitation spectrum has a gap), and for $g < -1$ they resemble those of an Ising ferromagnet. It is only in the range $-1 \leq g \leq +1$ that results are obtained similar to the gapless spectra we have seen here. For more details, the papers of *Orbach* [5.21] or *Walker* [5.22] or the more recent studies by *C.N. Yang* and *C.P. Yang* [5.25] or the generalization by *Baxter* [5.26] to the completely anisotropic linear chain can be consulted. The case $g > 1$ has been analyzed in [5.27, 28].

Des Cloizeaux and Pearson's results for the one-magnon triplet ($S = 1$) excitation spectrum is shown in Fig. 5.8, together with some numerical results on chains of 48 spins and compared there to the spectrum of the XY model ($g = 0$). This comparison will have further significance when we study approximate theories of antiferromagnetism in the following sections. If we denote the triplet excitation spectrum ε_1, it is given by

$$\varepsilon_1 = \frac{1}{2}\pi \, |\sin q| \ . \tag{5.172}$$

Yamada [5.29], and *Bonner* et al. [5.30] noted ε_1 is most likely the lower edge of a continuum of triplet excitations, with a curve ε_2 defining the upper edge

$$\varepsilon_2 = \pi \, |\sin \frac{1}{2}q| \ . \tag{5.173}$$

This continuum is illustrated in Fig. 5.9. Undoubtedly, excitations belonging to $S = 0, 2, 3, \ldots$ lie in other continua that partly intersect this. However, in an experimental situation where one unit of angular momentum is transferred to the magnetic system, only the triplet continuum can contribute. For such processes, *Hohenberg* and *Brinkman* [5.31] have computed that the matrix elements peak near the lower, des Cloizeaux-Pearson, threshhold. This topic, and its nontrivial extension to the case of an applied magnetic field, is increasingly being studied theoretically [5.32] and experimentally [5.33].

One important result follows immediately from Fig. 5.9, viz., *the size of the Brillouin Zone (BZ) remains 2π for the linear chain antiferromagnet*. With only the spectrum of Fig. 5.8 one might suspect that the BZ could be folded in half, which is equivalent to the choice of 2 spins as the primitive unit cell. Physically, this is a very attractive idea: the couple of spin "up" and "down" nearest-neighbors make an ideal unit cell, and this is indeed the starting point of many approximate analyses [5.33]. With the *complete* excitation spectrum, however, one sees that the translational invariance of the ground state is preserved *and the larger BZ must be retained.*[6]

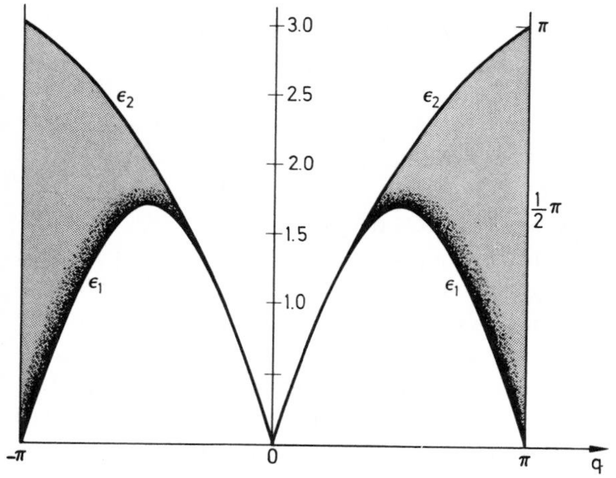

Fig. 5.9. The continuum of magnon ($S = 1$) excitations for the Heisenberg XYZ linear chain antiferromagnet, indicating the two bounding curves ε_1 and ε_2 given in the text

5.11 Linearized Antiferromagnetic Magnons

The quantization of antiferromagnons in the linear approximation is performed by analogy with the procedure in the theory of ferromagnetism. Such an approximate quantum theory, in which only terms quadratic in the P's and Q's are retained in the Hamiltonian, was originally proposed by *Anderson* [5.35], and was very successful in the early interpretation of the dynamics of antiferromagnets.[7] As can be expected, the frequencies are identically the same as found

---

[6] The general topic of correlation functions for the ground state (and also, at finite T) for this model has been much studied in the recent past. The interested reader may wish to study the papers in [5.34].

[7] See also [5.14].

in the classical linearized equations of motion of the antiferromagnet (discussed earlier); only the amplitudes are quantized as Bosons. Below, we shall indicate in what vital manner procedures which are satisfactory in ferromagnetism must be changed for the present problem.

The paramount difficulty is that the ground state of the Heisenberg antiferromagnet is not known, with the exception of the linear chain. Alas, it is quite apparent that the procedure which worked in one dimension is not generalizable to three because of the heavy reliance on ordering the spin deviates along a line. The somewhat simpler XY model is similar to the quantum-mechanical hard-sphere Boson problem, for which also no adequate intermediate- or high-density theory exists in two or three dimensions.

Fortunately, there exists a perturbation theory of sorts, which gives us a "handle" on an approximate, variational ground state. It is a series in powers of $1/s$. The off-diagonal parts of the Hamiltonian are linearized, and diagonalized as best can be, in the standard way. The starting, or zeroth, point of this perturbation procedure is the classical ground state: the configuration of lowest energy of the Hamiltonian with spins replaced by classical vectors of fixed length and position but variable orientations. The choice of this classical ground state is a soluble problem (see pp. 232, 233). We determine it to be a two-sublattice antiferromagnet with the following properties:

Antiferromagnetic coupling between nearest neighbors, and a lattice geometry such that two spins which are nearest neighbors of a third are not nearest neighbors of each other. These requirements are met, e.g., in the linear chain, simple square, simple cubic, and body-centered cubic lattices (but not in the face-centered cubic lattice). The classical ground state is then the *Néel state*: every spin *up* is surrounded by nearest neighbors which are *down*, and vice versa.

To give them a name, denote the spins down the *A sublattice*, and the spins up the *B sublattice*. Perform a canonical transformation on the B (but not on the A) spins: rotate them by 180° about the S^x axis, to obtain the so-called Néel state

$$S_j^{\pm} \rightarrow +S_j^{\mp}, \qquad S_j^z \rightarrow -S_j^z \qquad (j \text{ in } B)\,. \tag{5.174}$$

At first, assume that i is in A and j is a nearest neighbor of i in B to write the Hamiltonian in the form

$$\mathscr{H} = \sum_{i \leq A} \sum_{j \leq B} \left[\left(\frac{1}{2} S_i^+ S_j^+ + \text{H.c.}\right) - S_i^z S_j^z\right]. \tag{5.175}$$

It is convenient to take the exchange constant J to be the unit of energy, so that it need not be written explicitly. Now, because the last result is explicitly symmetric in the two sublattices, one allows i to range over the entire crystal and j over all nearest neighbors of i, denoted by $j(i)$, and divides by 2 so as not to double-count bonds.

$$\mathscr{H} = \frac{1}{2} \sum_{\text{all } i} \sum_{j(i)} \left[\frac{1}{2}(S_i^+ S_j^+ + \text{H.c.}) - S_i^z S_j^z \right] \tag{5.176}$$

In the "linear" approximation, the Hamiltonian becomes

$$\begin{aligned} \mathscr{H}_{\text{lin}} &= \frac{1}{2} \sum_i \sum_{j(i)} [s(a_i^* a_j^* + \text{H.c.}) - s^2 + s(a_i^* a_i + a_j^* a_j)] \\ &\to -\frac{1}{2} Nzs^2 + zs \sum_k a_k^* a_k + \frac{s}{2} \sum_k \left(a_k^* a_{-k}^* \sum_\delta e^{ik\cdot\delta} + \text{H.c.} \right). \end{aligned} \tag{5.177}$$

After the transformation to running-wave Boson operators, it is not yet diagonal. We eliminate the pair-creation terms by a method first due to *Holstein* and *Primakoff* [3.6]. Subsequently rediscovered by Bogolubov in his theory of Bose condensation, and now generally known as the *Bogolubov transformation*, it is:

$$\begin{aligned} a_k &\to (\cosh u_k) a_k + (\sinh u_k) a_{-k}^* \\ a_k^* &\to (\cosh u_k) a_k^* + (\sinh u_k) a_{-k} \qquad \text{with} \qquad u_k = u_{-k} = u_k^* . \end{aligned} \tag{5.178}$$

One must choose the function u_k so that the terms $a_k^* a_{-k}^*$ are eliminated. It is a matter of some simple algebra to determine that what is required is

$$\tanh 2u_k = \frac{-1}{z} \sum_\delta \cos \boldsymbol{k}\cdot\boldsymbol{\delta} \tag{5.179}$$

with $\boldsymbol{\delta}$, a vector connecting any spin with any of its z nearest neighbors. This reduces the linearized Hamiltonian to the diagonal form,

$$\mathscr{H}_{\text{lin}} = -\frac{1}{2} Nzs(s+1) + \sum_k \left(n_k + \frac{1}{2} \right)(zs)\sqrt{1 - \tanh^2 2u_k} . \tag{5.180}$$

Before discussing the magnon spectrum it is not unwise to digress somewhat and examine the ground-state energy. In (5.180), it is the energy eigenvalue when all $n_k = 0$, that is, the vacuum energy. Conventionally, one writes it in the following form:

$$E_0 = -\frac{1}{2} Ns^2 z \left(1 + \frac{\gamma}{zs} \right). \tag{5.181}$$

The parameter γ may be shown to be bounded between 0 and 1. Indeed, Hulthen's exact result for the linear chain of (5.171) ($z = 2, s = \frac{1}{2}$) yields $\gamma = 0.7726$, which is perhaps the largest attainable value. The results in Table 5.1 were computed by the linear theory [5.35] (5.180), for nearest-neighbor interactions. But it is interesting to see the good agreement (within $\approx$6 percent) with the exact result for the linear chain, where the spin-wave theory might have been expected to fail badly. Although for the one-dimensional magnon spectrum the

Table 5.1 γ for various lattices

Lattice	z	γ
Linear chain	2	0.726 (exact: 0.7726 . . for spins 1/2)
square	4	0.632
Simple cubic	6	0.58
Body-centered cubic	8	0.58

agreement will not appear so favorable, it is not farfetched to conceive that the ground state and excited states of the simple theory may be in good agreement with the exact result for lattices with higher coordination numbers.

Again for one dimension, using (5.179, 180), we find for the magnon energy,

$$\varepsilon(k) = (2s) \sin ka \tag{5.182}$$

with a = lattice spacing. For spins $\frac{1}{2}$ the result is similar to the exact result for the XY model, but is too low by a factor $2/\pi$ for spins $\frac{1}{2}$ in the Heisenberg model, which the spin-wave theory is supposed to approximate. The discrepancy is seen in Fig. 5.8.

In any number of dimensions, the magnon spectrum

$$\varepsilon_{\boldsymbol{k}} = s\sqrt{z^2 - (\sum_{\delta} \cos \boldsymbol{k}\cdot\boldsymbol{\delta})^2} \tag{5.183}$$

is *doubly* degenerate, and is linear in k for small k[It should be noted that, properly speaking, magnons are *triplet*, not *doublet* excitations ($S = 1$)].

Problem 5.3. Derive the magnon spectrum (5.183) from (5.179, 180) evaluated for nearest-neighbors. Generalize (5.183) to arbitrary range interactions.

The existence of two distinct modes of zero energy, one at $\boldsymbol{k} = \boldsymbol{0}$ and the other at the edge of the Brillouin zone, means that magnons belonging to these wave vectors can be emitted at no cost of energy, and indicates a high degeneracy of the approximate ground state. This is a manifestation of the rotational invariance of the spin Hamiltonian. By the emission of zero energy magnons, a state can also be reached in which the A and B sublattices have been interchanged. However, the *true* ground state is *non*degenerate [5.11].

The diagonalization of $\mathscr{H}_{\text{lin}}$ leads to a reduction of the S^z spin components from their saturation magnitude in the Neel state. Calculating this reduction we find the reduction in sublattice magnetization due to this:

$$\langle \delta S_i^z \rangle \equiv \left\langle \frac{1}{N} \sum_i (s - |S_i^z|) \right\rangle = \frac{1}{2N} \sum_{\boldsymbol{k}} \left(\frac{1}{\sqrt{1 - \tanh^2 2u_{\boldsymbol{k}}}} - 1 \right) \approx 0.078 \quad \text{sc.} \tag{5.184}$$

The numerical value represents an estimate by Anderson of the integral, for the simple cubic (sc) structure. Note that for the linear chain this formula yields

$$\langle \delta S_i^z \rangle = \infty \qquad \text{lin. chain} \tag{5.185}$$

due to the divergence of the integral at *long* wavelengths. This implies an absence of long-range order in one dimension even in the spin-wave approximation, a result in accord with other analyses.

We now compare the results of linear spinwave theory (LSW) with other calculations, including the most accurate cellular method of *Betts* and *Oitmaa* in a table adapted from one of their recent works [5.37]. LSW is in good agreement with the last, and presumably best, result.

Table 5.2. Comparison of various calculations of the ground state energy parameter $e_0 \equiv 1 + \frac{1}{2}\gamma$ and of $\langle \delta S_i^z \rangle$ for the $s = \frac{1}{2}$ Heisenberg antiferromagnet on a square lattice ($z = 4$)

Method	Reference	e_0	$\langle \delta S_i^z \rangle$	γ
LSW	[5.35]	1.32	0.20	0.64
perturbation theory	[5.36]	1.33	0.12	0.66
Variational	[5.38]	1.29	0.07	0.58
Variational	[5.39]	1.32	0.10	0.64
Cellular	[5.37]	1.31	0.25	0.62

The method by which Betts and Oitmaa have been obtaining the ground state parameters in 2 and 3 dimensions deserves comment. Let us illustrate with the square lattice, then quote their results for the XY antiferromagnet and for both Heisenberg and XY models in 3D.

Figure 5.10 illustrates finite cells of 4, 8, 10, 16 and 18 spins on the square lattice. In each case the infinite lattice can be filled by periodic extensions of the cell; thus, the wavefunction in the cell is subjected to periodic boundary conditions. It consists of linear combinations of states belonging to the symmetry appropriate to the ground state. For example, starting with the Néel state, one overturns a pair of n.n. spins in all possible ways. One then overturns a pair in any of these states, etc., until all the states which can mix in the ground state are found. For the 16 spin diamond illustrated in Fig. 5.10 there are 6 generations of spin flips before the total of 153 basis states is achieved (a great deal less than the total number of 2^{16} basis states)! For $N = 18$, 398 states are generated, and the resulting matrices are easily diagonalized to yield the lowest eigenvalue, the ground state energy. When plotted vs $1/N$, as in Fig. 5.11, these cellular ground state energies typically lie on a straight line that extrapolates to $1/N = 0$.

From the ground state energy one can readily extract the nearest neighbor correlation function $(\phi_0 | S_i^x S_j^x | \phi_0) = (\phi_0 | S_i^y S_j^y | \phi_0) = (\phi_0 | S_i^z S_j^z \phi_0 |)$, with i, j n.n.

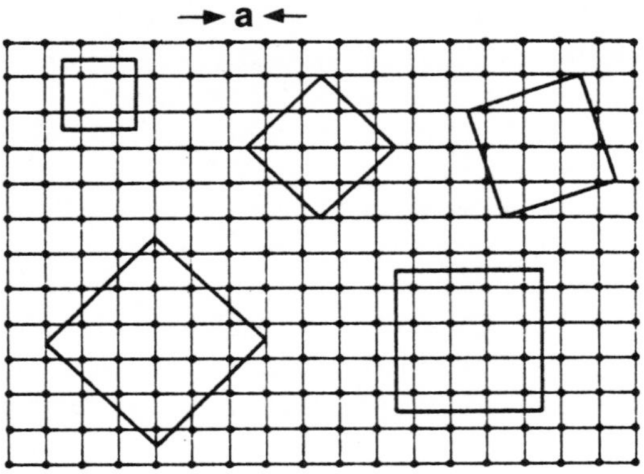

◁ **Fig. 5.10.** Finite cells of 4, 8, 10, 16, 18 spins on the square lattice. In each case the infinite lattice is filled by periodic repetition of the cells

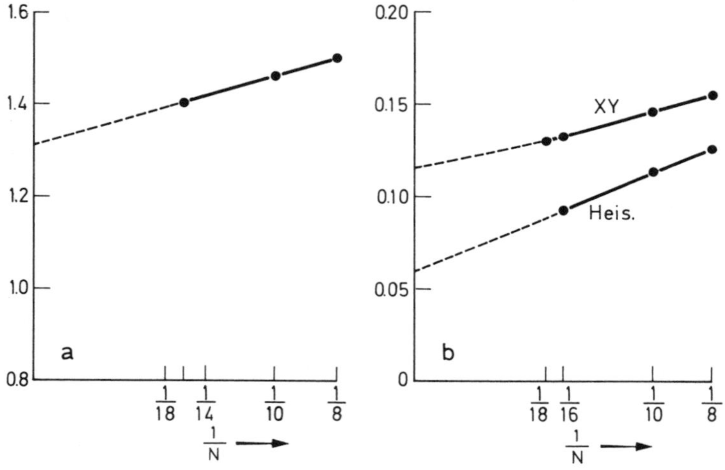

▽ **Fig. 5.11.(a)** Ground state energy of the $s = \frac{1}{2}$ Heisenberg antiferromagnet on a square lattice **(b)** Long range order parameter for XY and Heisenberg antiferromagnets, showing the existence of LRO in the ground state [5.37]

Using the computed ground states, Betts and Oitmaa also obtain the correlation functions for i, j *not* n.n. Two spins correlate positively when on the same sublattice, but negatively when on different sublattices. For the antiferromagnetic XY model, the X and Y correlations are of course equal, but the Z correlation function is distinct. Calculation of both $(\phi_0 | S_i^x S_j^x | \phi_0)$ and $(\phi_0 | S_i^z S_j^z | \phi_0)$ show them to decrease surprisingly slowly with distance. These results are evident in Table 5.3 taken with suitable modification from [5.37]. Betts and Oitmaa concluded that the long range order persists in $2D$ in the ground states; plotting $2/N \sum_{i,j} (-1)^{i-j} \langle S_i^x S_j^x \rangle$ vs $1/N$, they obtained the lines in Fig. 5.11 which extrapolate to LRO = 0.116 for the XY model and 0.059 for the Heisenberg antiferromagnet (compared to a possible maximum of 0.25).

Table 5.3. Ground state correlations for the 16 spin cell on the square lattice, $s = 1/2$

	XY		Heis. A.-F.
R_{ij}/a	$4\langle S_i^z S_j^z \rangle$	$4\langle S_i^x S_j^x \rangle$	$4\langle S_i^z S_j^z \rangle$
1	0.182	− 0.562	− 0.468
$2^{1/2}$	0.027	0.487	0.285
2	0.027	0.487	0.285
$5^{1/2}$	0.023	− 0.468	− 0.270
$2 \times 2^{1/2}$	0.018	0.458	0.240

LSW theory on the 2D and 3D XY models is extremely simple, although highly inaccurate for $s = 1/2$ (see, however, the following section). Starting with the exact Hamiltonian

$$\mathscr{H} = \frac{1}{4} \sum_{i,j} (S_i^+ S_j^- + \text{H.c.}) \tag{5.186a}$$

we use the linear approximation to the Holstein-Primakoff operators, and write the above as

$$\mathscr{H} = \frac{1}{4} s \sum_{i,j} (a_i^* a_j + \text{H.c.}) \tag{5.186b}$$

which, after the phase change $a_i, a_i^* \to -a_i, a_i^*$ on the B sublattice alone, yields

$$\mathscr{H} = -\frac{1}{4} s \sum_{i,j} (a_i^* a_j + \text{H.c.}) \tag{5.186c}$$

and after the transformation to running wave operators, (5.112, 113), yields

$$\mathscr{H} = -\frac{1}{4} s \sum_{k} (\cos k_x a + \cos k_y a) \mathfrak{n}_k \tag{5.186d}$$

in 2D, with an analogous expression in 3D.

In the ground state, a symmetry principle requires the eigenvalue of the operator S_{tot}^z to be zero [5.40], thus for $s = \frac{1}{2}$ the magnon population is:

$$\mathfrak{n}_{\text{op}} = \sum_{k} \mathfrak{n}_k = \frac{1}{2} N \tag{5.187}$$

in the ground state. Evidently, the ground state is achieved by populating the $\boldsymbol{k} = \boldsymbol{0}$ mode only, thus $n_0 = \frac{1}{2} N$ and $n_k = 0$ for $\boldsymbol{k} \neq \boldsymbol{0}$. This yields for the $s = \frac{1}{2}$ correlation functions

$$4(\phi|S_i^x S_j^x|\phi) = (-1)^{i-j}(\phi|(a_i^* a_j + a_j^* a_i + a_i a_j + a_i^* a_j^*)|\phi)$$
$$= (-1)^{i-j}\frac{1}{N}\sum_k (2\cos k\cdot R_{ij})\,\mathbf{n}_k \tag{5.188}$$

in any eigenstate ϕ. In the ground state ϕ_0, this sum is N, so (5.188) is ± 1 depending whether the two spins are on the same sublattice or not. This result is twice as large as the computed values in Table 5.3 and does not exhibit the rather mild, but quite definite fall-off with distance of the computed results. Similarly, the other $s = \frac{1}{2}$ correlation function is

$$4(\phi_0|S_i^z S_j^z|\phi_0) = 4(\phi_0|a_i^* a_j|\phi_0)(\phi_0|a_j^* a_i|\phi_0) = 1, \tag{5.189}$$

regardless of sublattice, having the correct sign but some 5 times too large! It seems that LSW is fortuitously good in the Heisenberg model, for the accuracy evaporates in the XY model, although it *does* explain LRO.

Turning to 3D, we compare LSW with other methods (in Table 5.4, adapted from [5.37]) and in Table 5.5, present the correlation functions in the XY and Heisenberg antiferromagnets computed by the cell method [5.41]. The LSW correlation functions for the XY model are the same in 3D as in 2D and remain in quantitative disagreement with the tabulated values. A striking result of Table 5.4 is that the spin deviation is larger by a factor of 3 or 4 in the cell method than in the other calculations. Experimental tests to establish the correct value of this parameter in Nature have not yet been devised.

Table 5.4. Ground state energy parameter γ and spin deviation in 3D Heis. A.-F. spin one-half

		s.c.($z = 6$)		b.c.c.($z = 8$)	
Method	Reference	γ	$\langle \delta S_i^z \rangle$	γ	$\langle \delta S_i^z \rangle$
LSW	[5.35]	0.58	0.078	0.58	0.059
Perturbation theory	[5.36]	0.60	0.064	0.60	0.047
Variational	[5.38]	0.54	0.045	0.54	0.033
Variational	[5.39]	0.60	0.059	0.60	0.051
Cellular	[5.41]	0.4 ± 0.4	0.20 ± 0.02	0.4 ± 0.4	0.19 ± 0.02

5.12 Nonlinearities in Antiferromagnetism

In this section, we discuss corrections to the linear spinwave theory (LSW) of the preceding section.

As previously noted, the only systematic small parameter in which to expand is $1/s$. For large s, LSW is capable of yielding the correct answers, the error being estimated by computation of the next terms in an expansion in $1/s$. Although this procedure might seem hopeless for $s = \frac{1}{2}$, it is possible to make progress there also. We illustrate by means of the XY model treated above, for which LSW gave such poor results.

Consider the LSW substitution

$$(S_i^x)^2 \to \frac{1}{4}(a_i + a_i^*)^2 = \frac{1}{4}(1 + 2\mathbf{n}_i + a_i^2 + a_i^{*2})$$

$$(S_i^y)^2 \to -\frac{1}{4}(a_i - a_i^*)^2 = \frac{1}{4}(1 + 2\mathbf{n}_i - a_i^2 - a_i^{*2})\ . \tag{5.190}$$

The ground state expectation values are $\frac{1}{2}$ for each, rather than the correct value $\frac{1}{4}$. Thus, if to correct this discrepancy we took "renormalized" operators:

$$S_i^+ \to 2^{-1/2} a_i^* \qquad \text{and} \qquad S_i^- \to 2^{-1/2} a_i \tag{5.191}$$

retaining $S_i^z = (a_i^* a_i - \frac{1}{2})$ as the counting operator, we would obtain $\pm 1/2$ for the ground state expectation value of (5.188) instead of ± 1. This is now in reasonable agreement with the second column of Table 5.3. But this theory still does not explain the discrepancies between the Tables 5.3 for 2D and 5.5 for 3D, for the ground state correlations we compute by spinwaves in this model are independent of dimensionality; nor does it explain the observed decay of correlations with distance R_{ij}.

Table 5.5. Ground state correlations for $s = \frac{1}{2}$ antiferromagnets in 3D, for a 16 spin cell in two distinct lattices [5.41]

	XY		Heis.
R_{ij}/a	$4\langle S_i^z S_j^z \rangle$	$4\langle S_i^x S_j^x \rangle$	$4\langle S_i^z S_j^z \rangle$
Simple cubic (s.c.)			
1	0.1230	−0.5464	−0.4327
$2^{1/2}$	0.0297	0.5054	0.3141
$3^{1/2}$	0.0246	−0.4926	−0.3107
2	0.0347	0.5174	0.3333
Body-centered cubic (b.c.c.)			
$(3/4)^{1/2}$	0.0965	−0.5395	−0.4167
$1, 2^{1/2}, 3^{1/2}$	0.0326	0.5163	0.3333

5.13 Ferrimagnetism

Ferrimagnets are generally insulators containing localized spins which are antiferromagnetically coupled. One example pointed out by Néel, who coined the term *ferrimagnetism* (because of its existence in the *ferrites*, of which the lodestone is an example) occurs if the spins on the A and B sublattices of the previous section are of unequal magnitude $s_A \neq s_B$. Other, vastly more complicated examples of ferrimagnetism exist in theory and in nature,[8] but their study is a complex and specialized field. In the present section, we want to accomplish two goals: to display the spin-wave Hamiltonian for the simple two-sublattice model of ferrimagnetism, including leading nonlinear terms, and to show the leading nonlinear terms in antiferromagnetism, obtained from the former by setting $s_A = s_B$. We also show how the same expression, which for unequal spins gives a magnon energy $\sim k^2$ at long wavelengths, will yield the antiferromagnon energy $\sim k$. The mathematics is based on work by *Nakamura* and *Bloch* [5.43] and consists of a straightforward expansion of the square roots in the Holstein-Primakoff representation. For large spins, agreement with the classical equations of motion gives some confidence in this procedure, for which there is no other formal mathematical justification.

Let $s_A \geq s_B$, let there be N of each, and put

$$s_A = (1 + \alpha)s \qquad s_B = (1 - \alpha)s \,. \tag{5.192}$$

Except for the unequal spins, the Hamiltonian is precisely that of (5.175) in the preceding section. We keep the first three terms in an expansion of the Hamiltonian in powers of s:

$$\mathscr{H} = \mathscr{H}_0 + \mathscr{H}_1 + \mathscr{H}_2 + O(s^{-2}) \,. \tag{5.193}$$

They are,

$$\begin{aligned}\mathscr{H}_0 = \sum [&\gamma_0(s_B a_k^* a_k + s_A b_k^* b_k) \\ &+ \sqrt{s_A s_B}\;\; \gamma_k(a_k b_k + \text{H.c.})] - N\frac{z}{2} s_A s_B\end{aligned} \tag{5.194a}$$

$$\begin{aligned}\mathscr{H}_1 = -\frac{1}{4N\sqrt{s_A s_B}} &\sum_{k_1 k_2 k_3 k_4} (s_B \gamma_{k_1} b_{k_1} a_{k_2}^* a_{k_3} a_{k_4} + s_A \gamma_{k_1} a_{k_1} b_{k_2}^* b_{k_3} b_{k_4} \\ &+ \sqrt{s_A s_B}\;\; \gamma_{k_1 - k_3} a_{k_1}^* a_{k_3} b_{k_4}^* b_{k_2}) \Delta(\boldsymbol{k}_1 + \boldsymbol{k}_2 - \boldsymbol{k}_3 - \boldsymbol{k}_4) + \text{H.c.}\end{aligned} \tag{5.194b}$$

with $\Delta(\boldsymbol{k})$ defined in the text following (5.124). Finally,

$$\begin{aligned}\mathscr{H}_2 = -\frac{1}{2(4N)^2(s_A s_B)^{3/2}} &\sum_{k_1 \cdots k_6} (s_A^2 \gamma_{k_1} a_{k_1}^* b_{k_4}^* b_{k_5}^* b_{k_2} b_{k_6}^* b_{k_3} \\ &- 2 s_A s_B \gamma_{k_1 + k_2 - k_4} a_{k_1}^* a_{k_2}^* a_{k_4} b_{k_5}^* b_{k_6}^* b_{k_3} \\ &+ s_B^2 \gamma_{k_6} a_{k_1}^* a_{k_2}^* a_{k_4} a_{k_3}^* a_{k_5} b_{k_6}^*) \Delta(\boldsymbol{k}_1 + \boldsymbol{k}_2 + \boldsymbol{k}_3 - \boldsymbol{k}_4 - \boldsymbol{k}_5 - \boldsymbol{k}_6) + \text{H.c.}\end{aligned} \tag{5.194c}$$

---

[8] For a comprehensive review and references to the literature, see, e.g., [5.42].

where

$$a_k = \frac{1}{\sqrt{N}} \sum_{j \le A} a_j e^{-ik\cdot R_j} \qquad \text{and} \qquad b_k = \frac{1}{\sqrt{N}} \sum_{i \le B} b_i e^{+ik\cdot R_i} \tag{5.195}$$

and

$$\gamma_k = \sum_\delta e^{ik\cdot\delta}, \qquad \gamma_0 = z\,. \tag{5.196}$$

Next, $\mathscr{H}_0$ is diagonalized by the Bogolubov transformation,

$$\begin{aligned} a_k &\to a_k \cosh u_k + b_k^* \sinh u_k \\ b_k &\to a_k^* \sinh u_k + b_k \cosh u_k \end{aligned} \tag{5.197}$$

which mixes operators of the A and B sites. The degeneracy of the magnon spectrum in antiferromagnetism is lifted for $\alpha \neq 0$, and the following two magnon branches are found:

$$\varepsilon_b(k) = \gamma_0 s(f_k - \alpha) \qquad \text{and} \qquad \varepsilon_b(k) = \gamma_0 s(f_k + \alpha) \tag{5.198}$$

where

$$f_k = \sqrt{1 - (1 - \alpha^2)\left(\frac{\gamma_k}{\gamma_0}\right)^2} \tag{5.199}$$

provided the transformation parameter, u_k, chosen so as to eliminate $a_k b_k$ + H.c. from the Hamiltonian, takes the value

$$\tanh 2u_k = -\frac{\gamma_k}{\gamma_0}\sqrt{1 - \alpha^2}\,. \tag{5.200}$$

In the long wavelength approximation, the magnon energies are

$$\begin{aligned} \varepsilon_a(k) &\approx (\sqrt{\alpha^2 + (1 - \alpha^2)(ka)^2} - \alpha)\gamma_0 s \propto \frac{1 - \alpha^2}{2\alpha}(ka)^2 \\ \varepsilon_b(k) &\approx [\sqrt{\alpha^2 + (1 - \alpha^2)(ka)^2} + \alpha]\gamma_0 s \end{aligned} \tag{5.201}$$

from which it is easy to see the range dependent on α, over which the lower branch is quadratic, as in ferromagnets, before becoming approximately linear, as in antiferromagnets.

The ground-state energy, in this linear approximation, is

$$E_0 = -\frac{z}{2}(2N)s^2(1 - \alpha^2)\left(1 + \frac{\gamma}{zs}\right) \tag{5.202}$$

(it must be noted that there is a total of $2N$ spins in the present calculation) with

$$\gamma = zs\left(1 - \frac{1}{N}\sum f_k\right). \tag{5.203}$$

This quantum-mechanical correction is smaller in ferrimagnets than in antiferromagnets, which is not too surprising in view of the resemblance with ferromagnets, for which $\gamma \equiv 0$.

The nonlinearities are handled in the following manner. *First*, $\mathscr{H}_1$ and $\mathscr{H}_2$ are both transformed by the rules of (5.197, 200). *Then*, the diagonal terms in $\mathscr{H}_1$ are combined with the now entirely diagonal $\mathscr{H}_0$ to give a first-order diagonal Hamiltonian which we may denote $\mathscr{H}_D$, by analogy with the treatment of ferromagnetism. *Finally*, the off-diagonal matrix elements of $\mathscr{H}_1$ are eliminated by a canonical transformation, such as T in the ferromagnetic case. The resulting additional diagonal terms, combined with the diagonal parts of $\mathscr{H}_2$, form the second-order correction to $\mathscr{H}_D$. The remaining nondiagonal terms are discarded, for their contribution is $O(s^{-2})$. These calculations are undertaken in [5.43] but are too lengthy to be reproduced here. They also analyze the temperature dependence of the magnon energies (a subject which is treated here for ferromagnets only, in a separate chapter). Some of their results at $T = 0$ K are as follows:

1) There is a small s-independent shift in the magnon energies as given in (5.198, 200).

2) There is a nonlinear diagonal magnon energy, of the form

$$\sum_{\boldsymbol{k},\boldsymbol{k}'}\left[\frac{1}{2}\Gamma_{aa}(\boldsymbol{k},\boldsymbol{k}')(a_{\boldsymbol{k}}^* a_{\boldsymbol{k}})(a_{\boldsymbol{k}'}^* a_{\boldsymbol{k}'}) - \Gamma_{ab}(\boldsymbol{k},\boldsymbol{k}')(a_{\boldsymbol{k}}^* a_{\boldsymbol{k}})(b_{\boldsymbol{k}'}^* b_{\boldsymbol{k}'}) + \frac{1}{2}\Gamma_{bb}(\boldsymbol{k},\boldsymbol{k}')(b_{\boldsymbol{k}}^* b_{\boldsymbol{k}})(b_{\boldsymbol{k}'}^* b_{\boldsymbol{k}'})\right] \tag{5.204}$$

which, together with the linear magnon terms, makes up $\mathscr{H}_D$. One significant property of the nonlinear Hamiltonian is that the coefficient of $a_{\boldsymbol{k}}^* a_{\boldsymbol{k}}$ turns out to vanish for $\boldsymbol{k} = \boldsymbol{0}$, regardless of the occupation numbers of the other modes. This expresses the rotationally invariant nature of the approximations leading to $\mathscr{H}_D$. But differences with the analogous treatment of the ferromagnet can be noted, due to the presence of two spin-wave branches with different properties; for example, an increase in the occupation numbers of either branch *increases* the magnon energies in that branch and *decreases* the magnon energies of the other branch. [However, this appears to be a consequence of the nearest-neighbor model, and is not a law of universal validity; see the discsusion follow ing (5.120)].

5.14 Effects of Surfaces on Spin-Wave Amplitudes [5.44]

The role of surfaces is dominant in the observed properties of magnetic substances: domains nucleate there, spinwaves are "pinned" there by impurities or stray fields, and phase transitions at a surface may foreshadow those in the bulk. To study these effects one appeals to a scattering theory which has some similarities with the two-magnon problem studied above. For definiteness, we consider only a single surface (in *extremely* thin films where the two surfaces interact, this treatment will have to be modified) perpendicular to one of the principal crystal axes in a simple cubic structure. The interactions are ferromagnetic, and J differs from the bulk value 1 only for the surface spins. A surface anisotropy field $h_A > 0$ is also included, but the bulk ferromagnet is otherwise isotropic. With the surface located at $\boldsymbol{R}_i = (X_i, Y_i, 0)$ and all interactions restricted to nearest-neighbors, we have the Hamiltonian

$$\mathscr{H} = -\frac{1}{2}\sum_{Z_i, Z_j \neq 0}\sum \boldsymbol{S}_i \cdot \boldsymbol{S}_j - J \sum_{Z_i=0,\, Z_j \neq 0}\sum \boldsymbol{S}_i \cdot \boldsymbol{S}_j$$
$$-\frac{1}{2}J \sum_{Z_i=Z_j=0}\sum \boldsymbol{S}_i \cdot \boldsymbol{S}_j - h_A \sum_{Z_i=0} S_i^z \,. \tag{5.205}$$

The ground state $|0)$ of all spins "down" has energy

$$E_0 = E_0(\text{bulk}) - 3N_z J s^2 - h_A N_z s \tag{5.206}$$

in which we define the bulk energy to be that part which does not depend on the surface exchange parameters (J, h_A) and we note that each of the N_z surface spins has 4 n.n. in the surface (i.e., two bonds per spin), plus one in the neighboring plane, for a net total of 3 J-dependent bonds *per* surface spin. (This should not be confused with the obvious fact that each surface spin has 5 n.n. rather than 6 in the bulk).

Absent the translational symmetry that enabled us to guess the plane wave form of the one-magnon states in the bulk, we shall have to construct the set of excited one-magnon eigenstates, in the form

$$|\psi) = (2sN)^{-1/2} \sum_j f_j S_j^+ |0) \tag{5.207}$$

and extract from it a set of coupled equations for the amplitudes f_j by projecting as follows:

$$(2s)^{-1/2} N^{1/2} (0|S_i^- \mathscr{H}|\psi) = (E_0 + \varepsilon) f_i \tag{5.208}$$

ε is the excitation energy, to be compared with $\hbar\omega(\boldsymbol{k})$ for the translationally invariant case. Evaluating the l.h.s. of (5.208) we readily obtain

$$(E_0 + 6s) f(\boldsymbol{R}_i) - s \sum_{\boldsymbol{\delta}} f(\boldsymbol{R}_i + \boldsymbol{\delta}) \tag{5.209a}$$

for $\boldsymbol{R}_i$ in the bulk; for a spin in the plane adjacent to the surface, the l.h.s. of (5.208) is, instead,

$$(E_0 + 6s)f(\boldsymbol{R}_i) - s\sum_{\boldsymbol{\delta}} f(\boldsymbol{R}_i + \boldsymbol{\delta}) + (J-1)s[f(\boldsymbol{R}_i) - f(X_i, Y_i, 0)] \quad (5.209\text{b})$$

and finally, for a surface spin

$$(E_0 + h_A + 5Js)f(X_i, Y_i, 0) - sJ[f(X_i + a, Y_i, 0) + f(X_i - a, Y_i, 0) + f(X_i, Y_i + a, 0) + f(X_i, Y_i - a, 0) + f(X_i, Y_i, a)] \,. \quad (5.209\text{c})$$

There remains translational invariance of sorts, i.e., in the X_i, Y_i plane. Therefore let us seek a solution of the form

$$f_i = C\, e^{i(k_x X_i + k_y Y_i)} F(n) \quad (5.210)$$

where C is a normalization constant, $Z_i = na$ and the surface plane is located at $n = 0$ (the first interior plane at $n = 1$, the bulk at $n = 2, 3, \ldots$). Without loss of generality, we write for the excitation energy of the continuum states

$$\varepsilon = 2s(3 - \cos k_x a - \cos k_y a - \cos q) \quad (5.211)$$

and for the three regions considered above, the Schrödinger equation becomes,

$$2s(2 - \cos k_x a - \cos k_y a)F(n) + s[2F(n) - F(n+1) - F(n-1)] = \varepsilon F(n) \quad (5.212\text{a})$$

for $n \geq 2$. At $n = 1$ we obtain

$$2s(2 - \cos k_x a - \cos k_y a)\, F(1) + s[2F(1) - F(2) - F(0)] + (J-1)s[F(1) - F(0)] = \varepsilon F(1) \quad (5.212\text{b})$$

and at the surface,

$$[h_A + 2Js(2 - \cos k_x a - \cos k_y a)]F(0) + sJ[F(0) - F(1)] = \varepsilon F(0) \,. \quad (5.212\text{c})$$

As the bulk eigenstates take the form $\exp(\pm iqn)$, we try a linear combination: let

$$F(n) = \cos(qn + \theta), \qquad \text{for} \quad n \geq 1 \quad (5.213)$$

in which $\theta(k_x, k_y, q)$ is a *phase shift* independent of n. However, this form is not suitable at the surface, $n = 0$. The substitution (5.213) into (5.212a) solves that set of (bulk) equations exactly. We are left with (b) and (c), 2 equations in the 2 unknowns, θ and $F(0)$:

$$F(0) = \frac{\cos\theta}{J} + \frac{(J-1)}{J}\cos(q+\theta) \tag{5.214a}$$

and

$$F(0) = \frac{J}{D}\cos(q+\theta) \tag{5.214b}$$

where,

$$D(q) \equiv h_A/s + J + 2(J-1)(2 - \cos k_x a - \cos k_y a) - 2(1 - \cos q). \tag{5.214c}$$

$D(q)$ is a function of q alone, at fixed h_A, s, J, k_x and k_y but its variation, especially with k_x, k_y may be important. Solving the coupled equations for $F(0)$, we find it to be a function of q and indicate this by a subscript

$$F_q(0) = \frac{J \sin q}{\{(J^2 - DJ + D)^2 \sin^2 q + [(J^2 - DJ + D)\cos q - D]^2\}^{1/2}} . \tag{5.215}$$

Note that as $q \to 0$, this surface amplitude vanishes

$$F_q(0) \underset{(q\to 0)}{\longrightarrow} \frac{\sin q}{|J - D(q)|} \to 0$$

and thus, long *wavelength magnons have small amplitudes* at the surface, unless $J = D(0)$. [For this special case, (5.215) yields $F_q(0) \underset{q\to 0}{\longrightarrow} 1$] .

When J exceeds $D(0)$, bound states can form below the continuum. Then, the form (5.213) is not suitable, and one uses $F(n) = C\exp(-\gamma n)$. Results by *Ilisca* and *Gallais* [5.44] indicate that when $h_A < 0$ the bound state even acquires negative energy, $\varepsilon < 0$, and that the surface can no longer be magnetized in the same way as the bulk.

Recapitulating, we have seen that the discontinuity at the surface, and the possible surface-specific parameters (change in coordination number, in J, anisotropy field, etc.) affect the surface magnon amplitudes in a major way. The same must therefore also be true for crystalline defects: magnons are expected to be strongly scattered at dislocations, grain boundaries, etc.

5.15 Vortices

Just as dislocations in solids cannot be studied using the theory of small amplitude vibrations, neither can the topological disorder known as a vortex in the magnetic problem be described in terms of spinwaves or magnon excitations. Vortices have to date been studied only in the classical ($s \to \infty$) limit, and were introduced into magnetism by *Kosterlitz* and *Thouless* [5.45] in the context of the 2D classical XY model.

Following those authors, let us consider classical vectors with ferromagnetic nearest-neighbor interactions, on a square lattice as illustrated in Fig. 5.12. There we picture a single counter-clockwise vortex of charge +1. (A similar but clockwise vortex has charge −1.) It extends throughout the lattice and because spins are practically parallel in any neighborhood far removed from the core, a vortex cannot be "cured" by any finite number of small re-orientations of the spins, nor can it be caused by them. Once introduced, it is a stable state, although one of higher energy than the ground state. The vortex represents an extra degree of freedom of the spin system, independent of those associated with the magnons.

The excitation energy of a single vortex is easily estimated. In 2D one reasons as follows: the misalignment of 2 neighboring spins is a small quantity, say $\delta\theta$. Starting at some spin at a radial distance R from the core of the vortex and moving along the circumference in a ccw direction, we must wind through a total angle $\theta = (2\pi R/a)\delta\theta$ which is required to be an integer multiple of 2π, say $2\pi n$; for the case illustrated in Fig. 5.12, $n = +1$. Thus,

$$\delta\theta = na/R\,. \tag{5.216}$$

The energy of misalignment at each bond is of the order of $J(\delta\theta)^2$, with J the exchange parameter. Thus the total excitation energy or "self-energy" of the vortex is

$$\begin{aligned}\Delta E &= 2\pi \int_a^L dRRa^{-2}J(na/R)^2 \\ &= 2\pi n^2 J \ln(L/a)\end{aligned} \tag{5.217}$$

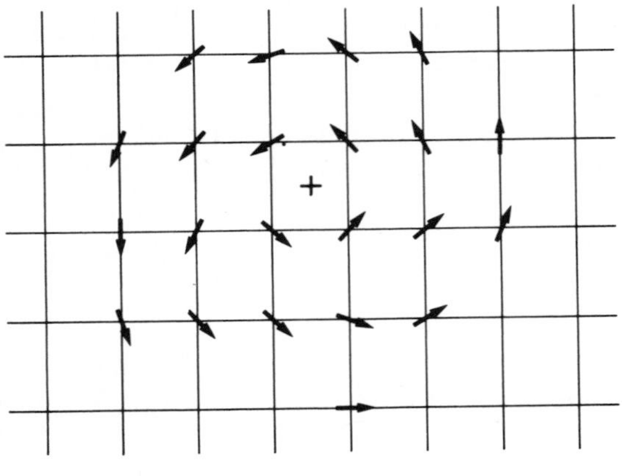

Fig. 5.12. A single vortex of "charge" +1. At large distances from the core, the spins are locally almost parallel but there is still circulation: a spin far to the left and one far to the right are antiparallel. The self-energy is infinite

where we take the lattice to be circular of radius L and area πL^2, centered about the core. Because the ln is a slowly varying function one does not expect any dependence on sample geometry and the most convenient one is used.

Two vortices of opposite sign (say a pair $+n$, $-n$) at a distance $\boldsymbol{R}_{ij}$ apart have an energy which is independent of L. Their fields cancel at large distances as shown in Fig. 5.13; and all that remains is the dipolar near-fields, which result in

$$\Delta E_{ij} = 2\pi n^2 J \ln\left(\frac{|\boldsymbol{R}_{ij}| + a}{a}\right). \tag{5.218}$$

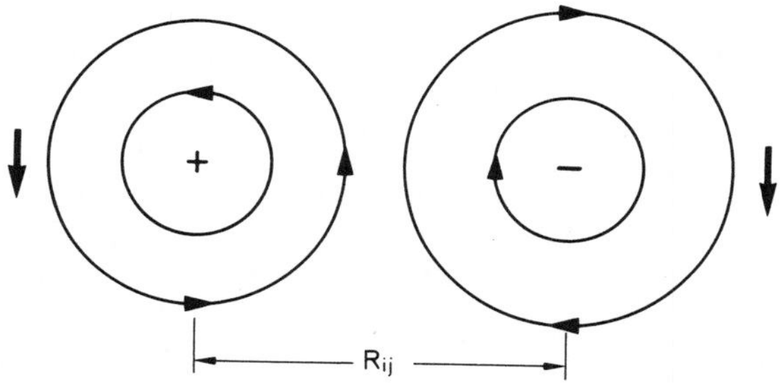

Fig. 5.13. A vortex dipole: effects on spins far from the cores is much reduced: the energy is finite

Such vortex-antivortex (dipole) pairs can have a strong effect on the finite temperature thermodynamics [5.45, 46].

One might wonder whether the effect depends on dimensionality. In 1D, for a chain of N spins we have

$$N\delta\theta = n2\pi \tag{5.219}$$

and a self-energy

$$\Delta E = JN(2\pi n/N)^2 = J(2\pi n)^2/N \tag{5.220}$$

a negligible value. Moreover, the interaction energy of a vortex-antivortex pair separated by m sites is

$$\Delta E_{i,j} = |m|(2\pi n/N)^2 J\ . \tag{5.221}$$

What is interesting here is the similarity to the interaction energy of two sheets of charge in 1D, and of course, the energetics of the vortices is analogous to the flux calculus of classical electrostatics. Thus, in any number of dimensions, the study of magnetic vortices parallels that of charged fluids.

5.16 Solitons: Introductory Material

It has been observed throughout this chapter that various problems in magnetism become simpler as s, the magnitude of the individual spin, becomes larger and the quantum fluctuations become correspondingly smaller. A century of progress has recently culminated in the exact solution of many problems in classical nonlinear dynamics in one spatial dimension, including those of greatest relevance to this chapter. A common thread is the identification of at least two different types of propagating excitations: the ordinary wave (here, *magnon*) and the solitary wave, or *soliton*. The soliton is a particle-like manifestation which preserves its shape and identity after a long course of travel or period of time, and after collisions with stationary defects or other solitons. To quote its discoverer [5.47]:

> "I was observing the motion of a boat which was rapidly drawn along a narrow channel by a pair of horses, when the boat suddenly stopped—not so the mass of water in the channel which it had put in motion—it accumulated round the prow of the vessel in a state of violent agitation, then suddenly leaving it behind, rolled forward with a great velocity, assuming the form of a large solitary elevation, a rounded, smooth and well-defined heap of water, which continued its course along the channel apparently without change of form or diminution of speed. I followed it on horseback, and overtook it still rolling on a rate of some eight or nine miles an hour, preserving its original figure some thirty feet long and a foot to a foot and a half in height. Its height gradually diminished and after a chase of one of two miles I lost it in the windings of the channel. Such, in the month of August 1834, was my first chance interview with that singular and beautiful phenomenon. . ."

To relate this remarkable phenomenon to magnetism, some concepts will have to be developed. Chief among these are Heisenberg's equations of motion for the spin dynamics, such as

$$\frac{\partial}{\partial t}\boldsymbol{S}_n = \frac{\mathrm{i}}{\hbar}[\mathscr{H}, \boldsymbol{S}_n] \tag{5.222}$$

with $\mathscr{H}$ the Hamiltonian of (5.1). Abbreviating $\partial/\partial t$ by $(\cdot)$, we find

$$\dot{\boldsymbol{S}}_n = -\frac{\mathrm{i}}{\hbar}\sum_m J_{mn}[\boldsymbol{S}_m\cdot\boldsymbol{S}_n, \boldsymbol{S}_n] - \frac{\mathrm{i}}{\hbar}\, g\mu_\mathrm{B}[\boldsymbol{H}\cdot\boldsymbol{S}_n, \boldsymbol{S}_n]\,. \tag{5.223}$$

The Zeeman contribution is easy to compute by the familiar commutation relations, [Chap. 3, (3.5, 6)]. The bracket coefficient of J_{mn} has been previously calculated in [Chap. 3, (3.55)]. Combining the results, we find

$$\dot{\boldsymbol{S}}_n = \boldsymbol{S}_n \times \left(\sum_m J_{mn}\boldsymbol{S}_m + g\mu_\mathrm{B}\boldsymbol{H}\right)\,. \tag{5.224}$$

Summing over all spins, we see that the cross terms cancel and the total spin merely precesses about the external field

$$\dot{\boldsymbol{S}}_T = \boldsymbol{S}_T \times g\mu_B \boldsymbol{H} \,. \tag{5.225}$$

As the quantum unit of action $\hbar$ does not appear explicitly in the equations of motion (5.224), therefore they are valid without modification in the classical limit $s \to \infty$. (In proceeding to this limit, it is necessary to scale J and μ_B as $1/s$ to keep the results finite, while setting $\hbar = 1$ for dimensional convenience). We can now represent a spin by a classical unit vector: $\boldsymbol{S}_n = s\boldsymbol{u}_n$. Assuming a simple lattice [s.c., or sq., or n.n. l.c. with primitive unit vectors $\boldsymbol{\delta}$ in the notation of (5.16), with $\delta = a$ the lattice parameter], the equations of motion take the form of difference equations

$$\dot{\boldsymbol{u}}(\boldsymbol{R}_n) = \boldsymbol{u}(\boldsymbol{R}_n) \times \{Js \sum_{\boldsymbol{\delta}} [\boldsymbol{u}(\mathbf{R}_n + \boldsymbol{\delta}) - \boldsymbol{u}(\boldsymbol{R}_n)] + \boldsymbol{h}\} \tag{5.226}$$

with $\boldsymbol{h} \equiv g\mu_B \boldsymbol{H} = (0, 0, h)$. For all long wavelength phenomena, whether linear or not, a lattice may be approximated by a continuum. In (5.226) we have subtracted terms such as $\boldsymbol{u}_n \times \boldsymbol{u}_n$ which vanish in the classical limit, to obtain a form in which the continuum limit is easily taken. One expands in a Taylor series

$$\boldsymbol{u}(\boldsymbol{R}_n + \boldsymbol{\delta}) = \boldsymbol{u}(\boldsymbol{R}_n) + \boldsymbol{\delta}\cdot\nabla\boldsymbol{u}(\boldsymbol{R}_n) + \frac{1}{2}(\boldsymbol{\delta}\cdot\nabla)^2\boldsymbol{u}(\boldsymbol{R}_n) + \dots \,. \tag{5.227}$$

Terms in δ_α or in $\delta_\alpha\delta_\beta(\alpha \neq \beta)$ vanish in (5.226) by symmetry, leaving, in leading order, only the following:

$$\dot{\boldsymbol{u}}(\boldsymbol{R}) = Jsa^2\, \boldsymbol{u}(\boldsymbol{R}) \times [\nabla^2\boldsymbol{u}(\boldsymbol{R})] + \boldsymbol{u}(\boldsymbol{R}) \times \boldsymbol{h} \,. \tag{5.228}$$

This is better written in terms of $u^\pm = u_x \pm \mathrm{i}u_y$ and u_z:

$$\dot{u}^+ = \mathrm{i}Jsa^2(u_z\nabla^2u^+ - u^+\nabla^2u_z) - \mathrm{i}u^+h \tag{5.229a}$$

with u^- just the complex conjugate of u^+, and:

$$\dot{u}_z = -\mathrm{i}\frac{1}{2}Jsa^2(u^-\nabla^2u^+ - u^+\nabla^2u^-) \,. \tag{5.229b}$$

From our earlier studies we already know the form of a spinwave: it is a solution of these equations with constant amplitude and plane wave character. Thus, $u^\pm = A\exp[\pm\, \mathrm{i}(\boldsymbol{k}\cdot\boldsymbol{R} - \omega t)]$ and (5.229) yield

$$\omega = Jsa^2u_z\boldsymbol{k}^2 + h \tag{5.230a}$$

and

$$\dot{u}_z = 0, \qquad \text{hence} \qquad u_z = p_A, \quad \text{a constant.} \tag{5.230b}$$

As $\hat{\boldsymbol{u}}$ is a unit vector, the amplitude A must be $\sqrt{(1 - p_A^2)}$. These results agree with (5.16) for $A \to 0$, $p_A \to 1$. At finite amplitude there is a "softening" of the frequency, in agreement with our expectations on the basis of nonlinear magnon interactions studied earlier in this chapter, e.g., (5.120). But at finite amplitude, the excitation energy of the spinwave is extensive, i.e., it is infinite. A finite amplitude spinwave is the superposition of $O(N)$ magnons. Thus the momentum and angular momentum are also infinite. This contrasts with the soliton, a localized excitation whose energy, momentum and angular momentum are all finite (1D) or proportional the cross-sectional area of the wave front (2 or 3D); see Fig. 5.14. To study it further, we shall need explicit expressions for the excitation energy, momentum and angular momentum. It will also be convenient to specialize to one dimension.

The Hamiltonian (5.1) can be written

$$\mathscr{H} = \frac{1}{2} J \sum_{n=1}^{Na} (\boldsymbol{S}_{n+1} - \boldsymbol{S}_n)^2 - h \sum_{n=1}^{N} (S_n^z - s) + E_0$$
$$\to \frac{1}{2} J s^2 a \int_0^{Na} dx \left[\frac{d\hat{\boldsymbol{u}}(x)}{dx}\right]^2 + (hs/a) \int_0^{Na} dx [1 - u_z(x)] + E_0 \qquad (5.231)$$

proceeding to both classical and continuum limits. As unit vectors need only two parameters to specify their direction, we can introduce the two, canonically conjugate, dynamical variables p, q through the Villain representation [see

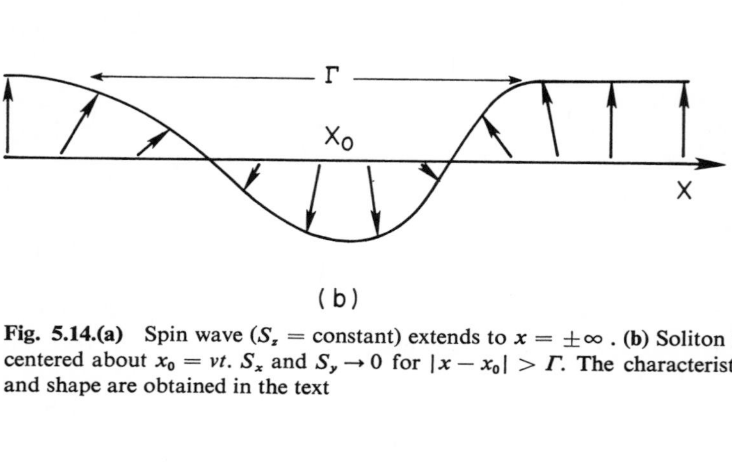

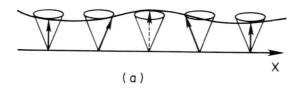

Fig. 5.14.(a) Spin wave (S_z = constant) extends to $x = \pm\infty$. **(b)** Soliton is a moving pulse centered about $x_0 = vt$. S_x and $S_y \to 0$ for $|x - x_0| > \Gamma$. The characteristic pulse width Γ and shape are obtained in the text

Chap. 3, (3.87–89)] in the classical limit

$$u^+ = e^{iq}\sqrt{1-u_z^2}, \qquad u^- = (u^+)^+, \qquad \text{and } u_z = p\ . \tag{5.232}$$

Thus, the excitation Hamiltonian, in the natural units ($Js^2a = 1$) is

$$\begin{aligned} \boldsymbol{H} = \mathscr{H}_0 - E_0 = \frac{1}{2}\int dx\left[\frac{1}{1-p^2}(dp/dx)^2 + (1-p^2)(dq/dx)^2\right] \\ + h'\int dx(1-p) \end{aligned} \tag{5.233}$$

writing hs/a as h'. The equations of motion (5.229) now take on the aspect

$$\begin{aligned} -\dot{q}(x,t) = \frac{1}{1-p^2}(d^2p/dx^2) \\ + \frac{p}{(1-p^2)^2}(dp/dx)^2 + p(dq/dx)^2 + h' \end{aligned} \tag{5.234a}$$

and

$$\dot{p}(x,t) = (1-p^2)(d^2q/dx^2) - 2p\frac{dp}{dx}\frac{dq}{dx} = \frac{d}{dx}\left[(1-p^2)\frac{dq}{dx}\right]. \tag{5.234b}$$

Along with the Hamiltonian, we identify two conserved quantities [5.48] as, the total momentum,

$$P = \frac{s}{a}\int dx(1-p)\frac{dq}{dx} \tag{5.235}$$

and the z component of the angular momentum carried by the excitations,

$$\mathrm{M}_z = \frac{s}{a}\int dx(p-1) \tag{5.236}$$

respectively. The latter is naturally negative, as the maximum spin of a ferromagnet occurs in the ground state.

5.17 Solitary Wave Solution

Here we set out to construct the most general solutions permitted by the equations of motion, having a "permanent profile" character, $p(x-vt)$ and $q(x-vt)$. To qualify as *solitons* they must moreover be stable against small perturbations, against collisions with other solitons, etc., i.e., they must possess some quality of permanence.

With time dependence arising only through $x - vt$ and v an adjustable parameter, (5.234) simplifies

$$v\frac{dq}{dx} = \frac{1}{1-p^2}\frac{d^2p}{dx^2} + \frac{p}{(1-p^2)^2}(dp/dx)^2 + p(dq/dx)^2 + h' \tag{5.237a}$$

and

$$v(dp/dx) = \frac{d}{dx}\left[(p^2-1)\frac{dq}{dx}\right] \tag{5.237b}$$

(b) is integrated at once, yielding

$$dq/dx = v\frac{p_0 - p}{1-p^2} \tag{5.238}$$

in which p_0 is a constant of integration. Insertion into (a) now yields a nonlinear, second-order differential equation for p

$$v^2\frac{p_0 - p}{1-p^2} = \frac{1}{1-p^2}\frac{d^2p}{dx^2} + \frac{p}{(1-p^2)^2}(dp/dx)^2 + pv^2\frac{(p_0-p)^2}{(1-p^2)^2} + h' \,. \tag{5.239}$$

This can nonetheless be solved, by the device of setting $(dp/dx)^2 \equiv F(p)$, so that $d^2p/dx^2 = \frac{1}{2}dF/dp$. The above then turns into a linear, first-order differential equation in the new unknown, F. It is readily solved, yielding

$$(dp/dx)^2 = F(p) = 2h'p(p^2-1) - v^2(1 + p_0^2 - 2p_0p) - p_1(p^2-1). \tag{5.240}$$

Here p_1 is the arbitrary constant of integration. It should be remarked that this equation is that of a classical particle of mass $\frac{1}{2}$ and energy E, in a potential well $V = E - F$. There exists the additional constraint, that $|p| \leq 1$. Aside from this, the motion is confined to the positive region of F, which is a cubic polynomial with asymptotic behavior $2h'p^3$. The constants $p_{0,1}$ must thus be adjusted so that the solution is physically allowed. It is possible to integrate (5.240) and analyze the resulting elliptic functions. It is, however, much simpler to study F for the behavior between the turning points, from which one concludes that $p_{0,1}$ must be adjusted to allow one of three distinct patterns shown as (a), (b) and (c) in Fig. 5.15.

The first curve, (a), satisfies the conditions appropriate to a spinwave of amplitude p_A, previously analyzed. Only $p = p_A$ is allowed, hence (5.237) has the solution $q = k(x - vt) + q_0$.

The second curve describes the soliton superposed onto an otherwise perfect ferromagnetic background. The asymptotic value of p is $p_C = 1$, and the largest deviation from the asymptotic value is at p_A which can lie anywhere in the range $-1 \leq p_A < +1$, by suitable adjustment of the parameters $p_{0,1}$.

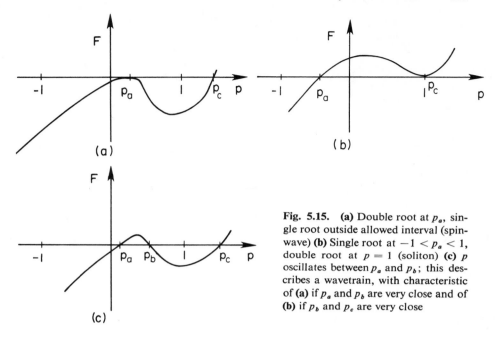

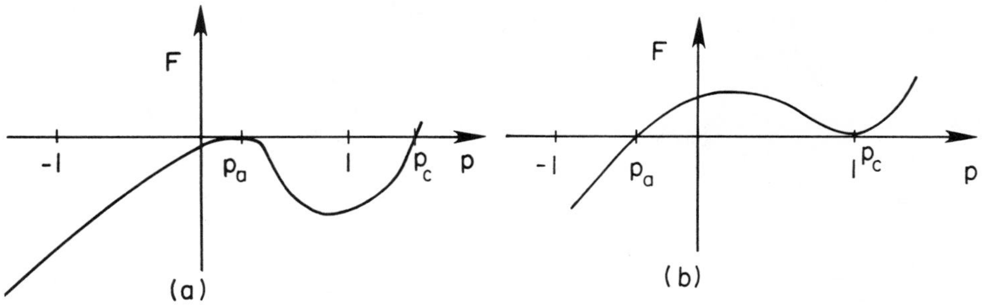

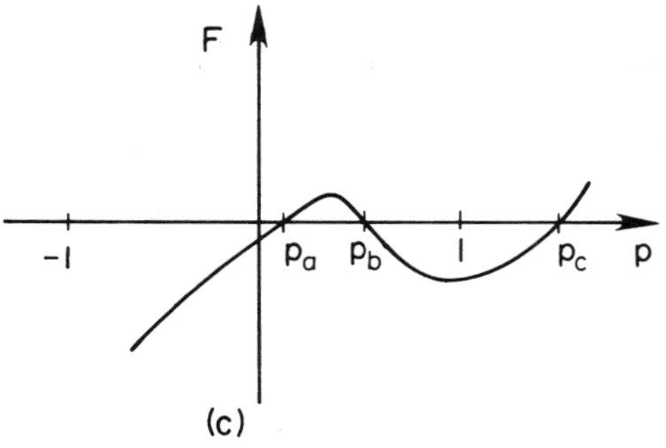

Fig. 5.15. **(a)** Double root at p_a, single root outside allowed interval (spinwave) **(b)** Single root at $-1 < p_a < 1$, double root at $p = 1$ (soliton) **(c)** p oscillates between p_a and p_b; this describes a wavetrain, with characteristic of **(a)** if p_a and p_b are very close and of **(b)** if p_b and p_c are very close

The final curve (c) leads to a periodic repetition of the soliton pulse, i.e., to a sort of wavetrain. With p_B and p_C very close and straddling 1, each pulse in the infinite train resembles the solution (b), although the period is finite rather than infinite. With p_A and p_B very close, the solution reduces to the spinwave of (a). We thus see the spinwave and the soliton as the two rather opposite limiting cases of the general behavior exemplified by (c).

The soliton solution of this problem was first obtained by *Nakamura* and *Sasada* [5.49], later by *Lakshmanan* et al. [5.50], and had its stability tested by *Tjon* and *Wright* [5.48]. The construction of an F having the shape (b) in Fig. 5.15 requires the right-hand side of (5.240) to have a double root at $p = 1$. $F(1) = 0$ implies $p_0 = 1$; then $(dF/dp)_{p=1} = 0$ yields $p_1 = v^2 + 2h'$. Writing $p = 1 - 2\sin^2 \frac{1}{2}\theta$ and defining a new variable $y = (1 - v^2/4h')^{-1/2} \sin \frac{1}{2}\theta$, and a new independent variable $t = x(h' - \frac{1}{2}v^2)^{1/2}$, we obtain

$$(dy/dt)^2 = y^2(1 - y^2) \tag{5.241}$$

a well known ("canonical") equation, known to have the solution $y = \operatorname{sech} t$. It is then a simple matter to substitute the original variables

$$p = 1 - 2\left(1 - \frac{1}{4}v^2/h'\right) \operatorname{sech}^2\left(\frac{x - vt - x_0}{\Gamma}\right) \tag{5.242}$$

where $\Gamma \equiv (h' - \frac{1}{2}v^2)^{-1/2}$, is the width of the pulse. This pulse is centered at x_0 (an arbitrary constant, chosen to satisfy initial conditions) at $t = 0$, travels to the left ($v < 0$) or right ($v > 0$) with the given speed in the range $0 \leq |v| < 2(h')^{1/2}$. Integration of the equation for q proceeds by similar substitutions, to yield

$$q = \phi_0 + \frac{1}{2}v(x - vt - x_0) + \tan^{-1}\left[\left(\frac{4h'}{v^2} - 1\right)^{1/2} \tanh\left(\frac{x - vt - x_0}{\Gamma}\right)\right]. \tag{5.243}$$

The distortions associated with the soliton are maximal at $v = 0$, yielding the narrowest possible pulse. Conversely, at $v^2 \to 4h'$, the pulse is infinitely broad and the local distortions infinitesimal.

Several quantities are of interest: the phase shift, the momentum, angular momentum, and energy. From (5.243) we find, for the total phase shift,

$$\Delta q = 2\tan^{-1}\left(\frac{4h'}{v^2} - 1\right)^{1/2} \tag{5.244}$$

which varies from a maximum of π at $v = 0$ to zero as $v^2 \to 4h'$.

Using (5.233) we find the energy density to be

$$\boldsymbol{H}(x, t) = \frac{4}{\Gamma^2} \operatorname{sech}^2\left(\frac{x - vt - x_0}{\Gamma}\right). \tag{5.245}$$

The total energy is time independent, of course. Integrating the above, we find it to be

$$\Delta E = 8/\Gamma .$$

Restoring the original units

$$\Delta E = 4Js^2a/\Gamma . \tag{5.246}$$

Similarly, the momentum is

$$P = \frac{4s}{a} \sin^{-1}\left(1 - \frac{1}{4}v^2/h'\right)^{1/2} \tag{5.247}$$

and

$$\mathbf{M}_z = -4s/h\Gamma a . \tag{5.248}$$

Combining these gives a more perspicuous relation

$$\Delta E = \frac{16Js^3}{|\mathbf{M}_z|} \sin^2(Pa/4s) . \tag{5.249}$$

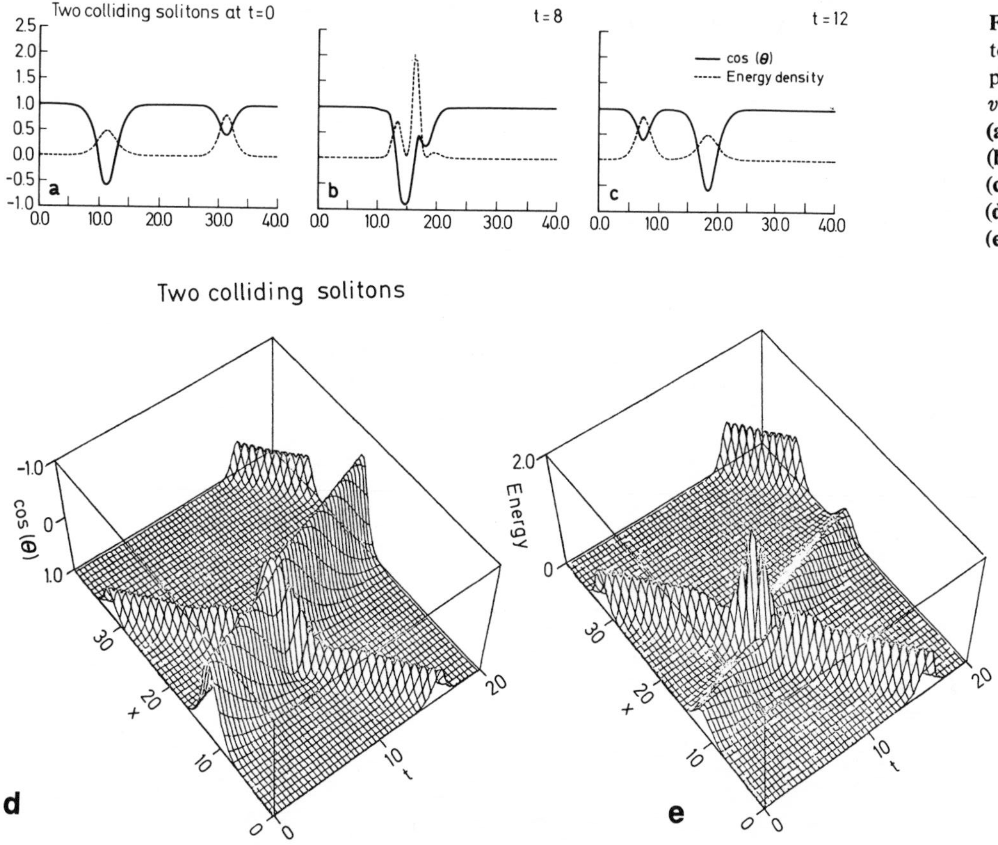

Fig. 5.16 a-b. Two colliding solitons: one has velocity $v = \frac{1}{2}$ and amplitude $b = \sin \frac{1}{2}\theta = 0.8$, the other $v = -2$, $b = 0.3$. [5.47]
(a) $t = 0$ (before collision)
(b) $t = 8$ (impact)
(c) $t = 12$ (after collision)
(d) $p = \cos\theta$ vs x and t.
(e) Energy density vs x and t

For the spin one-half two-magnon bound state, $|\mathbf{M}_z| = 2$, $s = \frac{1}{2}$, this formula agrees *precisely* with the energy previously derived in (5.85) by quantum-mechanical analysis. The above also demonstrates the further lowering of the energy upon binding 3, 4, or more magnons into a bound state "soliton" of a given total momentum.

Tjon and *Wright* [5.48] studied the collision of two solitons numerically, some of their results being displayed in Fig. 5.16. But the crucial progress has occurred only recently. First *Takhtajan* [5.51] and then *Fogedby* [5.52] have come to the realization that the nonlinear equations of motion (5.228) or (5.229) can be replaced by a set of coupled linear eigenvalue equations. Difficult though the linear problem may be to bring to a closed form solution, there is the fact that the spectrum of eigenfunctions and eigenvalues is complete. This spectrum maps, in a one-to-one correspondence, onto the magnons, solitons and wavetrains we have encountered. It is therefore no accident that two solitons, having a collision described in detail in Fig. 5.16, survive this collision with their identities unimpaired. This is a necessary consequence of the large number of constraints—constants of the motion—that the equations of motion imply and that a solution must satisfy.

The interested reader will find the analysis in [5.52] (also, in various stages, in [5.48–51, 53]), for the mathematics are beyond the scope of the present volume. Recent applications of this theory include the XY model, modified somewhat to allow slight oscillations out of the XY plane [5.53] as required for the experimental applications; future applications promise to explain the outstanding mysteries about one-dimensional magnets: their response to neutrons, to time-dependent forces and fields, etc.

Although there is no reason to exclude solitons from consideration in 2D and 3D systems, where indeed they must exist, their energy which is proportional to the area of the wave front (N^{D-1}) will be too large for them to be included among those elementary excitations that are found spontaneously at low T, although bound complexes of them (such as the vortex–antivortex pair of a preceding section) may be.

In a later chapter on thermodynamics, we shall see how much simpler is the equilibrium statistical mechanics of the linear chain when compared with its dynamics, of which the reader has had a glimpse in these pages, or compared with statistical mechanics in 2D and 3D.

6. Magnetism in Metals

Not all the electrons in a metal participate in the electrical conduction, nor in other physical, chemical, or magnetic processes; some are core electrons, belonging to filled shells tightly bound to the ion and unaware of the metallic environment. Electrons in unfilled shells have a range of behavior intermediate between such tightly bound localized electrons and that of quasifree particles experiencing the smooth periodic atomic potential and participating fully in the electrical conductivity.

It is easiest to discuss the theory of magnetism in such metals where the electrons can be clearly divided into distinct sets of tightly bound and quasifree particles. Hopefully, the results still have qualitative merit when the carriers have properties intermediate between these two extremes.

The simplest model of magnetism in metals is the following: electrons in well-localized magnetic d or f shells interact with one another via a Heisenberg nearest neighbor exchange mechanism, whilst an entirely distinct set of (quasi-free) electrons in Bloch states accounts for the metallic properties without partaking of the magnetic ones. *Unfortunately, this model is purely fictional;* x-ray data, optical experiments, measurements of specific heat all indicate that the d electrons in the iron transition series metals are largely conduction electrons, in bands an electron volt wide. Whereas in the rare earths, on the contrary, the magnetic f shells are *so* well localized ($\sim$0.3 Å) that the overlap between them (at a distance of approximately 3 Å) must be negligibly small, and there is therefore *no* Heisenberg nearest-neighbor exchange, to a good approximation. The observed magnetism in such a case must involve the s-band conduction electrons, which alone are capable of sustaining correlations over several interatomic distances: "indirect exchange".

For these and other reasons,[1] it is not possible to ignore the electrons' itineracy in any sensible theory of magnetism in metals.[2] In the present chapter we start with an elementary review of the band theory in the one-electron approximation with emphasis on *tight-binding*, the simplest approximation of any value in the investigation of magnetic properties. Proceeding from there, we shall see what gives rise to strong magnetic properties, such as ferromagnetic or antiferromagnetic (AF) behavior. We shall even show that the Heisenberg Hamiltonian for insulators can be derived on the basis of a band picture entirely

[1] See the indictment of the Heisenberg and Heitler-London theories in [6.1].

[2] For a comprehensive treatment see [6.2].

analogous to the band theory of metals. We shall also derive the theory of magnons in metals for the various models considered.

The case of a half-filled band is easiest. To the intratomic mechanisms (intra-atomic Coulomb repulsion, exchange corrections thereto, etc.) we can add the inter-atomic "transfer" matrix elements responsible for the motion of the electrons through the solid. One or the other of these mechanisms can be mathematically turned off, to recover the well-known atomic or free electron limits.

On the other hand, the case of partially-filled (low density) magnetic bands is far more intriguing. The electrons repel, and are not constrained by density to occupy the same atomic sites. Turning off the transfer matrix elements would result in an inhomogeneous localized array. We shall treat this case by a multiple scattering formalism that is exact in the dilute limit, and yields interesting criteria for the occurrence of magnetic moments in metals. We study it for clues to the eventual ferromagnetism or AF of the electron fluid.

We include in this chapter a thorough treatment of the magnetic impurity problem: how does an impurity atom acquire a magnetic moment, and how do these moments interact with one another.

6.1 Bloch and Wannier States

Consider the one-electron Hamiltonian

$$\mathscr{H} = \frac{\boldsymbol{p}^2}{2m} + \sum_i V(\boldsymbol{r} - \boldsymbol{R}_i) \tag{6.1}$$

where $V(\boldsymbol{r} - \boldsymbol{R}_i)$ is the averaged potential due to the nucleus and all other electrons except the one under consideration. In the simple cubic structure, lattice spacing a, the translation $\boldsymbol{r} \to \boldsymbol{r} + a(n_1, n_2, n_3)$ leaves $\mathscr{H}$ invariant, and so the translation operator can be used to provide the eigenfunctions of $\mathscr{H}$ with an important quantum number, the crystal momentum $\boldsymbol{k}$. In other lattices, the translations $\boldsymbol{R}_\alpha$ take a different form, but a crystal momentum can always be defined and, together with the band index t, it provides the quantum numbers for the *Bloch functions*,

$$\psi_{t,\boldsymbol{k}}(\boldsymbol{r}) = e^{i\boldsymbol{k}\cdot\boldsymbol{r}} u_{t,\boldsymbol{k}}(\boldsymbol{r}) \tag{6.2}$$

which are the eigenfunctions of $\mathscr{H}$. The function $u_{t,\boldsymbol{k}}(\boldsymbol{r})$ has periodicity of the lattice and satisfies the eigenvalue equation,

$$e^{-i\boldsymbol{k}\cdot\boldsymbol{r}} \mathscr{H} e^{+i\boldsymbol{k}\cdot\boldsymbol{r}} u_{t,\boldsymbol{k}}(\boldsymbol{r}) = \left[\frac{(\boldsymbol{p} + \hbar\boldsymbol{k})^2}{2m} + \sum_i V(\boldsymbol{r} - \boldsymbol{R}_i)\right] u_{t,\boldsymbol{k}}(\boldsymbol{r}) = E_t(\boldsymbol{k}) u_{t,\boldsymbol{k}}(\boldsymbol{r}) \tag{6.3}$$

subject to the boundary condition $u_{t,k}(\boldsymbol{r} + \boldsymbol{R}_\alpha) = u_{t,k}(\boldsymbol{r})$, with $\boldsymbol{R}_\alpha \equiv$ a translation vector of the lattice (see below), previously denoted $\boldsymbol{\delta}$ in Chap. 5.

The meaning of the band index t is best understood in connection with the Fourier transform of the Bloch functions, viz., the *Wannier functions*

$$\psi_{t,i}(\boldsymbol{r}) = \frac{1}{\sqrt{N}} \sum_{\boldsymbol{k}} e^{-i\boldsymbol{k}\cdot\boldsymbol{R}_i} \psi_{t,\boldsymbol{k}}(\boldsymbol{r}) = \psi_t(\boldsymbol{r} - \boldsymbol{R}_i) \tag{6.4}$$

which like the Bloch functions, form a complete, orthonormal set of functions in the Hilbert space of the Hamiltonian, (6.1). The sum over $\boldsymbol{k}$ is restricted to the *first Brillouin zone*, i.e., to N values of $\boldsymbol{k}$ in the region

$$|\boldsymbol{k}\cdot\boldsymbol{R}_\alpha| < \pi$$

where $\boldsymbol{R}_\alpha =$ any one of the smallest translation vectors of the lattice (primitive translation vectors). In the limit of infinite interactomic separation, the Wannier functions reduce to ordinary atomic orbitals. In that limit, i identifies the atom, and t the set of atomic quantum numbers (principal, orbital, azimuthal, spin; the use of a single index is for typographical simplicity). When atoms are brought close together, the atomic levels identified by t broaden into a band, unless, as in the f shell of the rare earths and the $1s$ helium core common to all metallic atoms, the electrons are still so tightly bound to the nucleus at the observed interatomic separation that the very concept of one-electron bands remains inapplicable. But this is not the case of the $3d$ states, and we note that the first metal to have a filled $3d$ band (Cu), and the elements immediately following it in the periodic table (Zn, Ga, etc.) are nonmagnetic; whereas the iron series just preceding these, noted for the unfilled d shell in the atom and d band in the metal, form materials with varied and interesting magnetic properties. One may rightly suspect the d-band electrons of being particularly important in the study of magnetism, and the unfilled d band of containing "magnetically active" electrons.

Before making these notions more precise it is necessary to review some of the properties shared by all electrons, including the nonmagnetic ones. Some of these can be studied in the "plane wave approximation," in which we set $V(\boldsymbol{r} - \boldsymbol{R}_i) = 0$ and $u_{t,\boldsymbol{k}} = 1$. But we do not wish to sacrifice the band structure, the qualitative features of which are retained in the tight-binding approximation which we study next.

6.2 Tight-Binding

The basic premise of this theory is that it is easier to estimate matrix elements involving Wannier functions (because of their supposed localization about specified atoms) than to solve the differential equations for the Bloch functions.

We illustrate this, using the Hamiltonian $\mathscr{H}$ defined in (6.1), and form the Wannier matrix elements within the nth band,

$$H(\boldsymbol{R}_{ij})_{n,n} \equiv \int \psi^*_{n,i}(\boldsymbol{r})\mathscr{H}\psi_{n,j}(\boldsymbol{r})\, d^3r \tag{6.5}$$

so that Schrödinger's equation reduces to the determinantal eigenvalue problem,

$$\det \|H(\boldsymbol{R}_{ij})_{n,n} - E\delta_{ij}\| = 0\,. \tag{6.6}$$

Evidently it does not matter which representation we solve Schrödinger's equation in, and the eigenvalues E will coincide exactly with the Bloch energies $E_n(\boldsymbol{k})$. Moreover, the eigenfunctions necessarily turn out to be precisely the proper linear combination,

$$\psi_{n,\boldsymbol{k}}(\boldsymbol{r}) = \frac{1}{\sqrt{N}}\sum_{i=1}^{N} e^{i\boldsymbol{k}\cdot\boldsymbol{R}_i}\psi_{n,i}(\boldsymbol{r}) \tag{6.7}$$

which make up the Bloch functions, (6.4).

If one uses *approximate* Wannier functions, however, the *interband* ($n \neq m$) matrix elements $H(\boldsymbol{R}_{ij})_{n,m}$ need not vanish. It is common practice to use atomic orbitals instead of Wannier orbitals as a first approximation, and therefore this method is often known as the LCAO method, for the initials of *linear combination of atomic orbitals*. The mixing of different orbitals to form the bands in the solid expresses the well-known fact that angular momentum is "quenched" (not conserved) due to conflict with the translational symmetry.

It is standard practice to limit the matrix elements $H(\boldsymbol{R}_{ij})_{n,m}$ to nearest-neighboring (in extreme cases, perhaps as far as second- and third-nearest-neighboring) atomic distances $\boldsymbol{R}_{ij}$. It is not sensible to consider more distant interactions, for if they become important the tight-binding procedure itself becomes unwieldy, and other methods such as the quasifree electron approximation are then simpler and more appropriate.

Consider the band structures derived in the following simple examples. The determinantal equation yields the energy at all points in $\boldsymbol{k}$-space, with only the constant parameters (overlap integrals) required to be numerically calculated. And if we do not know the atomic orbitals, nor trust them in the particular crystal under consideration, these constants may be taken as adjustable parameters to be fitted either by experiment or by comparison with a few calculated points given by more accurate band structure calculations. *Assume a simple cubic structure* and consider:

s bands: By symmetry, the matrix elements to the six nearest neighbors are all equal, so that only two parameters enter the problem:

$$A \equiv H(\boldsymbol{0}) = \int \psi_i{}^*(\boldsymbol{r})\mathscr{H}\psi_i(\boldsymbol{r})\, d^3r$$

and

$$-B \equiv H(0, 0, a) = \cdots = H(a, 0, 0) = \int \psi_i^*(|\boldsymbol{r} + (0, 0, a)|)\mathscr{H}\psi_i(\boldsymbol{r})\, d^3r.$$

In terms of these, the energy eigenvalues are

$$E(\boldsymbol{k}) = A - 2B(\cos k_x a + \cos k_y a + \cos k_z a). \tag{6.8}$$

Problem 6.1. (a) Assuming nearest-neighbor overlap, prove that in the body-centered cubic structure the s bands have the form

$$E(\boldsymbol{k}) = A - B \cos k_x a \cos k_y a \cos k_z a$$

and that in the face-centered cubic structure the appropriate formula is

$$E(\boldsymbol{k}) = A - B(\cos k_x a \cos k_y a + \cdots + \cos k_y a \cos k_z a)$$

(b) Derive the s-band structure for the hexagonal close-packed lattice.

---

p bands: Again in the simple cubic structure, with nearest-neighbor interactions only, the threefold degeneracy of the atomic orbitals is not lifted. We write the orbitals as

$$\psi_n(\boldsymbol{r}) = x\phi(r),\ y\phi(r),\ z\phi(r)$$

using $l = 1$ functions from Table 3.1. A first parameter,

$$A \equiv H(\mathbf{0})_{n,n} = \int x\phi^*(\boldsymbol{r})\mathscr{H}x\phi(\boldsymbol{r})\, d^3r$$

is the same for all three bands. A second parameter,

$$-B = \int x\phi^*(|\boldsymbol{r} + (0, 0, a)|)\mathscr{H}x\phi(\boldsymbol{r})\,d^3r = \int y\phi^*(|\boldsymbol{r} + (0, 0, a)|)\mathscr{H}y\phi(\boldsymbol{r})\,d^3r$$

and finally a third one

$$-C = \int (z + a)\phi^*(|\boldsymbol{r} + (0, 0, a)|)\mathscr{H}z\phi(\boldsymbol{r})\, d^3r$$

are required. All other integrals may be obtained from the above, except those for $m \neq n$; these vanish by symmetry, subject to the restriction to nearest-neighbor overlap. When the Hamiltonian eigenvalue equation (6.6) is finally solved, we find three degenerate bands:

$$E_1(\boldsymbol{k}) = A - 2B(\cos k_x a + \cos k_y a) - 2C(\cos k_z a) \tag{6.9a}$$

$$E_2(\boldsymbol{k}) = A - 2B(\cos k_y a + \cos k_z a) - 2C(\cos k_x a) \quad \text{and} \tag{6.9b}$$

$$E_3(\boldsymbol{k}) = A - 2B(\cos k_z a + \cos k_x a) - 2C(\cos k_y a)\ . \tag{6.9c}$$

Table 6.1. Matrix elements of (6.10) and (6.11) in tight-binding (LCAO) approximation

(s/s)	$H(000)_{s,s} + 2H(100)_{s,s}(X+Y+Z) + 4H(110)_{s,s}(XY+XZ+YZ) + 8H(111)_{s,s}XYZ$
(s/x)	$2iH(100)_{s,x}\tilde{X} + 4iH(110)_{s,x}\tilde{X}(Y+Z) + 8iH(111)_{s,x}\tilde{X}YZ$
(s/xy)	$-4H(110)_{s,xy}\tilde{X}\tilde{Y} - 8H(111)_{s,xy}\tilde{X}\tilde{Y}Z$
(s/x^2-y^2)	$\sqrt{3}H(001)_{s,3z^2-r^2}(X-Y) + 2\sqrt{3}H(110)_{s,3z^2-r^2}(Y-X)Z$
$(s/3z^2-r^2)$	$H(001)_{s,3z^2-r^2}(2Z-X-Y) - 2H(110)_{s,3z^2-r^2}(-2XY+XZ+YZ)$
(x/x)	$H(000)_{x,x} + 2H(100)_{x,x}X + 2H(100)_{y,y}(Y+Z) + 4H(110)_{x,x}X(Y+Z) + 4H(011)_{x,x}YZ + 8H(111)_{x,x}XYZ$
(x/y)	$-4H(110)_{x,y}\tilde{X}\tilde{Y} - 8H(111)_{x,y}\tilde{X}\tilde{Y}Z$
(x/xy)	$2iH(010)_{x,xy}\tilde{Y} + 4iH(110)_{x,xy}X\tilde{Y} + 4iH(011)_{x,xy}\tilde{Y}Z + 8iH(111)_{x,xy}\tilde{Y}XZ$
(x/yz)	$-8iH(111)_{x,yz}\tilde{X}\tilde{Y}\tilde{Z}$
(x/x^2-y^2)	$\sqrt{3}iH(001)_{z,3z^2-r^2}\tilde{X} + 2\sqrt{3}iH(011)_{z,3z^2-r^2}(\tilde{X}Y+\tilde{X}Z) + 2iH(011)_{z,x^2-y^2}\tilde{X}(Y-Z) + 8iH(111)_{x,x^2-y^2}\tilde{X}YZ$
$(x/3z^2-r^2)$	$-iH(001)_{z,3z^2-r^2}\tilde{X} - 2iH(011)_{z,3z^2-r^2}\tilde{X}(Y+Z) + 2\sqrt{3}iH(011)_{z,x^2-y^2}\tilde{X}(Y-Z) - (8/\sqrt{3})H(111)_{x,x^2-y^2}\tilde{X}YZ$
$(z/3z^2-r^2)$	$2iH(001)_{z,3z^2-r^2}\tilde{Z} + 4iH(011)_{z,3z^2-r^2}\tilde{Z}(X+Y) + (16/\sqrt{3})iH(111)_{x,x^2-y^2}XY\tilde{Z}$
(xy/xy)	$H(000)_{xy,xy} + 2H(100)_{xy,xy}(X+Y) + 2H(001)_{xy,xy}Z + 4H(110)_{xy,xy}XY + 4H(011)_{xy,xy}(X+Y)Z + 8H(111)_{xy,xy}XYZ$
(xy/xz)	$-4H(011)_{xy,xz}\tilde{Y}\tilde{Z} - 8H(111)_{xy,xz}X\tilde{Y}\tilde{Z}$
(xy/x^2-y^2)	Zero
$(xy/3z^2-r^2)$	$-4H(110)_{xy,3z^2-r^2}\tilde{X}\tilde{Y} - 8H(111)_{xy,3z^2-r^2}\tilde{X}\tilde{Y}Z$
(xz/x^2-y^2)	$2\sqrt{3}H(110)_{xy,3z^2-r^2}\tilde{X}\tilde{Z} + 4\sqrt{3}H(111)_{xy,3z^2-r^2}\tilde{X}Y\tilde{Z}$
$(xz/3z^2-r^2)$	$2H(110)_{xy,3z^2-r^2}\tilde{X}\tilde{Z} + 4H(111)_{xy,3z^2-r^2}\tilde{X}\tilde{Z}Y$
(x^2-y^2/x^2-y^2)	$H(000)^a + \frac{3}{2}H(001)^a(X+Y) + 2H(001)^b(\frac{1}{4}X+\frac{1}{4}Y+Z) + 3H(110)^a(X+Y)Z + 4H(110)^b(XY+\frac{1}{4}XZ+\frac{1}{4}YZ) + 8H(111)^a XYZ$
$(3z^2-r^2/3z^2-r^2)$	$H(000)^a + 2H(001)^a(\frac{1}{4}X+\frac{1}{4}Y+Z) + \frac{3}{2}H(001)^b(X+Y) + 4H(110)^a(XY+\frac{1}{4}XZ+\frac{1}{4}YZ) + 3H(110)^b(XZ+YZ) + 8H(111)^a XYZ$
$(x^2-y^2/3z^2-r^2)$	$\frac{1}{2}\sqrt{3}H(001)^a(-X+Y) - \frac{1}{2}\sqrt{3}H(001)^b(-X+Y) + \sqrt{3}H(110)^a(X-Y)Z - \sqrt{3}H(110)^b(X-Y)Z$

[a] $H(LMN)_{3z^2-r^2,3z^2-r^2}$.
[b] $H(LMN)_{x^2-y^2,x^2-y^2}$.

Note: Assuming nearest-neighbor interactions only, in the simple cubic structure retain only terms in (100), (010), and (001). For face-centered cubic retain only (110), (011), and (101). For bodycentered cubic retain only (111).

Key: $X = \cos k_x a$ $\quad \tilde{X} = \sin k_x a$
$Y = \cos k_y a$ $\quad \tilde{Y} = \sin k_y a$
$Z = \cos k_z a$ $\quad \tilde{Z} = \sin k_z a$

The band parameter constants are the integrals $H(LMN)_{m,n} = \int \psi^*_m[\boldsymbol{r}_m + a(L,M,N)]\mathscr{H}\psi_n(\boldsymbol{r})d^3r$

Source: Based on Table II in [6.3].

The three p bands are related to one another by those rotations in $\boldsymbol{k}$-space, which permute the Cartesian components $k_{x,y,z}$. In each band, the contours of constant energy have less than cubic symmetry, and even at small $\boldsymbol{k}$, these contours are not spherical but are ellipsoids of revolution with principal nonequivalent axes along the $k_{x,y,z}$ directions.

Had we chosen the axes of quantization of the three p functions along other than a crystal axis, the interband matrix elements $H(0, 0, a)_{n,m}$ for $n \neq m$ would not have vanished so conveniently. The eigenvalue equation, (6.6), which held for the exact Wannier functions, must be replaced in the general case by the r-dimensional equation

$$\det \| H(\boldsymbol{k})_{n,m} - E\delta_{n,m} \| = 0, \tag{6.10}$$

where r = number of interacting bands; $n, m = 1, 2, \ldots, r$. Here we have taken advantage of translational invariance to form the Fourier transforms, and to define

$$(n/m) \equiv H(\boldsymbol{k})_{n,m} = \frac{1}{N} \sum_{i,j} e^{i\boldsymbol{k}\cdot\boldsymbol{R}_{ij}} \int \psi^*_{n,i} \mathscr{H} \psi_{m,j} \, d^3r \, . \tag{6.11}$$

In Table 6.1 are reproduced results by *Slater* and *Koster*, who have calculated these matrix elements out to third nearest neighbor in the simple cubic structure. This is sufficient also for obtaining nearest-neighbor interactions in face- and body-centered cubic structures (and with some manipulations, next nearest neighbors also, but we shall not be interested in these). Note that $H(\boldsymbol{k})_{n,m}$ is abbreviated (n/m) in Table 6.1 for typographical simplicity. The matrix elements $H(\boldsymbol{k})_{n,m}$ are given, abbreviated as (n/m), among s-like functions (s); p-like functions $(x, y,$ and $z)$; and d-like functions $(xy, xz, yz, 3z^2 - r^2,$ and $x^2 - y^2)$.

It is sensible to consider these matrix elements as empirical constants; they can be estimated by performing the appropriate two-center integrals.

Let us consider five d bands in the simple cubic structure, ignoring s and p bands as well as non-nearest-neighbor interactions. Looking up Table 6.1, we extract the following special case of the eigenvalue quation, (6.10):

$$\det \begin{array}{c} \\ (xy) \\ (xz) \\ (yz) \\ x^2 - y^2 \\ 3z^2 - r^2 \end{array} \begin{array}{c} \begin{array}{ccccc} (xy) & (xz) & (yz) & x^2 - y^2 & 3z^2 - r^2 \end{array} \\ \left\| \begin{array}{ccc:cc} F_1(k) - E & 0 & 0 & 0 & 0 \\ 0 & F_2(k) - E & 0 & 0 & 0 \\ 0 & 0 & F_3(k) - E & 0 & 0 \\ \hdashline 0 & 0 & 0 & F_4(k) - E & V(k) \\ 0 & 0 & 0 & V(k) & F_5(k) - E \end{array} \right\| \end{array} = 0 \tag{6.12}$$

The definitions of the F's and $V(k)$, and some features of the solutions, are discussed in Problem 6.2.

Problem 6.2. (a) In the example of (6.12) in the text, find $F(k)$ and $V(k)$ by referring to Table 6.1. Show that the solutions of the eigenvalue equation describe three degenerate d bands with ellipsoidal contours of constant energy much like the p bands of (6.9) and two nondegenerate s-like bands. Obtain the contours of energy of the latter near $k = 0$. Plot the energy as a function of k along the three principal directions, (100), (110), and (111).

(b) Calculate the first-order effect of an infinitesimal next-nearest-neighbor interaction on the band structures calculated in part (a).

Making some assumptions about the relative and absolute magnitudes of the band-structure parameters, Slater and Koster calculated the histogram *density of states* curve (number of eigenvalues per unit energy) which includes all five d bands in the body-centered cubic structure, such as in Fe. This is reproduced in Fig. 6.1. The lower peak belongs to the *bonding orbitals* in the chemical terminology, and, according to Slater and Koster, explains the anomalously great binding energies of some metals in the first half of the iron transition series.

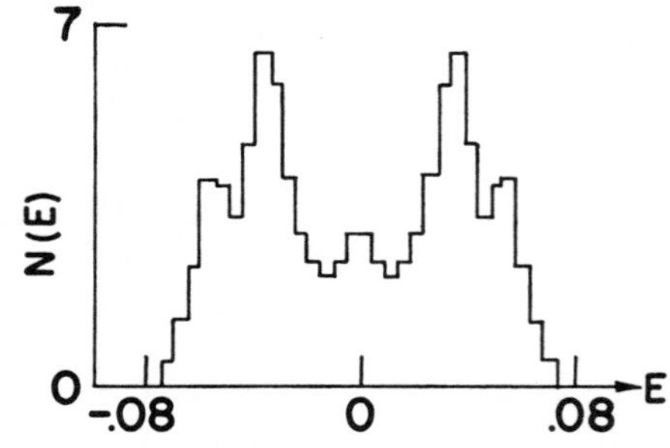

Fig. 6.1. Original tight-binding density-of-states histogram for d bands in bcc structure. $N(E)$ is plotted E (Rydbergs), and is equal to the number of eigenvalues $E_n(k)$ within $\pm$ 1.025 Ry of the energy E, normalized such that the total area under curve equals 5. [6.3]

What the histogram could not show are the so-called *Van Hove singularities.* Whenever an $E(\boldsymbol{k})$ curve has a minimum, maximum, or simply a saddle point, its contribution to the density of states becomes excessive. [For example, a totally constant $E(\boldsymbol{k}) = E_0$ contributes a delta-function singularity to the density of states function.] Some years after the Slater-Koster work, *Wohlfarth* and *Conwell* published the curve reproduced in Fig. 6.2, giving the density of states

$$N(E) \propto \sum_{n,k} \delta[E - E_n(\boldsymbol{k})] \tag{6.13}$$

as the output of a computer calculation in which the Van Hove singularities were scrupulously preserved. The peaky nature of this new curve is evident, with the principal maxima occurring whenever an energy band touches the Brillouin zone. Note also the revision upwards of the estimated effective width of the bands over the earlier work, by a factor of approximately 3.

As we shall see subsequently, it is of crucial importance in the theory of magnetism in metals[3] whether the density of states is high or low, particularly in the vicinity of the Fermi level μ.

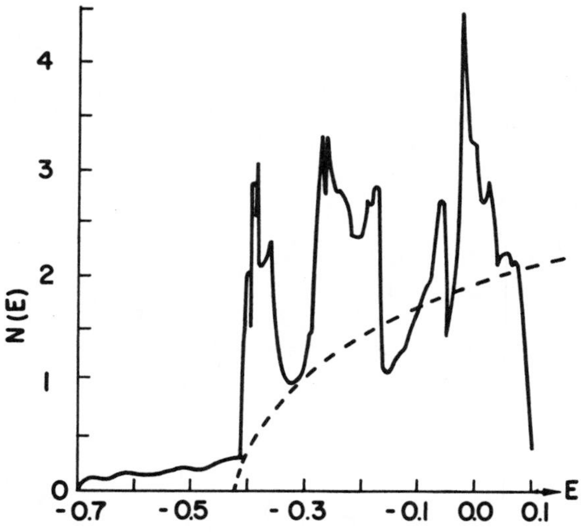

Fig. 6.2. More accurate density-of-states curve $N(E)$ *vs* E (Rydbergs) for bcc iron. Dashed line is average free-electron approximation $N(E) \propto \sqrt{E}$; and note how at several points (for example, $E = 0$ or -0.4) the computed density of states can exceed the average curve by a large factor [6.4]

6.3 Weak Magnetic Properties

All metals share some weak magnetic characteristics. The study of these has enjoyed great vogue lately, because of the very detailed information on the band structure and Fermi surface parameters which it divulges. As some examples of important dynamic properties, we list the Hall effect, magnetoresistance, and cyclotron resonance. Some important static properties of the electrons include Pauli spin paramagnetism, Landau diamagnetism, and the De Haas-Van Alphen effect. All these are well known and abundantly discussed in standard texts on solid-state or metal physics[4]. Therefore, here we shall be content with a qualita-

[3] The band structure of some magnetic metals is now known fairly well, such as nickel. See [6.5].

[4] Less known, but also interesting, is the subject of high fields in metals, treated by *Fawcett* [6.6].

tive discussion of the physical basis of the static phenomena listed, without emphasizing the mathematics which can become rather complicated. The principal purpose is to display the important role of the density of states function $N(E)$ in the magnetic properties of electrons, and to introduce the concepts of Fermi energy and Fermi distribution.

In the ground state of a normal metal ($T = 0$ K) all the one-electron states of energy less than μ are occupied, all those above are empty. μ is the chemical potential or Fermi energy. Moreover, in the absence of magnetic fields or spin-orbit coupling to lift the Kramers' degeneracy, every state of given $(n, \boldsymbol{k})$ below the Fermi level is *doubly* occupied, by an electron with spin up and one with spin down.

At finite temperature, states within $\pm kT$ of the Fermi level are partly occupied, as may be seen from the Fermi distribution function

$$f(E_{\boldsymbol{k}}) = \frac{1}{\mathrm{e}^{(E_{\boldsymbol{k}}-\mu)/kT} + 1} \tag{6.14}$$

which gives the thermal-average probability that the state of energy $E_{\boldsymbol{k}}$ is occupied (absorbing band and spin indices into $\boldsymbol{k}$). In a weak magnetic field, the spins of the electrons within the $\pm kT$ neighborhood of the Fermi energy will be free to orient themselves parallel to the field; and according to the laws of Langevin and Curie, each will contribute a magnetization proportional to the applied field, to Curie's constant C, and to the inverse temperature, viz.,

$$\delta\mathcal{M} \sim H\frac{C}{T}\,.$$

The number of participating electrons is $\sim 2kT\,N(\mu)$, and therefore the total paramagnetic spin susceptibility is

$$\mathcal{M}/H \equiv \chi_{\mathrm{p}} = 2CkN(\mu)\,. \tag{6.15}$$

This agrees with more rigorous derivations of Pauli's spin paramagnetism of free electrons and is correct to $O[(kT/\mu)^2]$ at finite temperatures. ($kT/\mu \ll 10^{-2}$ at room temperature for most nonmagnetic metals). Note the dependence on $N(\mu)$, the density of states *at* the Fermi energy. This susceptibility is smaller by a factor $2kTN(\mu)/\mathcal{N}$ than that of $\mathcal{N}$ free spins.

The De Haas-Van Alphen effect is not so easy to explain nor to understand; nevertheless it also reflects the dependence of the thermodynamic properties of the metal on the density of states. In this case, the *density of states is affected* by a magnetic field, and therefore the thermodynamic functions will depend on the field. For illustrative purposes, consider the wavefunctions in the free-electron approximation,

$$\psi_{\boldsymbol{k}} = \mathrm{e}^{i\boldsymbol{k}\cdot\boldsymbol{r}} \qquad \mathcal{H} = \frac{\boldsymbol{p}^2}{2m^*} = -\frac{(\hbar\nabla)^2}{2m^*}\,. \tag{6.16}$$

The effective mass m^* may differ from the free-electron mass $m_0 = 9.1 \times 10^{-28}$ gram by one or more orders of magnitude (greater or smaller). The effective mass approximation used here may be quite successful for describing s bands, but it does not lead to a realistic density of states for the d bands, as seen in Fig. 6.2; so the following derivation is merely illustrative.

In a weak electromagnetic field, described by the vector potential $\boldsymbol{A}(\boldsymbol{r}, t)$, the electron momenta $\boldsymbol{p}$ become $\boldsymbol{p} - e\boldsymbol{A}/c$, and the time-dependent Schrödinger equation becomes

$$\frac{[\boldsymbol{p} - (e/c)\boldsymbol{A}]^2}{2m^*}\psi(\boldsymbol{r}, t) = \hbar \mathrm{i}\frac{\partial}{\partial t}\psi(\boldsymbol{r}, t)\,. \tag{6.17}$$

For a static magnetic field, $\boldsymbol{A}(\boldsymbol{r}) = (0, Hx, 0)$ does not depend on the time and satisfies the two equations

$$\nabla \times \boldsymbol{A} = (0, 0, H) \qquad \text{and} \qquad -\frac{1}{c}\frac{\partial \boldsymbol{A}}{\partial t} = \boldsymbol{E}(\boldsymbol{r}, t) = 0\,. \tag{6.18}$$

Therefore, writing $\psi(\boldsymbol{r}, t) = \exp[\mathrm{i}(k_y y + k_z z)]\ \phi(x)\exp(-\mathrm{i}Et/\hbar)$, we find that $\phi(x)$ obeys a harmonic-oscillator equation, so that the total energy is given by

$$E = \frac{\hbar^2 k_z^2}{2m^*} + \left(n + \frac{1}{2}\right)\hbar\omega_c \qquad n = 0, 1, 2, \ldots \tag{6.19}$$

with the "cyclotron frequency" ω_c defined by

$$\omega_a = \frac{eH}{m^* c}\,. \tag{6.20}$$

Problem 6.3. Derive (6.19, 20) of the text, by solving Schrödinger's equation in the manner described.

The expression for the energy may be interpreted as the result of quantization on the classical circular motion of a charge in a magnetic field.

The density of states is obtained by differentiating the function which gives the total number of states lying below energy E. Thus, with a factor 2 for the two values of spin

$$\begin{aligned} N(E) &\propto 2\frac{d}{dE}\sum_{m=1}^{M(E)}\int_0^{\sqrt{E-(m+1/2)\hbar\omega_c}} \hbar dk_z (2m^*)^{-1/2} \\ &\propto \sum_{m=1}^{M(E)} \frac{1}{\sqrt{E - (m + 1/2)\hbar\omega_c}} \end{aligned} \tag{6.21}$$

where $M(E)$ = largest positive integer for which the radicand is positive. A plot of this function is given in Fig. 6.3a; it is similar to the free-electron function

$N(E) \sim E^{1/2}$ except for narrow, integrable, square-root singularities at half-odd-integer multiples of the cyclotron energy $\hbar\omega_c$.

The most interesting behavior occurs in the neighborhood of the Fermi energy μ as the magnitude of H is increased. Whenever $(\mu - \frac{1}{2}\hbar\omega_c)$ becomes an integer multiple of $\hbar\omega_c$, the above sums acquire a new integer $M(\mu)$, and the density of states $N(\mu)$ itself acquires a square-root singularity. This is shown in Fig. 6.3b. Clearly, this must lead to fluctuating, quasiperiodic behavior of all the thermodynamic properties of the metal: the specific heat, magnetic susceptibility, electrical resistance, etc., must all be oscillatory functions of the maximum integer, $M(\mu) \propto 1/H$.

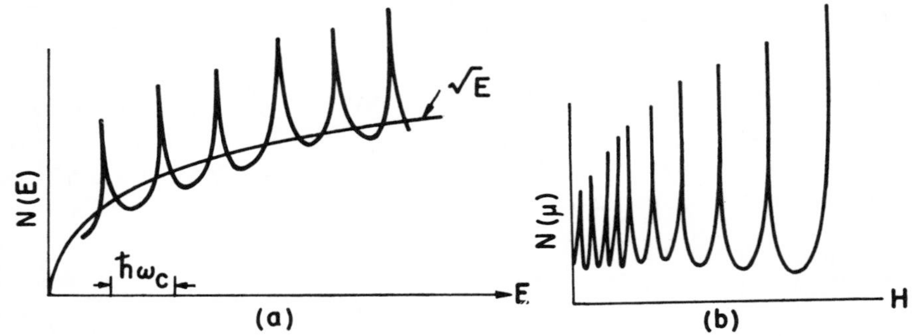

Fig. 6.3. (a) Density-of-states $N(E)$ vs E in constant magnetic field. Averaging the singularities would *lower* curve *below* its zero-field value $\sqrt{E}$. Therefore the energy of a constant number of electrons in weak magnetic fields is *higher* than in zero field, and the Landau motional susceptibility is diamagnetic. (b) Density of states at the Fermi energy, $N(\mu)$, as function of strong applied magnetic field. The oscillatory behavior results in De Haas-Van Alphen effect. Note that μ is itself a function of H, determined by the requirement $\mathcal{N} = \int^{\mu} dEN(E) =$ constant

The weak-field limit occurs either when $\hbar\omega_c \ll kT$ or when scattering results in a mean free path smaller than the radius of the cyclotron orbit. In either case it is permissible to expand the free energy and other thermodynamic quantities in powers of H.

A useful method due to *Peierls* [6.7] is based on the Poisson summation formula,

$$N(E) \propto \sum_{p=-\infty}^{+\infty} (-1)^p \int_0^{E/\hbar\omega_c} \frac{e^{2\pi ipx}}{\sqrt{E - x\hbar\omega_c}} dx$$

and some partial integrations to evaluate the leading terms in the free energy.[5] As an example, the internal energy to leading order is

[5] See also [6. 8,9].

$$E_{tot}(H) = \int dEN(E)Ef(E) \cong E_{tot}(0) + \frac{1}{2}\chi_d H^2 \qquad (6.22)$$

with χ_d a positive quantity. Because the increase in energy results in a force tending to repel the material from an applied field, this is a diamagnetic susceptibility.

Finally, the total susceptibility, combining the Pauli spin paramagnetism with the Landau orbital diamagnetism can be shown to have the value

$$\chi = 2CkN(\mu)\left[1 - \frac{1}{3}\left(\frac{m_0}{m^*}\right)^2\right] \qquad (6.23)$$

which explains the weak net paramagnetism of most metals (where $m^* \sim m_0$) and, on the other hand, the strong diamagnetism of bismuth, in which $m^* = O(0.01m_0)$, $(m_0/m^*)^2 = O(10^4)$.

In common with other magnetic phenomena studied in this book, the Landau diamagnetism is a purely quantum mechanical effect, which disappears in the correspondence limit by virtue of the oft-invoked Bohr-Van Leeuwen theorem. Also the diamagnetic increase in energy, (6.22), is extensive, i.e., every unit volume of the material contributes equally to the diamagnetic current density.

Note: It is possible to view the nonvanishing diamagnetism as a direct consequence of the uncertainty principle; for if the electrons have perfectly sharp momenta $\boldsymbol{p}$, the vector potential $\boldsymbol{A} = (0, Hx, 0)$ cannot be simultaneously specified nor removed by a gauge transformation and vice versa. Therefore for small H, the energy is raised above the ground state value it had in the absence of the field. This point of view has been carried through by an expansion of the free energy in powers of $\hbar$ in a review by *Van Vleck* [6.10] of the weak magnetic properties of metals and of the effects of exchange and correlation. Many-body effects have been thoroughly explored by *Isihara* and *Kojima* [6.10].

An inaccurate but frequently heard explanation of the Landau diamagnetism is that it is caused by the inability of surface currents to cancel volume currents, due to quantum mechanical effects. But this is only a half-truth; for it obscures the physically significant fact that χ_d depends only on the bulk properties, and is independent of the surface geometry, boundary conditions, scattering, etc.

6.4 Exchange in Solids: Construction of a Model Hamiltonian

In solids as in atoms, the really strong magnetic phenomena are *electrostatic* in origin, the powerful Coulomb forces being "triggered" by the spins of the electrons under the regulation of the Pauli principle. This is well demonstrated in second quantization, introduced in Chap. 4. Specific applications to insulators and metals will be the subject of the remainder of the chapter. In second quantization, it is possible for a unique Hamiltonian to apply to all the various sorts of solids, with only numerical parameters and the occupation of the various bands remaining to be specified. Thus there is no need to deal differently with insulators or metals at the present stage.

First, let us reformulate the theory of noninteracting electrons, starting with the operator $c_{j,n,m}$ which which destroys an electron at the jth Wannier site, in the nth band, with spin index $m(=\uparrow \text{ or } \downarrow)$. The operator which creates an electron in precisely the same state is the Hermitean conjugate operator $c^*_{j,n,m}$. The band Hamiltonian of (6.1) can be written in terms of these operators as

$$\mathscr{H}_0 = \sum_{i,j,n,m} H(\boldsymbol{R}_{ij})_{n,n} c^*_{i,n,m} c_{j,n,m} \,. \tag{6.24}$$

This represents quite graphically the "hopping" or transfer of an electron from site j to site i, with the matrix element previously calculated in (6.5), and displays the conservation laws obeyed by true Wannier functions in the present (one-electron) approximation. These are the conservation of the spin index, and of the band index (which may also be considered as an "isotopic" spin), ensuring that the one-electron bands are well defined.

The Fermion operators above obey the usual *anti*commutation relations

$$\begin{aligned} &c_r c_s + c_s c_r \equiv \{c_r, c_s\} = 0 \qquad \{c_r^*, c_s^*\} = 0, \qquad \text{therefore} \\ &(c_r)^2 = (c_r^*)^2 = 0 \qquad \text{and} \qquad \{c_r, c_s^*\} = \delta_{r,s} \end{aligned} \tag{6.25}$$

and the occupation number operator $\mathfrak{n}_r = c_r^* c_r$ has eigenvalues 0, 1 only; here, r or s stand for any *set* of quantum numbers, including the electron's spin, m.

The band Hamiltonian is diagonal in the Bloch representation. We show this by means of a canonical transformation, which in turn is equivalent to choosing the plane wave combination of Wannier operators, as follows:

$$c_{\boldsymbol{k},n,m} = \frac{1}{\sqrt{N}} \sum_{i=1}^{N} \mathrm{e}^{-\mathrm{i}\boldsymbol{k}\cdot\boldsymbol{R}_i} c_{i,n,m}$$

and (6.26)

$$c^*_{\boldsymbol{k},n,m} = \frac{1}{\sqrt{N}} \sum_{i=1}^{N} \mathrm{e}^{+\mathrm{i}\boldsymbol{k}\cdot\boldsymbol{R}_i} c^*_{i,n,m} \,.$$

The reader can easily verify that the $c_{\boldsymbol{k}}$'s and $c_{\boldsymbol{k}}^*$'s also satisfy anticommutation relations, (6.25). The inverse linear combinations are simply

$$c_{i,n,m} = \frac{1}{\sqrt{N}} \sum_{\substack{\boldsymbol{k} \text{ in} \\ \text{first BZ}}} \mathrm{e}^{\mathrm{i}\boldsymbol{k}\cdot\boldsymbol{R}_i} c_{\boldsymbol{k},n,m}$$

and (6.26a)

$$c^*_{i,n,m} = \frac{1}{\sqrt{N}} \sum_{\substack{\boldsymbol{k} \text{ in} \\ \text{first BZ}}} \mathrm{e}^{-\mathrm{i}\boldsymbol{k}\cdot\boldsymbol{R}_i} c^*_{\boldsymbol{k},n,m} \,.$$

Therefore let us substitute these expressions into $\mathscr{H}_0$, and obtain

$$\begin{aligned}\mathscr{H}_0 &= \frac{1}{N}\sum_j \mathrm{e}^{\mathrm{i}(\boldsymbol{k}-\boldsymbol{k}')\cdot\boldsymbol{R}_j} \sum_i H(\boldsymbol{R}_{ij})_{n,n} \mathrm{e}^{i\boldsymbol{k}\cdot\boldsymbol{R}_{ij}} \sum_{n,m} c^*_{\boldsymbol{k}',n,m} c_{\boldsymbol{k},n,m} \\ &= \sum_{\boldsymbol{k},n,m} E_n(\boldsymbol{k}) c^*_{\boldsymbol{k},n,m} c_{\boldsymbol{k},n,m} = \sum_{\boldsymbol{k},n,m} E_n(\boldsymbol{k}) \mathfrak{n}_{\boldsymbol{k},n,m}\end{aligned} \tag{6.27}$$

using the definition of the energy of a Bloch electron $E_n(\boldsymbol{k})$ given previously.

Because $\mathscr{H}_0$ is diagonal in the Bloch operator representation ($n_{\boldsymbol{k},n,m} = 0$ or 1), as eigenstate of this Hamiltonian can be specified merely by stating which $\boldsymbol{k}$, n, m are occupied, and which are not. For example, the no-particle *vacuum* state $|0)$ is annihilated by *every* $c_{\boldsymbol{k}}$: $c_{\boldsymbol{k}}|0) = 0$ therefore $\mathfrak{n}_{\boldsymbol{k}}|0) = 0$; and $\mathscr{H}_0$ must also have zero eigenvalue in this state. A more important eigenfunction is the *Fermi sea*, defined to be the state of lowest energy among all the eigenstates containing precisely $\mathscr{N}$ electrons. In terms of the Fermi energy μ (below which there are precisely $\mathscr{N}$ one-electron states $\boldsymbol{k}, n, m$) the Fermi sea can be written as

$$|F) \equiv \prod c^*_{\boldsymbol{k},n,m}|0) \tag{6.28}$$

where the product extends over all $\boldsymbol{k}$, n, m for which $E_n(\boldsymbol{k}) < \mu$.

The eigenvalue of $\mathscr{H}_0$ in this state will be the "unperturbed" ground state energy W_0,

$$W_0 = \sum E_n(\boldsymbol{k}) = \int_{-\infty}^{\mu} dEN(E)E \qquad \text{where } \mathscr{N} \equiv \int_{-\infty}^{\mu} dEN(E) \tag{6.29}$$

and where the sum over $\boldsymbol{k}$, n, m again extends only over the states in the Fermi sea. It is the principal object of the theory of magnetism in metals to explain precisely how the electronic interactions modify the Fermi sea, and perturb the ground state energy.

One possible result of the interactions, and of thermal excitations as well, is to create any number of *elementary excitations*. These are constructed by removing a single electron from the Fermi sea and placing it above in one of the unoccupied states. For example, letting b stand for a set of labels $\boldsymbol{k}$, *n,m within* $|F)$, and a for a set of such labels *outside* $|F)$, the eigenfunction and eigenvalue of a single elementary excitation are

$$\psi^+_{ab} = c^*_a c_b|F) \qquad \text{and} \qquad W_{ab} = W_0 + E_a - E_b\,. \tag{6.30}$$

As an alternative to the above description, we may conceive the elementary excitation of (6.30) as the creation of two *quasiparticles*: both a quasielectron of energy $E_a - \mu$, and a quasihole of energy $\mu - E_b$ are added to the ground state. The energy of each quasiparticle, and of the elementary excitation as well, must be positive (by definition of the *ground* state).

The elementary excitations occupy a continuum (in the limit $L \to \infty$, naturally) of energy levels even when restricted to a specific total momentum. In the free

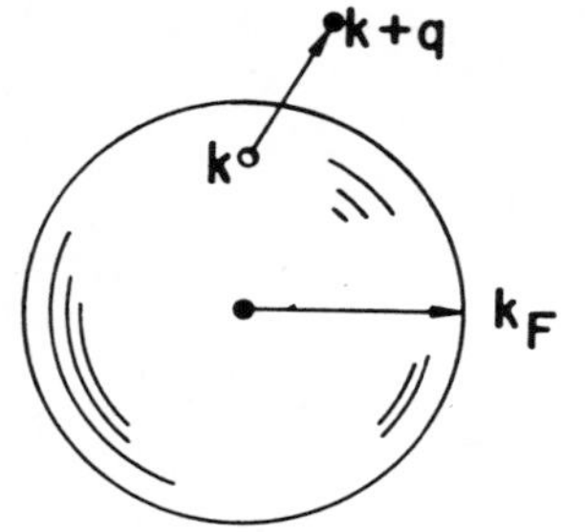

◁ **Fig. 6.4.** Fermi sphere of radius k_F with an elementary excitation indicated: electron taken to $(\boldsymbol{k} + \boldsymbol{q})$, leaving hole at $\boldsymbol{k}$

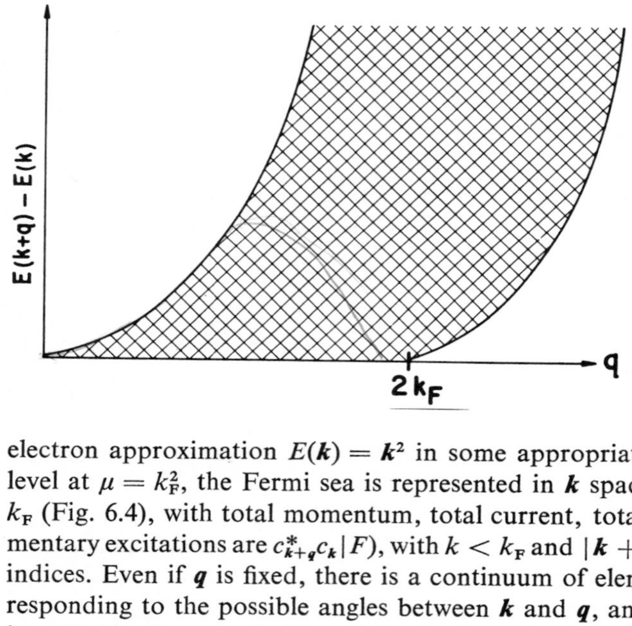

Fig. 6.5. Continuum of elementary excitations in nonmagnetic Fermi sea of noninteracting electrons

electron approximation $E(\boldsymbol{k}) = \boldsymbol{k}^2$ in some appropriate units, with the Fermi level at $\mu = k_F^2$, the Fermi sea is represented in $\boldsymbol{k}$ space as a sphere of radius k_F (Fig. 6.4), with total momentum, total current, total spin all zero. The elementary excitations are $c^*_{k+q} c_k |F)$, with $k < k_F$ and $|\boldsymbol{k} + \boldsymbol{q}| > k_F$, omitting spin indices. Even if $\boldsymbol{q}$ is fixed, there is a continuum of elementary excitations corresponding to the possible angles between $\boldsymbol{k}$ and $\boldsymbol{q}$, and the energy of these is bounded by two parabolas and the horizontal axis, as shown in Fig. 6.5. Brillouin zone and magnetic field effects on the spectrum of elementary excitations are discussed in Problem 6.4.

Problem 6.4. (a) Discuss the double spectrum of elementary excitations + spin flip,

$$\psi^+_{k,q} = c^*_{k+q\uparrow} c_{k\downarrow} |F_H) \text{ and } c^*_{k+q\downarrow} c_{k\uparrow} |F_H)$$

for free electrons whose spins, only, are interacting with a magnetic field (e.g., an exchange field); that is, for which $E_m(\boldsymbol{k}) = \boldsymbol{k}^2 + m\mu_B H$, where $m = \pm 1$, and $|F_H)$ is the Fermi sea appropriate to this situation. Plot the continua in the manner of Fig. 6.5, and notice that since $|F_H)$ is the ground state, all excitation energies are required to be positive. Consider H both small, *and* large.

(b) Neglecting spin (as in the text above), show the effects of the Brillouin zone by plotting qualitatively the elementary excitation spectrum of a half-filled s band in the simple cubic structure. Pay special attention to the effects of *umklapp* ($\boldsymbol{k}$ is necessarily in first B.z. but $\boldsymbol{k}+\boldsymbol{q}$ is not), and to the maximum energy cutoff in the spectrum.

Besides all one-body potentials, the band-theoretic Hamiltonian $\mathscr{H}_0$ can include the averaged effects of two-body forces. Let us see why this is the case; the most general two-body matrix element is, in the Wannier representation,

$$\mathscr{H}' = \sum V(i,j,n',n;i',j',t',t)c^*_{i't'm'}c^*_{in'm}c_{jnm}c_{j'tm'} \tag{6.31}$$

where

$$V(i,j,n',n;i',j',t',t) = \frac{1}{2}\int d^3r \int d^3r' \psi^*_{t',i'}(\boldsymbol{r}')\,\psi_{t,j'}(\boldsymbol{r}')\frac{e^2}{|\boldsymbol{r}-\boldsymbol{r}'|}\psi^*_{n',i}(\boldsymbol{r})\psi_{n,j}(\boldsymbol{r}). \tag{6.32}$$

These matrix elements will connect states which differ only by a change in the quantum numbers of two electrons, from jnm and $j'tm'$ to $in'm$ and $i't'm'$, and so long as we use orthogonal Wannier functions and two-body potentials (such as the physically important Coulomb repulsion), *there are no further matrix elements*, and *the total Hamiltonian consists of* $\mathscr{H}_{\text{tot}} = \mathscr{H}_0 + \mathscr{H}'$. As stated above, some of the terms in $\mathscr{H}'$ can be incorporated in $\mathscr{H}_0$; consider as one possible example the terms with $i' = j'$ and $t' = t$,

$$\left[\sum V(i,j,n',n;i',j',t,t)\mathbf{n}_{i'tm'}\right]c^*_{in'm}c_{jnm}\,. \tag{6.33}$$

Although the factor which multiplies $c^*_{in'm}c_{jnm}$ is an operator, its average value in the Fermi sea serves as a useful estimate. Thus, we should incorporate into $\mathscr{H}_0$ the terms,

$$\delta\mathscr{H}_0 = \sum \delta H(\boldsymbol{R}_{ij})_{n,n'}c^*_{in'm}c_{jnm}$$

with

$$\delta H(\boldsymbol{R}_{ij})_{n,n'} \equiv \sum_{i'tm'} V\cdot(F|\mathbf{n}_{i'tm'}|F) \tag{6.34}$$

and subtract them from $\mathscr{H}'$. This is the sort of procedure which was already anticipated in (6.1) and those following, when it was stated that $V(\boldsymbol{r}-\boldsymbol{R}_i)$ is the *averaged* potential due to the nucleus, and all other electrons except the one under consideration. Thus the transfer of all the averaged effects of $\mathscr{H}'$ into $\mathscr{H}_0$ has the result that what remains of the former has vanishing expectation value in the Fermi sea

$$(F|\mathscr{H}' - (F|\mathscr{H}'|F)F) \equiv 0 \, .$$

This procedure "renormalizes" the band structure $E_n(\boldsymbol{k})$ in a self-consistent way. Note that the self-consistent one-electron band picture must be altered somewhat if we consider a different state, such as the ferromagnetic state instead of $|F)$; but the resultant changes should be small if charge neutrality, the prime consideration in the energy balance, is maintained. We may therefore imagine $\mathscr{H}_0$ to have constant parameters, to include a priori all the important electron-electron interactions (on the average), as well as all the interactions of the electrons with nuclei and their kinetic energy. From $\mathscr{H}'$ we shall extract for present consideration only a significant subset of two-body terms. Using as a guide, that we should consider only those interactions which matter most when the atoms are very far apart (the intraatomic terms) or when they are very close (the ubiquitous Coulomb interaction), we can extract from $\mathscr{H}_{\text{tot}} = \mathscr{H}_0 + \mathscr{H}'$ the features of greatest physical significance and provide a starting point for later more ambitious investigations.

The first of the interactions to be retained is the direct two-particle *Coulomb repulsion* obtained from $\mathscr{H}'$ by setting $in' = jn$ and $i't' = j't$. After a change of dummy indices, it is

$$\mathscr{H}_{\text{c}} = \sum_{\substack{i,j,n,t,\\ m,m'}} V(\boldsymbol{R}_{ij})_{n,t} \mathfrak{n}_{i,n,m}[\mathfrak{n}_{j,t,m'} - (F|\mathfrak{n}_{j't'm'}|F)] \tag{6.35}$$

where the simplified notation is used,

$$\begin{aligned} V(\boldsymbol{R}_{tj}) &\equiv V(i, i, n, n; j, j, t, t) \\ &= \frac{1}{2} \int d^3r \int d^3r' \psi^*_{t,j}(\boldsymbol{r}') \psi_{t,j}(\boldsymbol{r}') \frac{e^2}{|\boldsymbol{r} - \boldsymbol{r}'|} \psi^*_{n,i}(\boldsymbol{r}) \psi_{n,i}(\boldsymbol{r}) \end{aligned} \tag{6.36}$$

and the factor 1/2 prevents double-counting. At large distances the leading term in a multipole expansion of this integral is

$$V(\boldsymbol{R}_{ij})_{n,t} \sim \frac{1}{2} \frac{e^2}{R_{ij} + a} \tag{6.37}$$

modified to be accurate even as close as nearest-neighbor separation. The terms subtracted in (6.35) ensure that we have electrical neutrality,

$$(F|\mathscr{H}_{\text{c}}'|F) = 0$$

as required.

The next important class of terms are the *exchange interactions*, which are the subset of terms in (6.31) for which $i = j'$, $n' = t$ and $i' = j$, $t' = n$. Because $c^*_{in'm}$ and c_{jnm} *anticommute*, the interchange of these two operators to properly associate the pairs in the manner indicated, results in a minus sign. Except for the operator formalism, this then is the ordinary exchange interaction discussed repeatedly in this volume. It acts as a correction to the Coulomb repulsion of *two electrons in a relative triplet state*, and lowers the energy of this state as compared to the singlet configuration. By grouping pairs of Fermions into spin operators, we can include in our Hamiltonian the natural and obvious generalization to the exchange operator $\frac{1}{2}(1 + 4\boldsymbol{S}_i \cdot \boldsymbol{S}_j)$, including the variable occupation-number feature of the second quantization. Associating the indices as prescribed, one extracts from (6.31) the exchange Hamiltonian,

$$\mathscr{H}_{\mathrm{ex}} = - \sum_{\substack{n,n' \\ i,j \\ (n,i)\neq(n',j)}} J(\boldsymbol{R}_{ij})_{n,n'} \left\{ \boldsymbol{S}_{n,j} \cdot \boldsymbol{S}_{n',i} + \frac{1}{4} \mathfrak{n}_{n,j} [\mathfrak{n}_{n'i} - (F | \mathfrak{n}_{n',i} | F)] \right\} \tag{6.38}$$

where

$$\mathfrak{n}_{n,j} \equiv \mathfrak{n}_{n,j,\uparrow} + \mathfrak{n}_{n,j,\downarrow} \tag{6.39}$$

and the spin operators,

$$S_p^z = \frac{1}{2}(\mathfrak{n}_{p\uparrow} - \mathfrak{n}_{p\downarrow}) = \frac{1}{2}(c_{p\downarrow}^* c_{p\downarrow} - c_{p\downarrow}^* c_{p\downarrow})$$

$$S_p^+ = S_p^x + \mathrm{i} S_p^y = c_{p\uparrow}^* c_{p\downarrow} \qquad \text{and} \qquad S_p^- = S_p^x - \mathrm{i} S_p^y = c_{p\downarrow}^* c_{p\uparrow} \tag{6.40}$$

are recognized as one of the Fermion representations of the spin one-half operators, with $\hbar = 1$, and $p \equiv (n, j)$.

The exchange constant, in the simplified notation, is the *positive-definite* integral

$$\begin{aligned} J(\boldsymbol{R}_{ij})_{n,n'} &= 2V(i, j, n', n; j, i, n, n') \\ &= \int d^3r \int d^3r' \psi_{n,j}^*(\boldsymbol{r}') \psi_{n',i}(\boldsymbol{r}') \frac{e^2}{|\boldsymbol{r} - \boldsymbol{r}'|} \psi_{n',i}^*(\boldsymbol{r}) \psi_{n,j}(\boldsymbol{r}) \, . \end{aligned} \tag{6.41}$$

The largest such exchange integral is the familiar *intra*-atomic Hund's rule integral to which we assign a special symbol,

$$\begin{aligned} J_{n,n,}^{Hu} &\equiv J(\boldsymbol{0})_{n,n'} \\ &= \int d^3r \int d^3r' \psi_{n,i}^*(\boldsymbol{r}') \psi_{n',i}(\boldsymbol{r}') \frac{e^2}{|\boldsymbol{r} - \boldsymbol{r}'|} \psi_{n',i}^*(\boldsymbol{r}) \psi_{n,i}(\boldsymbol{r}) \end{aligned} \tag{6.42}$$

and in most cases it will not be necessary to consider any exchange contribution other than $J(0)$. The reason for this is that $J(\boldsymbol{R}_{ij})$ decreases *much* faster than the Coulomb integral $V(\boldsymbol{R}_{ij})$, so that unless (6.42) vanishes by symmetry or for any other reason, even the nearest-neighbor exchange will be small in comparison, and the non-nearest-neighbor contributions completely negligible. (An exception: for only one band, two electrons can be in a triplet state only if they are on different Wannier sites; and the leading exchange integral is (6.41) at nearest-neighbor distances, with $n = n'$).

The designation of the exchange forces has a certain degree of arbitrariness, so it is useful to conclude this section with a review of the choice which was exercised. One traditional approach which was implicitly discarded, was the one-band, free electron approximation, with Bloch electrons interacting solely via the long-range Coulomb repulsion, (6.35). While this idea might have appeared plausible at the time *Bloch* [6.11] first proposed it, it had to be discarded in the face of strong theoretical and experimental evidence. "... those causes of the magnetical motions ... we relinquish to the moths and the worms" (see [Ref. 1.2, p. 64]) Obviously, experiment cannot disprove a correct theory; the fact that metals with simple nondegenerate bands are never found to possess strong magnetic properties should merely have been a spur to the experimentalists to look harder, had it not been for wiser counsel, notably Wigner's calculation [6.12] showing that correlations among the charged particles keep electrons both of parallel spin *and* of antiparallel spin apart (see also Chap. 1). In fact when the Coulomb interaction is strong enough, the ground state has to be a singlet as in the hydrogen molecule. *The key to magnetic behavior must therefore be the existence of degenerate bands* permitting a large number of electrons to be on the same or neighboring atoms. In such cases they cannot escape the Coulomb repulsion even by elaborate correlations. The fact is that the same forces which are present in the individual atoms to produce Hund's rule "atomic magnetism" cannot suddenly be nullified in the solid state.[6] Compelling as these arguments may be, they are not rigorous. A mathematical proof is available only in one dimension, where we have shown ([6.14], see also Chap. 4) that the ground state of $\mathscr{H}_0$ + any interaction Hamiltonian $\mathscr{H}_c$, is *always* a nonmagnetic singlet state, regardless of the number of electrons or the nature of the potential energy. Our proof breaks down, of course, in face of the degeneracy of three-dimensional atoms, as symbolized by the exchange Hamiltonian of (6.38). This adds plausibility to the choice of a single effective Hamiltonian for use in all solids which includes (6.38), viz.,

$$\mathscr{H}_{\text{eff}} = \mathscr{H}_0 + \mathscr{H}_c + \mathscr{H}_{\text{ex}} \tag{6.43}$$

defined in (6.27, 35, 38). *The retention of that part of $\mathscr{H}'$ which is explicitly spin-dependent, namely $\mathscr{H}_{\text{ex}}$, enhances the possibility of magnetic behavior.*

Unfortunately, the eigenfunctions and eigenvalues of $\mathscr{H}_{\text{eff}}$ cannot be found exactly because the three constituent terms do not commute with one another and therefore cannot be simultaneously diagonalized. $\mathscr{H}_{\text{eff}}$ itself is only part of the total Hamiltonian,

$$\mathscr{H}_{\text{tot}} = \mathscr{H}_0 + \mathscr{H}' \tag{6.44}$$

[6] The credit for this idea belongs to *Slater* [6.13].

defined in (6. 27,31), which contains all manners of complicated interactions, some of which have spin-dependent consequences and some not. Although we cannot assess their importance in any specific metal, we discard them because they are not required in the high-density limit for which $\mathscr{H}_0$ suffices, nor in the low-density limit where $\mathscr{H}_c + \mathscr{H}_{ex}$ will do. Thus, (6.43) which includes all these essential terms, will suffice.

6.5 Perturbation-Theoretic Derivation of Heisenberg Hamiltonian

To illustrate the uses of $\mathscr{H}_{eff}$, we shall obtain the Heisenberg Hamiltonian valid in an insulating magnetic material.[7] This derivation is an extension to the many-body problem of the method used in Problem 1, Chap. 2; it suggests what is required in a general theory of magnetism in insulators, a complicated and rather specialized subject when investigated in appropriate detail. According to the theory, the larger contributions will mostly be the antiferromagnetic ones, in accordance with experiment; although there exist exceptions to this prevalent antiferromagnetism in insulators:

> ... a curopium oxide of the formula EuO becomes truly *ferromagnetic* at 77 K with a saturation moment of close to 7 Bohr magnetons. This is thus the first rare earth oxide to be found to become ferromagnetic, and with the exception of CrO_2 the only oxide to our knowledge that has true ferromagnetic coupling [6.15].

In the magnetic insulator, $\mathscr{H}_c + \mathscr{H}_{ex}$ is the *big* part of $\mathscr{H}_{eff}$, and the "hopping" part, $\mathscr{H}_0$ is the perturbation. However, $\mathscr{H}_{ex}$ may be further decomposed; the Hund's rule part is retained in lowest order, whereas the nearest-neighbor exchange is treated by first-order perturbation theory. The starting Hamiltonian is, therefore,

$$\mathscr{H} = \mathscr{H}_c + \mathscr{H}_{ex}^{Hu} \tag{6.45}$$

describing the internal dynamics of *noninteracting* Wannier "atoms," each of net spin $\boldsymbol{S}_i$ of magnitude $s_i = \frac{1}{2}$ times the number of electrons in unfilled shell. (It is assumed that a crystal field quenches the angular momentum, or else the appropriate quantities are $\boldsymbol{J}_i$ and j_i). The ground-state eigenfunctions of $\mathscr{H}$ are highly degenerate, in fact they number N_0,

$$N_0 \equiv \prod_i (2s_i + 1)$$

and it is the perturbation $\mathscr{H}_0$ which will lift this degeneracy. The perturbation $\mathscr{H}_0$ can be seen from (6.24) to transfer an electron from one site to a neighboring site, this virtual transition occurring in second-order perturbation theory with matrix element $H(\boldsymbol{R}_{ij})_{n,n}$ and an increase of energy $U > 0$ in the intermediate

---

[7] The nature of the "insulating state" is the subject of [6.16] by *Kohn*

state due to the creation of two ionized sites: a positively ionized ion at $\boldsymbol{R}_i$ and a negatively ionized one at $\boldsymbol{R}_j$. Let one of the neutral ground states be $|\alpha)$ and one of the excited ionized states be $|\beta)$; then $U \equiv (\beta|\mathscr{H}|\beta) - (\alpha|\mathscr{H}|\alpha)$ and the second-order perturbation theoretic change in the energy of $|\alpha)$ equals

$$\begin{aligned} \delta E_\alpha &= -\sum_{\beta\neq\alpha} \frac{|(\beta|\mathscr{H}_0|\alpha)|^2}{U} \\ &= \frac{(\alpha|\mathscr{H}_0|\alpha)^2 - (\alpha|\mathscr{H}_0^2|\alpha)}{U} \end{aligned} \tag{6.46}$$

with the second line obtained by closure $[\sum_\beta|\beta)(\beta| \equiv \mathbb{1}]$. $\mathscr{H}_0$ has vanishing ground state expectation value in the insulator (as opposed to a metal) and therefore $(\alpha|\mathscr{H}_0|\alpha)$ vanishes. In writing out the nonvanishing second term, one collects such terms as

$$-H^2(\boldsymbol{R}_{ij})_{n,n} c^*_{in\uparrow} c_{in\downarrow} c^*_{jn\downarrow} c_{jn\uparrow} = -H^2(\boldsymbol{R}_{ij})_{n,n} S^+_{i,n} S^-_{j,n} \tag{6.47}$$

and others corresponding to $S^z_i S^z_j$, as well as terms which do not depend on the relative orientation of the two spins. The latter have the same magnitude for all states $|\alpha)$ and contribute a constant shift in their energy δE_α. The degeneracy of the ground state and relative ordering of the energy levels is given entirely by the subset of spin-dependent terms we lump together into an effective Hamiltonian

$$\mathscr{H}_{\text{Heis}} = \sum_{i,}\sum_{j\neq i}\left[\frac{2}{U}H^2(\boldsymbol{R}_{ij}) - J(\boldsymbol{R}_{ij})\right]\boldsymbol{S}_i\cdot\boldsymbol{S}_j = -\sum_{i\neq j} J_{\text{eff}}\boldsymbol{S}_i\cdot\boldsymbol{S}_j\,. \tag{6.48}$$

It is important to note that the vectors $\boldsymbol{S}_i$ are the *total atomic* spin operators and *not* individual electron components such as $\boldsymbol{S}_{i,n}$, which cannot be specified in any of the N_0 ground states. Therefore the quantities $H^2(\boldsymbol{R}_{ij})$ and $J(\boldsymbol{R}_{ij})$ are appropriate averages of the corresponding quantities in the various bands. Note that the non-Hund's rule nearest-neighbor exchange $J(\boldsymbol{R}_{ij})$ has been reintroduced at this point by means of first-order perturbation theory. The Hund's rule parameter $J(\boldsymbol{0})$ is absent from (6.48). Its role has been to create the atomic spins $\boldsymbol{S}_i$. The total Heisenberg Hamiltonian in nonconducting media must be considered as the sum of two effects: an invariably antiferromagnetic interaction due to the virtual "hopping" of an electron from $\boldsymbol{R}_i$ to $\boldsymbol{R}_j$ (and back), which is a "one-body" or "kinetic" exchange mechanism. It is antiferromagnetic because the hopping is enhanced when the spins are antiparallel, the exclusion principle prohibiting certain hops when the spins are parallel. The second, ferromagnetic, contribution arises from the exchange of two electrons not on the same atom; its matrix elements among the degenerate ground states may be described by first-order perturbation theory; it is always ferromagnetic because $J > 0$.

The same arguments hold for the interactions between two distinct partly occupied shells on the *same* atom, e.g., the valence and magnetic shells on a transition or rare earth atom or ion. The net coupling, J_{eff} given by the square bracket in (6.48), gives the tendency of the two shells to have their spins parallel or antiparallel, with important consequences in the Kondo effect studied later in this chapter.

As in the Heitler-London theory, the net interaction is the *difference* between two positive contributions and can have either sign. (Estimates of the various integrals indicate that the result is usually antiferromagnetic). However the present derivation is free from some of the undesirable features of the H-L derivation discussed in an earlier chapter. The natural expansion parameter here is U^{-1}, a physically measurable quantity; whereas in the older theory, the parameter l is not, in fact, an observable. The use of orthogonalized functions removes an element of arbitrariness from the theory, and the usual mathematical analysis of Hamiltonian mechanics can now be applied systematically.

In real materials (e.g., magnetite) the hopping occurs via an intermediary nonmagnetic ion, such as O^{-2}. This is called *superexchange*, a mechanism first proposed by *Kramers* [6.17] 50 years ago. The detailed theory of magnetism in insulators, including the theory of superexchange, is discussed by *Anderson* [6.18], who has contributed much towards the development of this field.

6.6 Heisenberg Hamiltonian in Metals

The "indirect exchange theory" of magnetism in metals is another example where an effective interatomic Heisenberg Hamiltonian can be derived by second-order perturbation theory. The relative importance of the various energies is here completely reversed over the previous section, for in a metal $\mathscr{H}_0$ is the principal Hamiltonian, and $\mathscr{H}_c$ and $\mathscr{H}_{ex}$ the perturbations.

The theory developed below was first invented in connection with nuclear magnetic resonance by *Ruderman* and *Kittel* [6.19] and independently, by *Bloembergen* and *Rowland* [6.20]; these authors studied the effective long-ranged interaction between nuclear spins due to the hyperfine coupling with the common sea of conduction electrons. The extension of their analysis to the *s-d* or *s-f* interaction [6.21] permitted the explanation of some significant experiments by Zimmerman on the long-ranged interaction between Mn atoms[8] dissolved in Cu. Briefly, the Mn impurity atom retains part of its Hund's rule magnetization in the solute state and by J_{eff}, (6.48), polarizes the spins of the conduction electrons in its neighborhood. The conduction electrons, constrained by the Pauli exclusion principle, respond with a characteristic wavelength

$$\lambda_F = \pi/k_F \tag{6.49}$$

[8] See review of theory and experiments in [6.22–24], and discussions on the "spin glass" elsewhere in this volume.

and the resultant spin polarization is not well localized in the vicinity of the impurity but is oscillatory and long-ranged. A second manganese atom at an arbitrary distance from the first, suffers a ferromagnetic or an antiferromagnetic interaction with it, depending upon whether it is in the trough or on the crest of the polarization wave. The strength of the interaction gradually decreases with distance, in a manner we shall calculate.

Assume a pair of solute magnetic atoms at $\boldsymbol{R}_1$ and $\boldsymbol{R}_2$, in an otherwise ideal nonmagnetic metal characterized by an s-band Hamiltonian $\mathscr{H}_0$. Internal Hund's rule coupling maintains the magnitudes of the solutes' spins fixed at s_1 and s_2, respectively, but the *relative orientation* of the two spins will be governed by the interaction which is derived below. The exchange coupling of the localized electrons with the conduction electrons is the perturbation,

$$\mathscr{H}_{\text{ex}} = -J_{\text{eff}}[\boldsymbol{S}_1 \cdot \boldsymbol{s}_c(\boldsymbol{R}_1) + \boldsymbol{S}_2 \cdot \boldsymbol{s}_c(\boldsymbol{R}_2)] \tag{6.50}$$

with the conduction-band spin operators $\boldsymbol{s}_c(\boldsymbol{R}_i)$ given by (6.40). The substitution of Bloch operators for the Wannier operators, given in (6.26a), results in the following:

$$\boldsymbol{s}_c(\boldsymbol{R}_i)^z = \frac{1}{2N} \sum_{\boldsymbol{k},\boldsymbol{q}} \mathrm{e}^{-\mathrm{i}\boldsymbol{q}\cdot\boldsymbol{R}_i}(c^*_{\boldsymbol{k}+\boldsymbol{q}\uparrow}c_{\boldsymbol{k}\uparrow} - c^*_{\boldsymbol{k}+\boldsymbol{q}\downarrow}c_{\boldsymbol{k}\downarrow})$$

$$\boldsymbol{s}_c(\boldsymbol{R}_i)^+ = \frac{1}{N} \sum_{\boldsymbol{k},\boldsymbol{q}} \mathrm{e}^{-\mathrm{i}\boldsymbol{q}\cdot\boldsymbol{R}_i}(c^*_{\boldsymbol{k}+\boldsymbol{q}\uparrow}c_{\boldsymbol{k}\downarrow}) \quad \text{and} \quad \boldsymbol{s}_c(\boldsymbol{R}_i)^- = \frac{1}{N} \sum_{\boldsymbol{k},\boldsymbol{q}} \mathrm{e}^{-\mathrm{i}\boldsymbol{q}\cdot\boldsymbol{R}_i}(c^*_{\boldsymbol{k}+\boldsymbol{q}\downarrow}c_{\boldsymbol{k}\uparrow}). \tag{6.51}$$

With this definition in mind, let us calculate the eigenvalues and eigenfunctions of

$$\mathscr{H} = \mathscr{H}_0 + \mathscr{H}_{\text{ex}} = \sum_{\substack{m=\uparrow,\downarrow \\ \boldsymbol{k}}} E(\boldsymbol{k})\mathfrak{n}_{\boldsymbol{k},m} - J_{\text{eff}} \sum_{i=1}^{2} \left\{ s_c(\boldsymbol{R}_i)^z S_i^z + \frac{1}{2}[s_c(\boldsymbol{R}_i)^+ S_i^- + \text{h.c.}] \right\} \tag{6.52}$$

by ordinary perturbation theory. In particular, we wish to see how the perturbation lifts the degeneracy of the $r \equiv (2s_1 + 1) \times (2s_2 + 1)$ states of orientations of the two solute spins. The conduction electrons will be assumed to be in their ground state, except for the resultant polarization effects.

The first-order correction to the energy,

$$\delta E^{(1)} = (t; F|\mathscr{H}_{\text{ex}}|F; t) = 0 \tag{6.53}$$

is seen to vanish. We use the notation $|F; t)$ to indicate the product state of Fermi sea with the two solute spins, with the index t spanning the range $t = 1, \ldots, r$ of degenerate orientations of the two spins.

The conduction-band spin operators c^*c create elementary excitations of energy $E(\boldsymbol{k}+\boldsymbol{q}) - E(\boldsymbol{k})$; their matrix elements are unity if $k < k_F$ and $|\boldsymbol{k}+\boldsymbol{q}| > k_F$, and zero otherwise. Therefore, by second-order perturbation theory

$$\delta E_t^{(2)} = -\frac{J_{\text{eff}}^2}{2N^2} \sum_{\substack{t', k<k_F \\ |\boldsymbol{k}+\boldsymbol{q}|>k_F}} \frac{(t|\mathrm{e}^{\mathrm{i}\boldsymbol{q}\cdot\boldsymbol{R}_1}\boldsymbol{S}_1 + \mathrm{e}^{\mathrm{i}\boldsymbol{q}\cdot\boldsymbol{R}_2}\boldsymbol{S}_2|t')\cdot(t'|\mathrm{e}^{-\mathrm{i}\boldsymbol{q}\cdot\boldsymbol{R}_1}\boldsymbol{S}_1 + \mathrm{e}^{-\mathrm{i}\boldsymbol{q}\cdot\boldsymbol{R}_2}\boldsymbol{S}_2|t)}{E(\boldsymbol{k}+\boldsymbol{q}) - E(\boldsymbol{k})}$$

$$= -\frac{J_{\text{eff}}^2}{2N^2} \sum_{\substack{k<k_F \\ |\boldsymbol{k}+\boldsymbol{q}|>k_F}} \frac{(t|s_1(s_1+1) + s_2(s_2+1) + 2\boldsymbol{S}_1\cdot\boldsymbol{S}_2 \cos \boldsymbol{q}\cdot\boldsymbol{R}_{12}|t)}{E(\boldsymbol{k}+\boldsymbol{q}) - E(\boldsymbol{k})}. \tag{6.54}$$

The second line is the result of using closure on the intermediate states $|t')$. It is convenient to separate this formula into two parts: a self-energy

$$\delta E^{(2)} = -K \sum_i s_i(s_i+1) \tag{6.55}$$

with

$$K = \frac{J_{\text{eff}}^2}{2N^2} \sum_{\substack{k<k_F \\ |\boldsymbol{k}+\boldsymbol{q}|<k_F}} \frac{1}{E(\boldsymbol{k}+\boldsymbol{q}) - E(\boldsymbol{k})} \tag{6.55a}$$

and an interaction energy, which is the eigenvalue of the effective Hamiltonian

$$\mathscr{H}_{\text{IE}} = -\sum_{(i,j)} J(\boldsymbol{R}_{ij})_{\text{IE}} \boldsymbol{S}_i \cdot \boldsymbol{S}_j \tag{6.56}$$

where the indirect exchange coupling constant is

$$J(\boldsymbol{R}_{ij})_{\text{IE}} = \left(\frac{J_{\text{eff}}}{N}\right)^2 \sum_{\substack{k<k_F \\ |\boldsymbol{k}+\boldsymbol{q}|>k_F}} \frac{\cos \boldsymbol{q}\cdot\boldsymbol{R}_{ij}}{E(\boldsymbol{k}+\boldsymbol{q}) - E(\boldsymbol{k})}. \tag{6.56a}$$

If instead of only two impurities there are N_I, the sum in (6.55) runs over all N_I spins and the interaction, (6.56), over all $\frac{1}{2}N_I(N_I - 1)$ distinct pairs.

For an estimate of the indirect exchange coupling, one uses the effective mass approximation, $E(\boldsymbol{k}) = \hbar^2\boldsymbol{k}^2/2m^*$ and $\mu = \hbar^2 k_F^2/2m^*$, and introduces an imaginary part $\hbar\mathrm{i}/\tau$ to the denominator to account for a finite electronic mean free path.

$$J(\boldsymbol{R}_{ij})_{\text{IE}} = J_{\text{eff}}^2 \left(\frac{a_0}{2\pi}\right)^6 \int_{[E(k)<\mu]} d^3k \int d^3k' \frac{\cos(\boldsymbol{k}-\boldsymbol{k}')\cdot\boldsymbol{R}_{ij}}{(\hbar^2/2m^*)(\boldsymbol{k}'^2 - \boldsymbol{k}^2 + \mathrm{i}2m^*/\hbar\tau)}$$

$$= J_{\text{eff}}^2 \left(\frac{a_0}{2\pi}\right)^6 \frac{m^*}{2\hbar^2} \left(\frac{4\pi}{R_{ij}}\right)^2 \int_0^{k_F} dkk \int_{-\infty}^{\infty} dk'k' \frac{\sin kR_{ij} \sin k'R_{ij}}{k'^2 - k^2 + \mathrm{i}2m^*/\hbar\tau}$$

$$= \frac{-J_{\text{eff}}^2}{\mu} \frac{(k_F a_0/2)^6}{\pi^3} \left[\frac{\sin 2k_F R_{ij} - 2k_F R_{ij} \cos 2k_F R_{ij}}{(2k_F R_{ij})^4}\right] \mathrm{e}^{-R_{ij}/\lambda} \tag{6.56b}$$

where λ = mean free path = $\hbar k_F \tau/m^*$ and a_0^3 = volume of unit cell. This result (without the mean free path factor) was first published by *Ruderman* and *Kittel* [6.19] and is universally referred to as the *Ruderman-Kittel interaction.* The mean free path factor was introduced in the first edition of the present book, and has since been measured [6.25].

6.7 Ordered Magnetic Metals: Deriving the Ground State

Some of the most interesting applications of the indirect exchange theory are to metals containing elements in the gadolinium rare-earth series (*lanthanides*). Generally the *f*-shell radii of the rare-earth atoms are so small that even nearest neighboring atoms do not have significant direct overlap and the interaction is assumed to be principally the Ruderman-Kittel indirect exchange mechanism derived in the preceding section (with crystal field anisotropy and hybridization *i.e.*, band-mixing, the principal correction) [6.26]. Ordered alloys containing transition elements are also described by this theory if the magnetic atoms are sufficiently far apart for the *d*-shell overlaps to be unimportant.

Therefore we are led to consider the Hamiltonian

$$\mathscr{H}_{\mathrm{IE}} = -\sum J_{ij}\boldsymbol{S}_i\cdot\boldsymbol{S}_j \quad [J_{ij} \equiv J(\boldsymbol{R}_{ij})_{\mathrm{IE}}] \tag{6.57}$$

and to calculate its eigenstates and eigenvalues in cases where the spins $\boldsymbol{S}_i$ occupy points on a regular lattice. For convenience, we shall assume it to be one of the Bravais lattices, and particularly one of the three principal cubic lattices. There is no exact method known to obtain the ground state, but the following procedure avoids unnecessary complications, and appears reliable. First, we construct the product wavefunction

$$\psi = \Pi\phi_i \tag{6.58}$$

in which the ϕ_i are, as yet, unspecified but normalized states of the spins $\boldsymbol{S}_i$. The variational energy in this configuration is

$$E = (\Psi|\mathscr{H}_{\mathrm{IE}}|\Psi) = -\sum J_{ij}(\phi_i|\boldsymbol{S}_i|\phi_i)\cdot(\phi_j|\boldsymbol{S}_j|\phi_j) \qquad \text{where} \tag{6.59}$$

$$(\phi_i|\boldsymbol{S}_i|\phi_i)^2 \leq s_i^2\,. \tag{6.60}$$

In the remainder, assume all N_{I} magnetic atoms to belong to the same species, so that $s_i = s$. The *method of Luttinger and Tisza*, similar to the "spherical model" discussed later, can then be used to find the lowest energy attainable with a trial function of the type in (6.58). It consists mainly of relaxing the inequality above, requiring only

$$\sum_i (\phi_i|\boldsymbol{S}_i|\phi_i)^2 \leq N_\text{I}s^2 \tag{6.61}$$

which is a weaker constraint. *The lowest energy* (6.59) *subject to the weaker constraint* (6.61) *must be lower than the lowest energy* (6.59) *subject to the more rigorous constraint.* (6.60). E is calculated by Fourier transforms, letting $\boldsymbol{S}_k$ be defined by

$$\boldsymbol{S}_k \equiv \frac{1}{N_\text{I}} \sum_i \mathrm{e}^{-\mathrm{i}\boldsymbol{k}\cdot\boldsymbol{R}_i}(\phi_i|\boldsymbol{S}_i|\phi_i) \qquad \text{and} \qquad (\phi_i|\boldsymbol{S}_i|\phi_i) = \sum_{\boldsymbol{k}} \mathrm{e}^{\mathrm{i}\boldsymbol{k}\cdot\boldsymbol{R}_i}\boldsymbol{S}_k\,, \tag{6.62}$$

substituting it into (6.59) and obtaining,

$$E = -N_\text{I} \sum_{\boldsymbol{k}} J(\boldsymbol{k})|\boldsymbol{S}_k|^2 \tag{6.63}$$

where

$$J(\boldsymbol{k}) = \frac{1}{2N_\text{I}} \sum_{i,j} J_{ij}\mathrm{e}^{\mathrm{i}\boldsymbol{k}\cdot\boldsymbol{R}_{ij}}\,. \tag{6.64}$$

The weak inequality (6.61), by substitution of the Fourier transform, is found to be

$$\sum_{\boldsymbol{k}} |\boldsymbol{S}_k|^2 \leq s^2 \tag{6.65}$$

and if we define $\boldsymbol{q}_0$ to be the wave vector for which $J(\boldsymbol{q}_0)$ attains its largest value (or one of the wave vectors which have this property if there are more than one) then let $\boldsymbol{S}_{q_0} = \boldsymbol{s}$ and all other $\boldsymbol{S}_k = 0$, and

$$E(\boldsymbol{q}_0) = -N_\text{I}J(\boldsymbol{q}_0)s^2 \tag{6.66}$$

is certainly the lowest energy subject to the weak constraint. Now choose the wavefunctions ϕ_i sueh that

$$(\phi_i|\boldsymbol{S}_i|\phi_i) = s(\cos \boldsymbol{q}_0\cdot\boldsymbol{R}_i, \sin \boldsymbol{q}_0\cdot\boldsymbol{R}_i, 0) \tag{6.67}$$

which is called a *spiral* configuration of pitch $\boldsymbol{q}_0$. Inserting this variational *ansatz* into (6.59) leads to precisely the energy $E(\boldsymbol{q}_0)$ calculated above; and being an upper bound as well as a lower bound to the ground-state energy (in the Hartree product wavefunction approximation), $E(\boldsymbol{q}_0)$ must itself be the Hartree ground-state energy.

When $\boldsymbol{q}_0 = 0$, all spins are parallel and the ground state is ferromagnetic. When $\boldsymbol{q}_0$ is a wavevector on one of the points of symmetry of the Brillouin zone boundary, for example, $\pi/a(1\pm, \pm 1, \pm 1)$ in the simple cubic structure, the ground state is an antiferromagnetic configuration of some sort (the Néel state in the example given). If $\boldsymbol{q}_0$ is none of these special wave vectors, the ground state is a "spiral spin configuration," as these screw structures are known in general. The various possible configurations can be found from group theory or by matrix methods, even in non-Bravais lattices, particularly in insulators when $\boldsymbol{J}_{ij}$ is limited to the Heisenberg nearest-neighbor forces [6.27, 28]. Note that in present derivation the range of the interaction does not matter, therefore the proof holds equally well for the short-ranged Heisenberg interaction in insulators. In the case of the long-ranged Ruderman-Kittel interaction, numerical calculation is required to obtain $J(\boldsymbol{k})$, and to determine $\boldsymbol{q}_0$.

The numerical calculation of $J(\mathbf{0})$ is particularly interesting, because according to the theory developed in the next volume, *it is proportional to the paramagnetic Curie temperature* θ. A negative θ necessarily precludes ferromagnetism; a positive θ is a likely indication of ferromagnetism, but it is still possible for $\boldsymbol{q}_0 \neq 0$ to be the ground state solution, and for a spiral configuration to be the stable ground state at low temperatures. Therefore it is also necessary to study the spin-wave spectrum, or magnon energy

$$\hbar\omega(\boldsymbol{k}) = 2s[J(\mathbf{0}) - J(\boldsymbol{k})] \tag{6.68}$$

to determine that all $\hbar\omega(\boldsymbol{k})$ are positive, if the ferromagnetic state is to be stable. This is a necessary but *not sufficient* condition for ferromagnetism.[9] The minimizing wavevector $\boldsymbol{q}_0$ can also be found at the minimum of the function $\hbar\omega(\boldsymbol{q}_0)$.

As an example, we calculate $\hbar\omega(\boldsymbol{k})$ for the Ruderman-Kittel interaction, in the continuum limit $k_F a \to 0$, where the lattice sums can be replaced by integrals. Let us define ($x_i \equiv 2k_F R$, a = separation of n.n. *magnetic* atoms):

$$\varepsilon(\boldsymbol{k}) \equiv \sum a^3 \frac{\sin x_i - x_i \cos x_i}{5x_i R_i^3} \mathrm{e}^{-R_i/\lambda}(1 - \cos \boldsymbol{k} \cdot \boldsymbol{R}_i) \tag{6.69}$$

which is (6.68), the magnon energy, with unimportant constant factors eliminated, for tabular convenience. Also,

$$\theta \equiv \sum_{R_i \neq 0} a^3 \frac{\sin x_i - x_i \cos x_i}{5x_i R_i^3} \mathrm{e}^{-R_i/\lambda} \tag{6.70}$$

is the paramagnetic Curie temperature with similar constant factors removed. The limiting values are elementary integrals

$$\lim_{k_F \to 0} \theta = \frac{8\pi}{10}\left(1 + \frac{1}{32k_F\lambda} - \frac{1}{16\pi k_F\lambda} \tan^{-1} \frac{1}{2k_F\lambda}\right) \tag{6.71}$$

---

[9] Cf. [6.29], in which it is shown that in some exceptional instances the ground state may be nonferromagnetic, due to quantum fluctuations *even though* the ferromagnetic state is stable a gainst the emission of (any finite number of) spin waves. This analysis has been extended, see [6.30], and ongoing work. As for the practical task of fitting the formulas to experiment, we may cite [6.31] for the Heusler alloys.

and

$$\lim_{k_F \to 0} \varepsilon(\boldsymbol{k}) = \frac{4\pi}{10}\left\{1 - \frac{(2k_F\lambda)^2 + 1 - (k\lambda)^2}{(4k\lambda)(2k_F\lambda)} \ln \frac{(2k_F\lambda + k\lambda)^2 + 1}{(2k_F\lambda - k\lambda)^2 + 1} + \frac{1}{2k_F\lambda}\left[\tan^{-1}\frac{4k_F\lambda}{(2k_F\lambda)^2 - 1} - \tan^{-1}\frac{4k_F\lambda}{(2k_F\lambda)^2 - (k\lambda)^2 - 1}\right]\right\}. \quad (6.72)$$

These are the magnetic parameters for a "continuum" or "jellium" lattice or for a low density electron gas. Note that they are both positive, indicating ferromagnetism.

Closely associated with k_F is the dimensionless parameter $n_{c/a} \equiv$ number of conduction electrons per magnetic atom. (If there is one conduction electron per unit cell, but a magnetic atom is present only in every other cell, $n_{c/a} = 2$, etc. The "jellium" limit is equivalent to $n_{c/a} \to 0$.) As $n_{c/a}$ is raised, the paramagnetic Curie temperature goes through zero at approximately $n_{c/a} = 1/4$ for the three principal cubic lattices. There is then an antiferromagnetic region as $n_{c/a}$ is increased further, until it exceeds the value 3/2 for the bcc and fcc lattices, and 5/2 for the sc lattice. At this point the paramagnetic Curie temperature becomes positive once more until $n_{c/a}$ is approximately doubled, whereupon a new antiferromagnetic region is encountered, etc.

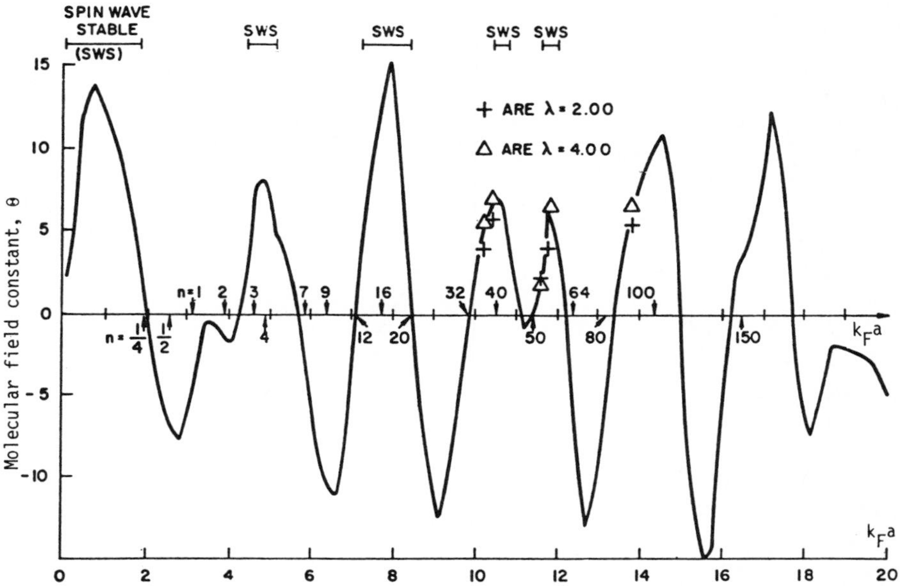

Fig. 6.6. Θ vs $k_F a$ in the sc lattice (bcc, fcc are similar). Values of electron density $n = (8\pi/3)(k_F a/2\pi)^3$ = number of conduction electrons *per* magnetic atom, are indicated by arrows. Main curve is for $\lambda = 3$ (some points at $\lambda = 2$ or 4 are shown to indicate the insensitivity of the results). Spinwave-stable regions are indicated, and are seen to correlate well with $\Theta > 0$

A plot of θ as a function of k_F and $n_{c/a}$, which displays the features just discussed, is given in Fig. 6.6 for the simple cubic lattice. In Fig. 6.6 are also shown the spin-wave stable regions where $\hbar\omega(\boldsymbol{k}) > 0$; these are seen to be somewhat smaller than the regions of positive paramagnetic Curie temperature. To scale the results so that they correspond to the Ruderman-Kittel interaction of (6.56) with constant J, it is necessary to multiply the plotted and tabulated values by appropriate functions of k_F, a, and a_0 = nonmagnetic metal lattice parameter. In the calculations, the value $a = 1$ is taken for convenience Therefore, comparing (6.56b, 68, 69), we obtain the physical parameters $\tilde{\theta}$ and $\hbar\omega(\boldsymbol{k})$

$$\frac{\tilde{\theta}}{\theta} = \frac{\hbar\omega(\boldsymbol{k})}{\varepsilon(\boldsymbol{k})} \propto J^2(k_F a_0)(k_F a)^3 \left(\frac{a_0}{a}\right)^6 \tag{6.73}$$

in terms of θ and $\varepsilon(\boldsymbol{k})$, the computed functions.

In Fig. 6.7 we reproduce experimental results of *Methfessel* and collaborators [6.32], on the paramagnetic Curie temperature of certain ordered Eu-Gd-Se

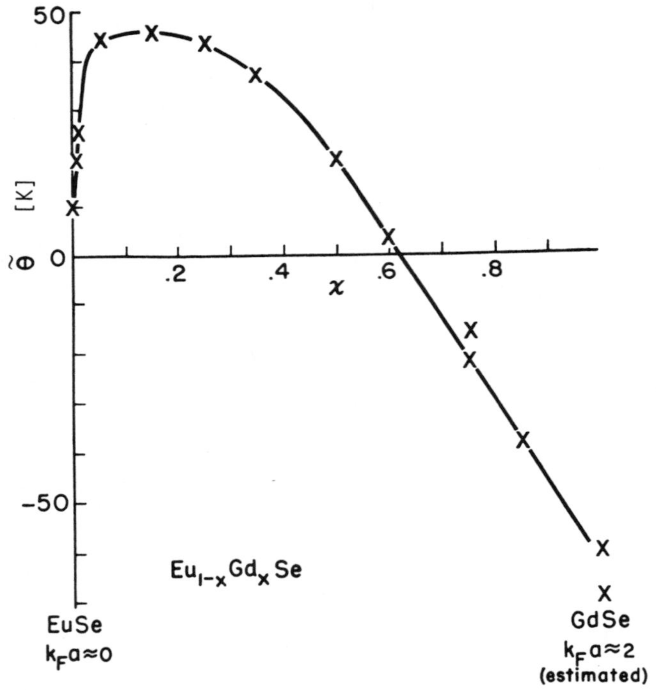

Fig. 6.7. Experimental values of $\tilde{\Theta}$ (paramagnetic Curie temperature) *vs.* composition (roughly, k_F^3) at constant a, obtained by *Methfessel* et al. [6.32] in a series of ordered rare-earth alloys. An order-of-magnitude estimate is that $k_F a$ varies in the range 0–2 over the range of compositions, and these results are in qualitative agreement with Ruderman-Kittel theory (cf. Fig. 6.6 over same range of k_F)

alloys, in which the electronic concentration could be varied from insulator to metallic, corresponding to the range $0 < k_F a < 2$ (the upper value is an order of magnitude estimate). These results are apparently in good qualitative agreement with the applicable portions of the theoretical curves.

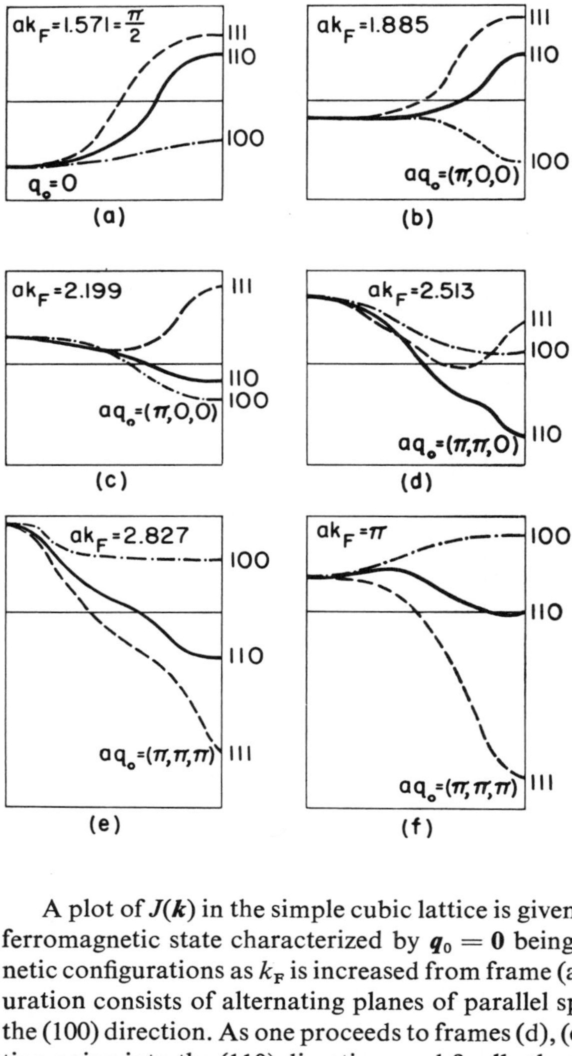

Fig. 6.8. Plot of $-J(\boldsymbol{k})$ for the Ruderman-Kittel interaction versus $\boldsymbol{k}$ in the three principal directions, sc lattice. In the range $0 < k_F a \leq \pi/2$ the ground state is ferromagnetic. However, as k_F is increased in frames **(b)** to **(f)** various antiferromagnetic states become stable, indicated by nonzero values of spiral pitch parameter $\boldsymbol{q}_0$. $J(\boldsymbol{k})$ is defined in (6.56, 64)

A plot of $J(\boldsymbol{k})$ in the simple cubic lattice is given in Fig. 6.8. One observes the ferromagnetic state characterized by $\boldsymbol{q}_0 = \boldsymbol{0}$ being succeeded by antiferromagnetic configurations as k_F is increased from frame (a) to (b). The resulting configuration consists of alternating planes of parallel spins, the alternation being in the (100) direction. As one proceeds to frames (d), (e), and (f), he sees the alternation going into the (110) direction, and finally the (111) direction, the last being the Néel state.

Note that if the Ruderman-Kittel interaction is a valid one to use for small k , then the statistical mechanics will be very well described by the molecular field theory which we develop in the next volume. For then, the interaction is long-ranged and practically nodeless (J_{ij} is ferromagnetic out to distances $\sim 1/k_F$ and is very small beyond) and the criteria of the molecular field theory are met at all but the lowest temperatures, where spin-wave theory is applicable.

Because the indirect exchange theory is used in the description of the magnetic properties of the rare-earth metals and alloys [6.33], it is interesting to note that in many cases the angular momentum of the f shell in these ions is not quenched, and the total $\boldsymbol{J}_i$ angular momentum must be specified and not just the total spin $\boldsymbol{S}_i$. That is, the magnetic degrees of freedom of each rare earth are described by $2j_i + 1$ eigenfunctions, and not by $2s_i + 1$. But this is easily taken into account by using the definition of the Landé g factor. In the subspace of the $2j_i + 1$ eigenfunctions, the following equality defines the Landé factor g_i,

$$\boldsymbol{M}_i = \boldsymbol{J}_i + \boldsymbol{S}_i = g_i \boldsymbol{J}_i \tag{6.74}$$

where $\boldsymbol{M}_i$ is the magnetic moment operator of the ion. Subtracting $\boldsymbol{J}_i$ from both sides of the equation, we obtain the prescription useful in the present case:

$$\text{replace } \boldsymbol{S}_i \text{ by } (g_i - 1)\boldsymbol{J}_i \tag{6.75}$$

in $\mathscr{H}_{\mathrm{IE}}$ for all rare earths *except* when $j = 0$, as in the case of Eu in some states. Because $(g_i - 1)$ can be positive or negative, a sort of "charge" is introduced

Table 6.2. g-factor and angular momenta of rare earths

Number of electrons in f shell	Symbol	s	l	j	$g - 1$
0	La	0	0	0	...
1	Ce	$\frac{1}{2}$	3	$\frac{5}{2}$	$-\frac{1}{7}$
2	Pr	1	5	4	$-\frac{1}{5}$
3	Nd	$\frac{3}{2}$	6	$\frac{9}{2}$	$-\frac{3}{11}$
4	Pm	2	6	4	$-\frac{2}{5}$
5	Sm	$\frac{5}{2}$	5	$\frac{5}{2}$	$-\frac{5}{7}$
6	Eu	3	3	0	...
7	Gd	$\frac{7}{2}$	0	$\frac{7}{2}$	$+1$
8	Tb	3	3	6	$+\frac{1}{2}$
9	Dy	$\frac{5}{2}$	5	15/2	$+\frac{1}{3}$
10	Ho	2	6	8	$+\frac{1}{4}$
11	Er	$\frac{3}{2}$	6	15/2	$+\frac{1}{5}$
12	Tm	1	5	6	$+\frac{1}{6}$
13	Yb	$\frac{1}{2}$	3	$\frac{7}{2}$	$+\frac{1}{7}$
14	Lu	0	0	0	...

Note: From the following formula: $g - 1 = [j(j + 1) + s(s + 1) - l(l + 1)]/2j(j + 1)$.

into the indirect exchange theory: ions with opposite signs of $(g_i - 1)$ will interact antiferromagnetically for ferromagnetic J_{ij}, and vice versa, so that this gives to mixed rare-earth alloys yet another degree of freedom.

The indirect exchange Hamiltonian is thus,

$$\mathscr{H}_{\mathrm{IE}} = -\sum J_{ij}(g_i - 1)(g_j - 1)\boldsymbol{J}_i \cdot \boldsymbol{J}_j \tag{6.76}$$

except when $j_i = 0$, when $\boldsymbol{S}_i$ is used. Table 6.2 lists the rare earths and their effective "spin charge" $g_i - 1$.

In transition metal ions the angular momentum is quenched because of strong crystal field effects on the relatively extensive d orbitals. This is reflected in experimentally measured g factors close to 2, the theoretical spin-only value. In those cases, the correct low-lying states are designated by m_s, and the correct vector operator is just $\boldsymbol{S}_i$.

6.8 Kondo Effect

In the preceding, a large number of spins arranged in a regular way in a metallic matrix were found to correlate their orientations via the Ruderman-Kittel interaction mediated by the nonmagnetic conduction electrons. The symmetry of the ground state depended on geometric factors, such as the ratio of distance between magnetic atoms to the deBroglie wavelength of an electron at the Fermi surface.

In this section, we give the reader a brief introduction to the case of random, dilute, magnetic alloys, in which the concentration of spins may be as few as parts per million (ppm). In the extremely dilute limit where the self-energy of each spin is the dominant factor, one finds the *Kondo* effect [6.34]: an anomaly in the conduction electrons in the vicinity of the impurity spin. In the moderately dilute case, when interactions among neighboring spins become more significant than the self-energy of each, the spin-glass phase—a condensed phase in the presence of a high degree of disorder—will yield the presumed ground state. In either case, the internal degrees of freedom peculiar to spins are the cause of highly unusual phenomena, which theorists, assisted by a variety of experimental facts, have been seeking to understand for the past two decades and to which we shall return more comprehensively in the companion volume.

One goes back to the 1930's for the first inklings of these phenomena. *Meissner* and *Voigt* [6.35] first noted an anomaly in the resistivity of nominally pure gold samples, in reality containing trace impurities of iron from the manufacturing process. These showed an increasing resistance as the temperature was lowered below some 10 K. Such behavior is in diametric opposition to that of ordinary metals and alloys, which have their minimal resistance at $T = 0$ when all thermal vibrations have ceased. A recent confirmation of the Meissner-Voigt discovery is shown in Fig. 6.9. The effects are highlighted in Fig. 6.10, from which the resistivity contributed by lattice vibrations has been subtracted out. A

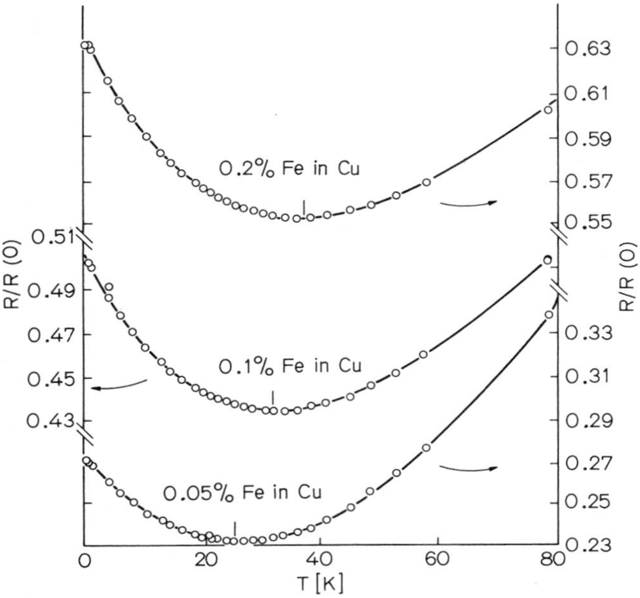

Fig. 6.9. Resistance as a function of temperature for dilute magnetic alloys; note various scales [6.36]

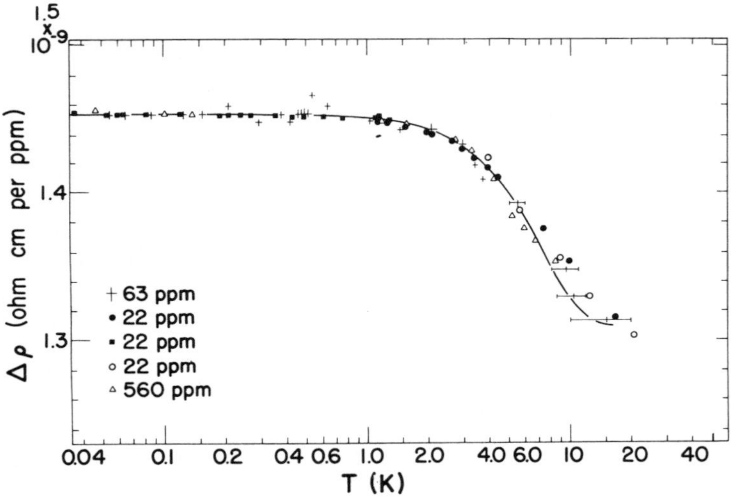

Fig. 6.10. Low temperature Resistivity of CuFe, *per* iron atom; with phonon contribution subtracted out [6.37] Note logarithmic scale

few ppm iron in Cu or Au are seen to be capable of substantially affecting the resistivity and, as it turns out, a variety of other thermodynamic properties as well. Between the low-temperature saturation of the resistivity (at 1K in the example of Fig. 6.10) and the high temperatures at which the magnetic contributions become negligible, there exists a wide range over which the resistivity follows an approximately *logarithmic* behavior. This is seen in Fig. 6.11 for several samples, which satisfy a universal law, once the temperature is expressed in terms of suitable units. The scaling temperature (the unit) is constant for a given alloy but varies from one to the other. It is most appropriately denoted the "Kondo temperature" T_K, given as:

$$T_K = (D/k_B)\exp(-D/2|J|) \tag{6.77}$$

with D = bandwidth, J the exchange parameter ($J > 0$ for ferromagnetic $s - d$ coupling, < 0 for AF coupling), k_B = Boltzmann's constant. Kondo first gave an expression for the resistivity [6.34]:

$$R = R_0(T) + c(J^2/D)s(s+1)A\left[1 - \frac{2J}{D}\ln(D/k_BT)\right] \tag{6.78}$$

in which R_0 is the nonmagnetic contribution, A is a lumped constant, s = spin of the impurity, and the other parameters have already been given; the Kondo temperature T_K signals the onset of peculiar behavior in (6.78) at $T \leq T_K$. In fact, (6.78), derived by perturbation theory, ceases to be valid at low temperature. Since the enunciation of the logarithmic law, it has been observed in a number of laboratories and Fig. 6.11 is typical Kondo behavior.

It is also generally accepted that this behavior is the consequence of a large "soft" cloud of conduction electrons' spin polarization resonating about the impurity's local moment, so that the role of temperature is to break up this cloud

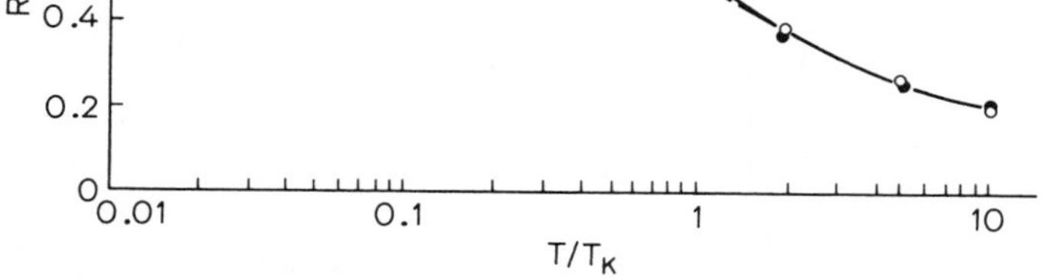

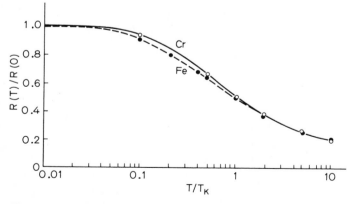

Fig. 6.11. The normalized resistivity $R(T)/R(0)$ for a variety of samples, plotted as function of $\log(T/T_K)$. The results suggest a universal curve [6.38]

and reduce the strong scattering that such a resonance produces. It should then be possible to predict at which concentration of magnetic impurities the overlap of neighboring clouds becomes more significant than the energy binding each to its local moment. At such a value of the concentration, the *spin glass* phase must become important; at yet higher concentrations, the *ordered magnetic alloys* studied in previous sections become relevant.

The one-spin interaction Hamiltonian appropriate to the very dilute Kondo limit (ppm) is precisely half of what we considered in (6.50, 51), viz.

$$\mathscr{H}_{ex} = -J\boldsymbol{S}\cdot\boldsymbol{s}_c(0), \qquad \text{with} \qquad J = J_{eff} \text{ of (6.48).} \tag{6.79}$$

The isolated spin is taken to be located at the origin, without loss of generality.

Now, however weak the interaction parameter J may be compared with the other parameters: Fermi energy ε_F, electron bandwidth D, etc., it cannot be considered a small perturbation because of a singularity at $J = 0$. Specifically, one can show [6.39]:

In any large region surrounding the magnetic defect, the ground state spin (impurity + conduction band) has the value $S_{tot} = s - \frac{1}{2}$ for antiferromagnetic interactions and $S_{tot} = s$ or $s + \frac{1}{2}$ for ferromagnetic interactions.

The proof of this follows the arguments in Chap. 4, by comparing the ground state of the present system with that of a reference system for which the ground state quantum numbers are known. It is a tricky proof, however, as the energy levels of the conduction band form a continuum and they are so dense that their fluctuations can overwhelm the impurity. *Cragg* and *Lloyd* [6.39] have investigated this problem by numerical means and use of the renormalization group, finding for $J > 0$ (ferromagnetic coupling) a dependence of the ground state symmetry on the number of iterative steps in their procedure. For $J < 0$ (AF coupling), however, the predicted $S_{tot} = s - \frac{1}{2}$ is obtained without ambiguity. In any event, the significant feature is the change in symmetry as the sign of J is varied; this indicates that perturbation theory in powers of J has a vanishing radius of convergence, as the point $J = 0$ is singular.

It is known from thermodynamics that the entropy $\mathscr{S}(T)$ relative to the value at high temperature $\mathscr{S}_\infty$ is given by

$$\mathscr{S}(T) = \mathscr{S}_\infty - \int_T^\infty dT' c(T')/T'$$

with $\mathscr{S}_\infty = k_B \ln 2$ for spin one-half. In Fig. 6.12 the specific heat curves of dilute copper-iron alloys are shown; the area under the suitable integral yields $k_B \ln 2$, implying $s = 1/2$ on the iron and confirming that the ground state entropy is zero. Such experimental evidence of a singlet ground state is paralleled by the variational estimates concerning the ground state of the system for antiferromagnetic coupling [6.40] and by the above-mentioned theorem [6.39] applied to

$s = 1/2$. A variety of approximate methods[10] have all shown the expected decrease in total moment as the temperature is lowered ($J < 0$) or the increase ($J > 0$).

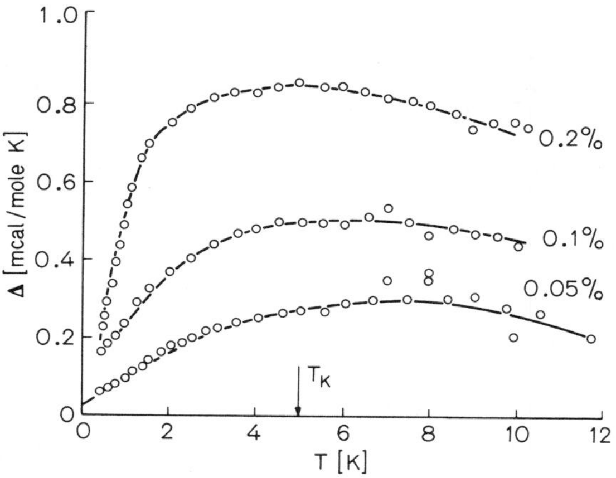

Fig. 6.12. The excess specific heat for the CuFe alloys, corresponding to a total entropy of precisely $k_B \ln 2$ per spin [6.36]

In (6.55) we have already obtained the self-energy of a magnetic ion in a metal, seen to be finite and symmetric in J. It is thus the higher-order terms that must lead to the expected anomaly, third- or higher-order in the energy or second- or higher-order in the Born expansion of the scattering amplitude. To pinpoint the origin of the difficulties, consider what occurs when one attempts to eliminate the perturbing Hamiltonian $\mathscr{H}_{\text{ex}}$ by a unitary transformation, even to lowest order in the "small parameter" J. First,

$$\begin{aligned} \mathscr{H} &= \mathscr{H}_0 + \mathscr{H}_{\text{ex}} \\ \mathscr{H}_0 &= \sum_{km} \varepsilon_k c_{km}^* c_{km} \\ \mathscr{H}_{\text{ex}} &= (-J/2N)\boldsymbol{S}\cdot\sum_{\substack{kk' \\ mm'}} c_{km}^* \boldsymbol{\sigma}_{mm'} c_{k'm'} \end{aligned} \tag{6.80}$$

in which $\boldsymbol{\sigma} = (\sigma_x, \sigma_y, \sigma_z)$ is a vector with 2×2 matrix components, the Pauli spin matrices inserted into (6.79); *cf.* (3.79). Following a common procedure,[11] pick an operator $\boldsymbol{\Omega}$ such that

$$[\boldsymbol{\Omega}, \mathscr{H}_0] = \mathscr{H}_{\text{ex}} \,. \tag{6.81}$$

It will be $O(J)$. A canonical transformation $\mathscr{H} \to \mathrm{e}^{-\Omega} \mathscr{H}\, \mathrm{e}^{\Omega}$ produces

---

[10] See the various chapters of [6.41].

[11] A simplified version of time-ordered perturbation theory.

$$\left.\begin{aligned} \mathscr{H}_0 &\to \mathscr{H}_0 - [\boldsymbol{\Omega}, \mathscr{H}_0] + \frac{1}{2!}\Big[\boldsymbol{\Omega}, [\boldsymbol{\Omega}, \mathscr{H}_0]\Big] + \cdots \\ \mathscr{H}_{\text{ex}} &\to \mathscr{H}_{\text{ex}} - [\boldsymbol{\Omega}, \mathscr{H}_{\text{ex}}] + \cdots \end{aligned}\right\} \tag{6.82}$$

and

$$\mathscr{H} \to \mathscr{H}_0 - \frac{1}{2}[\boldsymbol{\Omega}, \mathscr{H}_{\text{ex}}] + O(J^3)\,. \tag{6.83}$$

The operator linear in J has been transformed away (except for the small part "on the energy shell" which leads to scattering and thus cannot be physically removed). The $\boldsymbol{\Omega}$ which achieves this and solves (6.81) is,

$$\boldsymbol{\Omega} \equiv J\boldsymbol{S}\cdot\boldsymbol{\lambda}_{\text{c}}, \quad \boldsymbol{\lambda}_{\text{c}} = (1/2N)\sum_{\substack{\varepsilon\neq\varepsilon' \\ mm'}} \frac{c^*_{km}\boldsymbol{\sigma}_{mm'}c_{k'm'}}{\varepsilon_k - \varepsilon_{k'}} \tag{6.84}$$

It is well defined except on the energy shell ($\varepsilon = \varepsilon'$). Writing $\boldsymbol{S}\cdot\boldsymbol{s}_{\text{c}}(0)$ as $S_\alpha s^\alpha_{\text{c}}$ and $\boldsymbol{S}\cdot\boldsymbol{\lambda}_{\text{c}}$ as $S_\beta\lambda^\beta_{\text{c}}$ using a summation convention on repeated indices, we find, to new leading order in J, a Hamiltonian

$$\mathscr{H} = \mathscr{H}_0 - \frac{1}{2}J^2\{S_\alpha S_\beta[\lambda^\alpha_{\text{c}}, s^\beta_{\text{c}}] + [S_\alpha, S_\beta]s^\beta_c\lambda^\alpha_c\}\,, \tag{6.85}$$

in which we have used the fact that $\boldsymbol{S}$ commutes with all components $\boldsymbol{s}_{\text{c}}$ of and $\boldsymbol{\lambda}_{\text{c}}$. For classical spins, the components of $\boldsymbol{S}$ commute with each other and the second commutator vanishes. This commutator is also absent in ordinary potential scattering. The first commutator is representative of conventional scattering. $\boldsymbol{\lambda}_{\text{c}}$ and $\boldsymbol{s}_{\text{c}}$ are both quadratic forms in fermion operators; their commutator is, again, quadratic. The contribution from this term to the scattering amplitude (which, in first order, was J) is $O(J^2/D)$. It is negligible for small J.

For quantum mechanical spins, the second commutator does *not* vanish. In terms of components, the spin commutation relations of Chap. 3 are

$$[S_\alpha, S_\beta] = \mathrm{i}\varepsilon^{\alpha\beta\gamma}S_\gamma \tag{6.86}$$

(with $\hbar = 1$, $\varepsilon^{\alpha\beta\gamma} = \pm 1$ for even/odd permutations of x, y, z and 0 otherwise.) The operator $s^\alpha_{\text{c}}\lambda^\beta_{\text{c}}$ is *quartic* in the fermion operators. Among the contributions to electron scattering on the energy shell, we find such new terms as

$$\frac{J^2}{2N}\sum \frac{1 - 2f_{k'}}{\varepsilon_k - \varepsilon_{k'}} \tag{6.87}$$

in which $f_{k'} = \langle c^*_{k'}c_{k'}\rangle$ is the Fermi function. The principal part of (6.87) diverges logarithmically as $\varepsilon_k \to \varepsilon_{\text{F}}$ and $T \to 0$; the Fermi function discontinuity at ε_{F} is sharper at low temperature.

Indeed, the calculation by Kondo showed that the scattering amplitude of an electron ($\boldsymbol{k}$) including constructive interference between first and second orders became

$$\sim J\left(1 + \frac{J}{D}\ln\frac{|\varepsilon_k - \varepsilon_{\mathrm{F}}|}{D}\right) \tag{6.88}$$

at $T = 0$. The scattering cross section should then vary as

$$\sim J^2\left(1 + \frac{2J}{D}\ln\frac{|\varepsilon_k - \varepsilon_{\mathrm{F}}|}{D}\right) \tag{6.89}$$

to the stated order. A typical energy is $\varepsilon_k = \varepsilon_{\mathrm{F}} + kT$; then, the resistivity is:

$$R = R_0 + R_1\left[1 - \frac{2J}{D}\ln(D/kT)\right] \tag{6.90}$$

in which R_0 contains the temperature-independent contributions, R_1 is a lumped constant. The result is the Kondo formula quoted earlier, (6.78).

The magnitudes of T_{K} estimated from resistivity measurements range from less than 1K (MgMn, CuMn, CdMn, etc.,) to greater than 1000 K (AuTi, CuNi) [6.42] while typical conduction electron bandwidths D range from 10^3 to 10^4 $\times\, k_{\mathrm{B}}$. From this we may assume that even at the largest T_{K}'s the Kondo problem remains within the realm of "weak coupling", however peculiar the ground state might be. The analyses by which one extracts the leading divergences from an expansion in a "small" parameter are beyond the scope of this volume, but will be taken up again in the companion volume on finite temperature effects. See also, chaps. 7–9 in Vol. 5 of the *Magnetism* series [6.41]

6.9 Spin Glasses

As the polarization cloud surrounding a magnetic ion which is antiferromagnetically coupled to the conduction electrons effectively screens it from further interactions with other magnetic ions, there must be a phase transition to an interactive phase when the interaction energy between two neighboring spins becomes more favorable than the Kondo binding energy of each. For spins ferromagnetically coupled to the conduction sea, this question does not pose itself: the Ruderman-Kittel interaction will connect them at all concentrations c. We can estimate c for the AF coupled ions by recalling that the interaction energy is $\sim(J^2/D)\,(a/R_{ij})^3\cos(2k_{\mathrm{F}}R_{ij})$. With $c \sim (a/R_{ij})^3$, this yields

$$c_{\mathrm{crit}} = A(D/J)^2\exp(-\,D/2|J|) \tag{6.91}$$

where A is a constant. For J/D much less than 0.1, the critical concentration drops exponentially to zero. Much of the current research in this topic is centered on the distinction, and competition, between Kondo and spin glass phases. It is clear that the spin glass (SG) phase is the more common.

In the SG phase, the spins interactiong *via* the Ruderman-Kittel interactions, have either ferromagnetic or AF couplings depending on the phase of $\cos(2k_F R_{ij})$, an uncertain quantity at large distances R_{ij}. Later, we shall see that the principal requirements for the existence of a SG phase concern the AF bonds: the concentration of AF bonds, disposed at random, must exceed a critical value, the magnitude of which depends on the topology of the lattice and the effective range of the interaction. We shall return to these considerations in due course, but first, point out the results of numerical experiments by *Binder* and *Schröder*

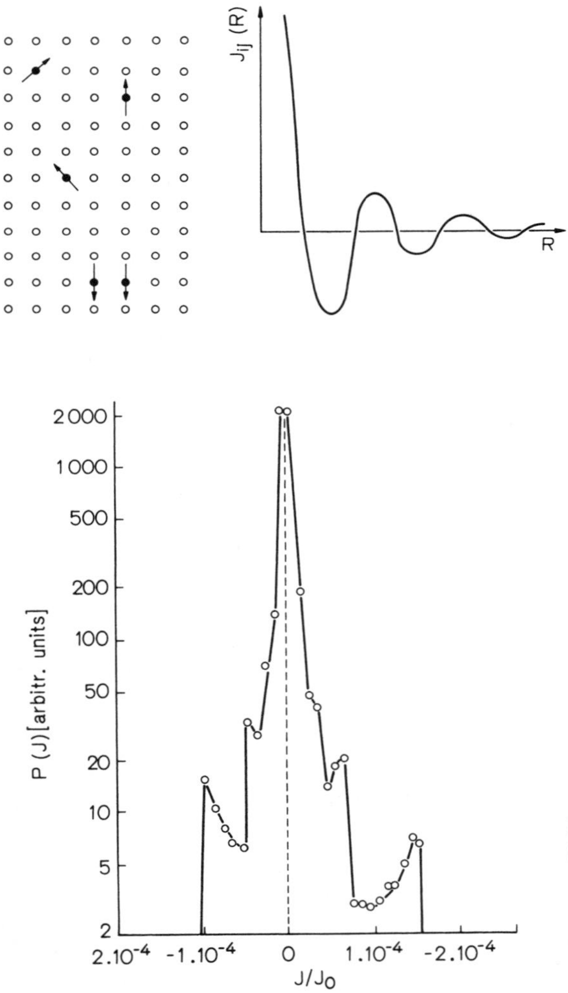

Fig. 6.13. Interactions leading to Spin Glass phase in dilute magnetic alloy, as computed in [6.43]. Top figures represent RKKY interactions among spins randomly distributed in dilute alloy. Lower figure shows computed distribution of J_{jl}

on the statistical properties of some half a million bonds J_{ij}, as summarized in Fig. 6.13. Note the near symmetry in the distribution of positive and negative bonds. [6.43].

Given a particular set of bonds, the spin orientations in the ground state have to minimize the energy. While presenting the appearance of total disorder, the spin-glass ground state is, in fact, highly correlated. It is parametrized by the internal fields (also called "molecular fields") felt by each spin. The distribution $P(h)$ of molecular fields h characterizes the ground state uniquely. On the other hand, the ground state itself is not unique, for there may be many different configurations yielding the same energy; all, however, with presumably the same $P(h)$.

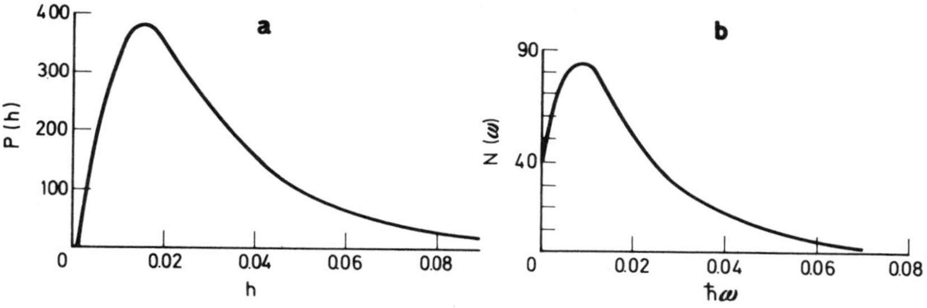

Fig. 6.14 a,b. Properties of computer simulation of RKKY spin-glass. (**a**) Distribution of internal fields and (**b**) Density of elementary excitations. Curves shown are smoothed approximations to the histograms in [6.44]

A computer simulation of the ground state and elementary excitation spectrum of the Ruderman-Kittel SG has been worked out by *Walker* and *Walstedt* [6.44], with the $\boldsymbol{S}_i$ taken to be classical vectors. The spectrum of elementary excitations is found by quantizing each spin along the direction in which the classical spin points, then diagonalizing leading terms in the Holstein-Primakoff LSW approximation, the quadratic Hamiltonian discussed in Chap. 5. The results of their computations, confirmed by semi-empirical theoretical considerations, lead to the curves of Fig. 6.14 obtained for 324 spins on a fcc lattice, $c = 0.003$. It should be noted that the elementary excitations (magnons) are extended states for the entire low-frequency spectrum, although the states in the high-frequency "tail" are localized at the various strongly interacting closest neighbor pairs. We return to this model in a later discussion of the finite temperature properties of the SG phase, but now fix our attention on other popular models.

The simplest "separable," model [6.45], takes the signs of the bonds to be separately random

$$J_{ij} = \varepsilon_i \varepsilon_j F(\boldsymbol{R}_{ij}) \tag{6.92}$$

with the $\varepsilon_i = \pm 1$ random variables, distributed independently according to a specified probability distribution. The $F(\boldsymbol{R}_{ij})$ can be chosen to mimic the Ruderman-Kittel long-range interaction, but two simpler choices can be made: $F(\boldsymbol{R}_{ij}) \neq 0$ for $\boldsymbol{R}_{ij}$ a nearest-neighbor vector only, or $F(\boldsymbol{R}_{ij}) = \text{constant}$, independent of distance. The former leads straightaway to the ordered n.n. models which have been extensively studied, and are found in the various chapters of the present book. The long-range version [6.46] is soluble in closed form by mean-field theory. The trick leading to the solution of the separable models is the transformation from spins $\boldsymbol{S}_i$ to new spin variables $\boldsymbol{\sigma}_i$, as follows: $\boldsymbol{S}_i \equiv \varepsilon_i \boldsymbol{\sigma}_i$ $(= \pm \boldsymbol{\sigma}_i)$. Only for a quantum Heisenberg model is this transformation *inadmissible*, for it is forbidden to invert all *three* components of a spin vector [recall the example following (3.6) of Chap. 3]. For classical spins, or quantum XY or Ising models, the transformation brings the Hamiltonian into a simpler form

$$\mathscr{H}_{\text{sep}} = -\sum_{i,j} J_{ij} \boldsymbol{S}_i \cdot \boldsymbol{S}_j = -\sum_{i,j} F(\boldsymbol{R}_{ij}) \boldsymbol{\sigma}_i \cdot \boldsymbol{\sigma}_j \; . \tag{6.93}$$

The latter is the type Hamiltonian we have already considered in Chap. 5. Now, obtaining correlated $\boldsymbol{\sigma}_i \cdot \boldsymbol{\sigma}_j$ in the ground state (e.g., parallel spins in the case $F > 0$) does not mean the physical spins $\boldsymbol{S}_i$ have an *obvious* correlation, for the quantity $\boldsymbol{S}_i \cdot \boldsymbol{S}_j = \varepsilon_i \varepsilon_j \boldsymbol{\sigma}_i \cdot \boldsymbol{\sigma}_j$ will have the sign of $\varepsilon_i \varepsilon_j$ and the *appearance* of a random variable. Nevertheless, it is obvious that the $\boldsymbol{S}$'s are as strongly correlated as the $\boldsymbol{\sigma}$'s, *for a given set* of ε_i. The long-range version of this model is remarkably simple; We illustrate for the ground state, replacing F by its average

$$\mathscr{H}_{\text{sep L.R.}} = -\frac{\langle F \rangle}{N} \Big(\sum_i \boldsymbol{\sigma}_i\Big)^2 \; . \tag{6.94}$$

For $\langle F \rangle < 0$, we have $\sum \boldsymbol{\sigma}_i = 0$, a condition satisfied by almost any random configuration of the $\boldsymbol{\sigma}$'s. The ground state is then totally disordered, just as for N noninteracting spins, regardless of the magnitude of F! This implies also the lack of an order-disorder phase transition at finite T, as we shall determine later. In the absence of correlations, the SG phase does not exist, so these spins are either perfectly free or are in their Kondo ground states. At the opposite extreme is $\langle F \rangle > 0$, for which the ground state $\boldsymbol{\sigma}_i$'s are *all* parallel. Although the corresponding $\boldsymbol{S}_i$'s would give the appearance of being uncorrelated, we already know this is not the case.

The physical problem studied numerically by Walker and Walstedt does not have all bonds positive or negative, but rather a mixture of both. There have been several attempts to introduce models that combine this element together with some simplification, sufficient to render the analysis tractable. Notably, *Edwards* and *Anderson* [6.47] and *Sherrington* and *Kirkpatrick* [6.48] have introduced models with such features, but these have also, to a great extent, resisted analysis. Although such studies have attracted much attention lately, the nature of the ground state is still not clear, not to mention the question of the existence of a thermodynamic phase transition at finite temperature! Recent studies indicate [6.49] that $T_c = 0$ in less than 4 dimensions (if the bonds are random with zero mean, and restricted to nearest-neighbors). Nevertheless, the study of ground state and low-lying states in such models is of the greatest interest.

Take, for example, a model in which Ising spins $S_i = \pm 1$ are regularly arrayed at the vertices of a square lattice with nearest-neighbor bonds $J_{ij} = \pm 1$. Let p be the probability of an AF bond, be in the range $0 \leq p \leq 1/2$ ($p > 1/2$ can be obtained by symmetry). The ground state energy, E_0, is easily found numerically: for p in the range $0 < p \leq p_c$ ($p_c \sim 1/4$) the energy increases monotonically from $-2N$ to approximately $-2^{1/2}N$. From p_c on, E_0 is independent of p and maintains the value $-2^{1/2}N$; this is suggestive of a distinct phase. And indeed, in 3D, E_0 starts at $-3N$ at $p = 0$ and rises to approximately $-3^{1/2}N$ at p_c, staying constant thereafter. In 1D, we of course have the trivial result: $-1N$ and $-1^{1/2}N$, with $p_c = 0$. It's not known whether in 4D the numbers are 4 and $4^{1/2}$, respectively, but this is certainly suggestive; the typical behavior being shown in Fig. 6.15. The ground state properties have been studied, in a variety of models, by a variety of techniques too numerous to list.[12]

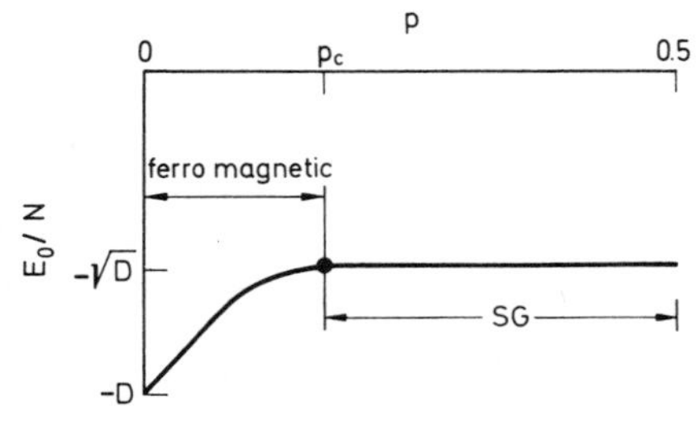

Fig. 6.15. Ground state energy per spin in n.n. model, bonds ± 1, with p = probability of AF bond. Energy is independent of p in the SG phase, and takes the values $-D^{1/2}$ for D= 1,2,3 (l. c., sq. and sc); p_c varies according to D

A somewhat more tractable model SG is based on the "spherical model"; each spin interacts with every other spin, the $N(N-1)/2$ bonds $J_{ij} = \pm 1/N^{1/2}$ being randomly selected. If the spins are treated in the spherical model, i.e., $\sum S_i^2 = N$ is required (and there is no constraint on the individual S_i^2) the ground state and thermodynamic properties can be calculated [6.53] from the eigenvalue spectrum of the J_{ij} matrix. Fortunately, this spectrum is known exactly [6.53, 54]. For example, the ground state energy is precisely $E_0 = -N$.

[12] Ground state properties, particularly the problems of "frustration" were pioneered in [6.50]. The quantum features are treated in [6.51] and also in [6.45]; the S-K model [6.48] in [6.52].

6.10 Magnetism Without Localized Spins: Preliminaries

The strongly magnetic properties of the iron transition series metals and of their alloys must be explained by band theory. This would be required for the following two reasons alone: the number of Bohr magnetons per atom is generally far from being an integer, and both calculated and observed band widths are of the order of electron volts. Neither of these facts could be accorded with a scheme based on localized spins, and together with such additional evidence as the abnormally high specific heat (which can only be explained by a continuum of states for the magnetic carriers) they point instead to the need for associating the uncompensated spins with Bloch electrons rather than with their Wannier counterparts in the first approximation.

The band theory of magnetism is currently a very active field of research, and it is not yet possible to predict the degree of accuracy and sophistication it may attain. At present, it is far less developed than the relatively straightforward Heisenberg theory for insulators; but the physical mechanisms are established and some of the consequences therefrom approximately understood. Let us start by visualizing, inaccurately perhaps, localized spins in the indirect exchange theory being gradually modified so as to allow some overlap and the formation "magnetic" electron bands. The energy gap against the excitation of electrons in the magnetic states disappears and a finite density of states appears at the Fermi surface. This now allows charge fluctuaitons, hence the conduction of electricity by magnetic electrons; it permits them to contribute to the electronic specific heat, microwave absorption, etc. On the other hand, there is no particular reason why the inherently *magnetic* properties—such as the magnon spectrum-should be affected to leading order. As with indirect exchange, there is no universally valid reason that the long-range order in the ground state be ferromagnetic, and spiral or other antiferromagnetic configurations, such as are actually observed in Mn and in Cr, are not difficult to reconcile with this approach.

One salient difference with the indirect-exchange theory, is that nonmagnetic *s*-like bands can be ignored in the first approximation. Doubtless some indirect exchange still goes on, but the dominant, primary interactions are now *among* the magnetic bands themselves. An interesting point of view has it that the *d* electrons (e.g., in Fe) effectively split into two sub-bands; one of which is narrow, so that the electrons it contains are well localized like the rare-earth *f* states, and the other is more like an ordinary conduction band. It is quite likely, that even were such a model describable by the indirect exchange Hamiltonian of (6.52), the perturbation-theoretic solution on which the Ruderman-Kittel interaction is based, would not be valid. The most spectacular feature of the band theory is that with neglect of the nonmagnetic electrons the remaining magnetic ones do not number any particular rational multiple of the number of atoms. Therefore, the occupation number of the magnetic bands is a vital, and often an adjustable, parameter in the theory.

For definiteness in this study, let us assume that whenever an ordered state exists, it is ferromagnetic. We then calculate the magnon spectrum, and if some magnon has negative energy, the assumption is then evidently false and the ground state must be some species of AF. This procedure is simpler than calculating the energy of every possible ordered state, although, as we saw in the indirect exchange theory, it leads to almost identical results. The first important physical fact we shall wish to demonstrate is the existence of a threshold magnitude for the interactions. That is, we shall show that *below* a certain magnitude of the Coulomb repulsion and Hund's rule coupling parameters, *there is no metallic magnetism.* To establish the possible orders of magnitude, we list in Table 6.3 the relative strengths of various plausible physical forces.

Table 6.3. Scale of energies

Order of magnitude	Mechanism
~10 eV	(a) Atomic Coulomb integrals U
↓	(b) Hund's rule exchange energy, J
	(c) Energy of electronic excitations violating Hund's rule
	(d) Electronic band widths, W (also denoted D)
~ 1 eV	(e) $\mathcal{N} \div$ (density of states at Fermi Surface)
0.1 – 1.0 eV	Crystal field splittings
10^{-2} – 10^{-1} eV	Spin-orbit coupling kT_C or kT_N
10^{-4} eV	Magnetic spin-spin coupling Interaction of a spin with external field 10 kG
10^{-6} – 10^{-5} eV	Hyperfine electron-nuclear coupling.

In the creation of magnetic moments in the metal, the competition is between the "kinetic energy" terms (d) and (e), and the potential energy terms (a)-(c), representing forces internal to each incomplete shell in each atom of the solid. For isolated atoms, the kinetic terms are zero and the magnetization of the incomplete shell proceeds according to Hund's rules, detailed and proved in an earlier chapter but what happens when kinetic energy is introduced?

And once the criteria for the existence of magnetic moments are satisfied, what is their alignment? We have already observed in connection with the indirect-exchange theory, that the geometric factor $k_F a$ plays an important role, i.e., the filling of the band. For an almost full or almost empty band, $k_F \to 0$ and long wavelengths predominate—i.e., ferromagnetism. For nearly half-filled bands, some species of AF are possible. The use of a variable axis of quantization for local spin polarization as an adjunct in the study of electron interactions in magnetic metals has become increasingly popular in the recent literature [6.55].

Much of the commonplace understanding about itinerant ferromagnets comes from the example of nickel, a ferromagnetic metal having 0.6 Bohr magnetons *per* atom, with 0.6 holes per atom in the *d* band. Alloying with copper, having a similar band structure but a higher Fermi level, "plugged up" the holes and quenched the ferromagnetism in direct proportion to the added number of electrons. Now, the important discovery of metallic ferromagnetic alloys made out of ordinarily nonmagnetic constituents [6.56]—e.g., $ZrZn_2$—has made the need for a reliable theory of itinerant electron ferromagnetism more urgent. Neutron diffraction has shown that some of the magnetic moment in such an alloy resides between the atoms and not on them, i.e., is identified with the bonding, mobile electrons [6.57].

The 1966 systematic review of itinerant electron magnetism theory by *Herring* [6.2], did much to place the topic in perspective. In the following sections, we deal with easily understood aspects of this knotty many-body problem: the exactly soluble low-density limit (appropriate to the few holes in Ni) and the case of a nearly half-filled band. We proceed to an elementary study of the magnon spectrum in magnetic metals, and suggest the type of calculation required to understand the magnon spectrum of Iron.

6.11 Low-Density Electron Gas

The low density limit of the electron gas can be calculated systematically. First, one recognizes that the long-range part of the electron-electron Coulomb interaction (the troublesome part) is of no interest in the magnetic problem: the high cost in energy to create a plasmon guarantees that charge neutrality is maintained throughout the metal. On the other hand, there is no requirement for charge neutrality on an atomic scale. A Hamiltonian incorporating the kinetic energy of electrons in a single nondegenerate band, with intra-atomic Coulomb repulsion U between electrons of opposite spin (electrons of parallel spin cannot simultaneously occupy the same atom) has been extensively studied by *Hubbard* [6.58] and a number of other workers, in the form

$$\mathscr{H} = \mathscr{H}_0 + U \sum_i \mathbf{n}_{i\uparrow} \mathbf{n}_{i\downarrow} \tag{6.95}$$

with $\mathscr{H}_0$ from (6.24) or (6.27). We shall solve it exoctly for two particles. The triplet eigenstates and eigenvalues are

$$|\boldsymbol{k}, -\boldsymbol{k}'; 1) = c^*_{\boldsymbol{k}\uparrow} c^*_{-\boldsymbol{k}'\uparrow} |0) \quad \text{and} \quad E_{\boldsymbol{k}\boldsymbol{k}'} \equiv E(\boldsymbol{k}) + E(\boldsymbol{k}') \tag{6.96}$$

for all $\boldsymbol{k}$, $\boldsymbol{k}'$ in the first BZ, with $E(\boldsymbol{k})$ the Bloch energy. Two projections of (6.96):

$$2^{-1/2}(c^*_{\boldsymbol{k}\uparrow} c^*_{-\boldsymbol{k}'\downarrow} + c^*_{\boldsymbol{k}\downarrow} c^*_{-\boldsymbol{k}'\uparrow}) |0) \quad \text{and} \quad c^*_{\boldsymbol{k}\downarrow} c^*_{-\boldsymbol{k}'\downarrow} |0) \tag{6.97}$$

are also, trivially, eigenstates of $\mathscr{H}$ belonging to the same eigenvalue. However, for the same two electrons in a *singlet* state, the energy involves U, and is obtained from scattering theory. For brevity define

$$b_{kk'} \equiv 2^{-1/2}(c^*_{k\uparrow}c^*_{-k'\downarrow} - c^*_{k\downarrow}c^*_{-k'\uparrow})|0) \tag{6.98}$$

having band ("kinetic") energy $E_{kk'}$ [defined in (6.96)]. Then let the eigenstate have energy $W_{kk'}$ and be of the form

$$B_{kk'} \equiv |k, -k'; 0) = b_{kk'} + \frac{1}{N}\sum_{q\neq 0} L_q b_{k+q,k'+q} \tag{6.99}$$

the usual form, an incoming wave and a scattered part. The Schrödinger equation consists of three parts:

$$\mathscr{H}_0 B_{kk'} = E_{kk'}b_{kk'} + \frac{1}{N}\sum_q L_q E_{k+q,k'+q}b_{k+q,k'+q} \tag{6.100a}$$

$$\mathscr{H}' B_{kk'} = \frac{U}{N}B_{kk'} + \frac{U}{N}\left(\sum_{q\neq 0} b_{k+q,k'+q} + \frac{1}{N}\sum\sum_{q',q\neq q'} L_{q'}b_{k+q,k'+q}\right) \tag{6.100b}$$

$$W_{kk'}B_{kk'} = W_{kk'}b_{kk'} + \frac{1}{N}\sum_q L_q W_{kk'}b_{k+q,k'+q}\ . \tag{6.100c}$$

Equating the coefficients of $b_{kk'}$ yields an expression for the energy eigenvalue

$$W_{kk'} = E_{kk'} + \frac{U}{N}\left\{1 + \frac{1}{N}\sum_{q\neq 0} L_q\right\}. \tag{6.101}$$

The scattering amplitudes L_q are obtained from the coefficients of the $b_{k+q,k'+q}$. After minor algebra, the result is

$$L_q = \frac{1}{W_{kk'} - E_{k+q,k'+q}} \times \frac{U}{1 + UG_0} \tag{6.102}$$

where

$$G_0 \equiv \frac{1}{N}\sum_q \frac{1}{E_{k+q,k'+q} - W_{kk'}}\ . \tag{6.103}$$

Consequently, the energy is

$$W_{kk'} = E_{kk'} + \frac{1}{N}\left(\frac{U}{1 + UG_0}\right) \equiv E_{kk'} + \frac{1}{N}t(W_{kk'}) \tag{6.104}$$

which also serves to define the scattering t matrix, which is the expression of the interaction energy between the two particles with all multiple scattering taken into account. For all practical purposes the $W_{kk'}$ can be replaced by the unperturbed $E_{kk'}$ from which it differs by only $O(1/N)$. (The exception is for bound states, i.e., the zeros of $1 + UG_0 = 0$ which, if they exist, lie above the highest $E_{kk'}$ and thus will not be relevant to the subsequent discussion. (In passing, one notes that for *negative* U, no matter how weak, such a pole in the t matrix could develop, *below* the lowest $E_{kk'}$; the lowest of these bound states belongs to $q = 0$, and is the zero-momentum Cooper pair, famous in the BCS theory of superconductivity [6.59]. Our interest, however, is in fairly large, *positive* U).

The modification of 2-body scattering required for the low density electron gas is straightforward: the range of $\boldsymbol{q}$'s in (100) et seq., especially G_0, (6.103), must be restricted to the unoccupied states: $\boldsymbol{k} + \boldsymbol{q}$, $-\boldsymbol{k}' - \boldsymbol{q}$ must lie outside the Fermi volume, i.e., $E(\boldsymbol{k} + \boldsymbol{q})$ and $E(-\boldsymbol{k}' - \boldsymbol{q})$ must exceed E_F. Denote the suitably modified G_0 by $\tilde{G}_0$ and similarly for $\tilde{t}(E_{kk'})$ to obtain the effective Hamiltonian for the singlet pairs

$$\mathscr{H}_{\text{eff}} = \sum_{\boldsymbol{k},m} E(\boldsymbol{k})\mathfrak{n}_{\boldsymbol{k}m} + \frac{1}{N}\sum_{\boldsymbol{k},\boldsymbol{k}'} \tilde{t}(E_{\boldsymbol{k}\boldsymbol{k}'})\mathfrak{n}_{\boldsymbol{k}\uparrow}\mathfrak{n}_{-\boldsymbol{k}'\downarrow}, \qquad \tilde{t} \equiv \frac{U}{1 + U\tilde{G}_0} \tag{6.105}$$

To proceed we need evaluate $\tilde{t}(E)$. Our approach is greatly simplified, yet retains essential features. In a dilute electron gas, the Fermi level is near enough to the bottom of the conduction band(s) (or, for holes, the top) that the effective mass approximation $E(k) = \hbar^2k^2/2m^*$ is appropriate. With this, all the integrals that go into the calculation of the ground state energy become simple.

First, we introduce the cutoff k_0 to retain, properly, the volume of the first BZ

$$\frac{1}{N}\sum_{k \leq \text{BZ}} 1 = 1 = \frac{4\pi}{3}\left(\frac{k_0 a_0}{2\pi}\right)^3. \tag{6.106}$$

The density parameter ρ is defined: *the number of electrons per atom per band in a given spin direction* ↑ or ↓. With a spherical Fermi surface at k_F, ρ is given as

$$\frac{1}{N}\sum_{k<k_F} 1 \equiv \rho = (k_F/k_0)^3 \text{ (by comparison with (6.106)).} \tag{6.107}$$

Similarly the kinetic energy (KE) per band per spin component is

$$\text{KE} = \sum_{k<k_F} E(k) = \frac{N\sum_{k<k_F} E(k)}{\sum_{\text{all}} 1} = N\left[\frac{\int_0^{k_F} dk k^4 \hbar^2/2m^*}{\int_0^{k_0} dk k^2}\right]$$

$$= \frac{9}{10} N \frac{\hbar^2 k_0^2}{3m^*}\rho^{5/3}. \tag{6.108}$$

We start with the exact expression for $\tilde{G}_0$:

$$\tilde{G}_0 = (a/2\pi)^3 \int_{\substack{|\boldsymbol{k}+\boldsymbol{q}|>k_F \\ |\boldsymbol{k}'-\boldsymbol{q}|>k_F \\ q<k_0}} d^3q \frac{1}{(\hbar^2/2m^*)[(\boldsymbol{k}+\boldsymbol{q})^2 + (\boldsymbol{k}'-\boldsymbol{q})^2 - \boldsymbol{k}^2 - \boldsymbol{k}'^2]}$$

and evaluate it, neglecting $\boldsymbol{k},\boldsymbol{k}' \ll k_0$, in the dilute limit. An approximate but very convenient expression, independent of $\boldsymbol{k}$ and $\boldsymbol{k}'$, is what results:

$$\tilde{G}_0 \approx (a/2\pi)^3 \int_{k_0>q>k_F} d^3q \frac{1}{(\hbar^2/2m^*)2\boldsymbol{q}^2} = \frac{3m^*}{\hbar^2 k_0^2}(1 - \rho^{1/3}) \tag{6.109}$$

using (6.107) to eliminate k_F. With this, we obtain a $\tilde{t}$ matrix which is independent of $\boldsymbol{k},\boldsymbol{k}'$ and depends explicitly on ρ. *Kanamori* [6.60] has made a different simplification, with basically similar final results.

As the combination $\hbar^2 k_0^2/3m^*$ occurs in the interaction and in the KE, we eliminate it. It is recognized as a measure of the bandwidth, in units appropriate to the effective mass approximation. We therefore denote it W as usual

$$W \equiv \hbar^2 k_0^2/3m^*. \tag{6.110}$$

With $\tilde{G}_0$ and $\tilde{t}$ suitably re-expressed in terms of W, and with the KE also simplified, we now have for the ground state energy of (6.105) the perspicuous expression

$$\begin{aligned} E_0 &= NW\left[2\cdot\frac{9}{10}\rho^{5/3} + \frac{U\rho^2}{W + U(1-\rho^{1/3})}\right] \\ &= NW\rho^{5/3}\left[\frac{9}{5} + \frac{U\rho^{1/3}}{W + U(1-\rho^{1/3})}\right]. \end{aligned} \tag{6.111}$$

Both KE and interaction energies have been expressed in terms of the sole physical parameters of interest: the bandwidth W, the coupling constant U, and the electron density ρ. Writing a similar expression for the *totally* ferromagnetic state at this stage, we do not bother with the added complications of partially magnetized states, so $\rho_\uparrow = 2\rho$, and $\rho_\downarrow = 0$; $\tilde{t}$ vanishes and we find

$$E_{\text{ferro}} = NW(2\rho)^{5/3}\left(\frac{9}{5}\right). \tag{6.112}$$

A criterion for the occurrence of ferromagnetism can be obtained by comparing the two expressions. If (6.112) is lower, then the ground state has spin $\mathcal{N}\frac{1}{2}\hbar$. After a few most elementary algebraic manipulations, this criterion is seen to be

$$\frac{U\rho^{1/3}}{U+W} > 0.5139\,, \tag{6.113a}$$

There is no solution for $\rho < (0.5139)^3 = 0.136$. Other wise, we find ferromagnetism when

$$\frac{U}{W} > \frac{1}{\rho^{1/3} - (0.136)^{1/3}} \quad \text{for} \quad \rho > 0.136\,. \tag{6.113b}$$

The phase diagram is plotted in Fig. 6.16, where the (uncertain) application to high-density (half-filled bands) is indicated by a dashed curve. Note the existence of a critical density $\rho_c = 0.136$. For densities lower than this no magnetism is possible, even at infinite interaction strengths $U = \infty$. This general feature is in excellent accord with a theorem proved in an earlier chapter, concerning the non-existence of ferromagnetism of 2 electrons, and extends it (for this particular interaction) to an infinite number of particles as long as the density does not exceed ρ_c. In Fig. 6.17 we anticipate the AF spatial ordering. As we have seen in the indirect exchange theory, the $\rho = 1/2$ density is always AF [see point labelled $n = (2\rho) = 1$ in Fig. 6.6], and we may combine this knowledge with (6.113)

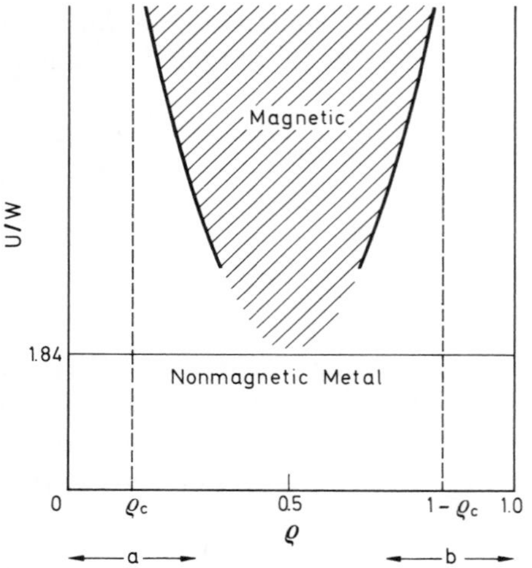

Fig. 6.16. Phase diagram of 1-band Hubbard model at $T = 0$, based on low-density theory (see text). Region (*a*) is low density for electrons and (*b*) for holes. The highdensity results ●-●-●- near 0.5 are unreliable, due to limitations of the theory. Magnetic region is shaded, and $\rho_c = 0.136$

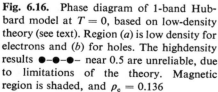

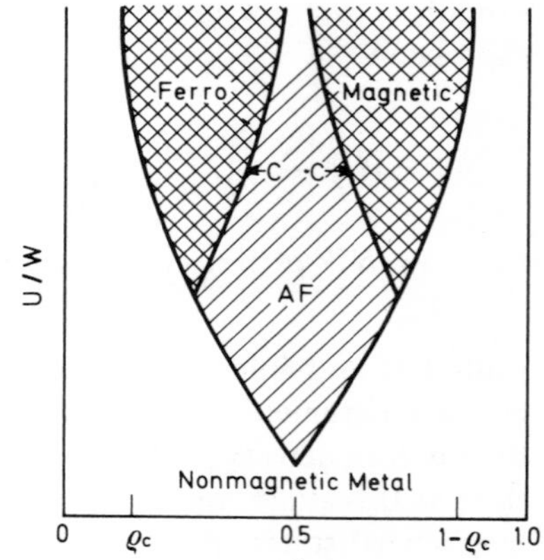

Fig. 6.17. Same as preceding, with magnetic ordering taken into account. The half-filled band is most easily susceptible to AF ordering (Examples: NiO, and also 1D Hubbard model [6.62]) but at strong enough coupling in 2D or 3D (not 1D) gives way to the ferromagnetic phase (cross-hatching). The equation of curve C separating the magnetic phases is derived in the next section

to produce a phase diagram of the type shown in Fig. 6.17. Similar diagrams have been obtained from many other theories, e. g., [6.61] or *Liu* [6.55], (note their $\boldsymbol{c}$ or $\boldsymbol{n}$ is *twice* our ρ). The exact solution by *Lieb* and *Wu* [6.62] of the one-dimensional Hubbard model for a half-filled band conclusively showed this to be an AF insulator for any finite U and to have a spectrum of low-lying excitations readily identifiable as antiferromagnons. This interesting work is the generalization to the interacting electrons of the Bethe solution for interacting spins, but involves mathematical techniques beyond the scope of the present text. It should be noted that the replacement of $\tilde{G}_0$ by its average value as in (6.109) is not permissible in 1D (nor perhaps in 2D), for the integral is very sensitive to the precise value of k,k' in this case. Indeed, for $k \sim k' \sim \pm k_F$, $\tilde{G}_0 \sim \infty$ in 1D and thus $\tilde{t} \to 0$. This has the interesting consequence that the well-known instability of a 1D "metal" against perturbations of wave vectors $2\boldsymbol{k}_F$-is compensated by the vanishing, at precisely the same wave vectors, of the interactions!

6.12 Quasi-Particles

Interactions affect the states of the system. At a given total momentum, only one state is absolutely stable; interactions cause higher lying states to decay. This is shown, by redefining the interacting Fermi sea, whether or not magnetic, to be the new "vacuum". Excited states are then described in terms of quasi-particles (fermions) or collective modes (bosons) in one-to-one correspondence with the familiar electrons (above the Fermi surface, FS) or holes (below the FS) of the noninteracting electron gas and the plasmons of the charged gas, as well as the magnons of the magnetic systems.

If one electron is added to those already present, the added energy is

$$E(\boldsymbol{k}) + \frac{1}{N}\sum_{k'<k_F} \tilde{t}(E_{kk'}) \equiv E(\boldsymbol{k}) + \Delta(\boldsymbol{k})\,. \tag{6.114a}$$

The quasi-particle energy must be measured from the FS, thus we define it to be

$$\begin{aligned}\varepsilon(\boldsymbol{k}) &= |E(\boldsymbol{k}) - E(\boldsymbol{k}_F) + \Delta(\boldsymbol{k}) - \Delta(\boldsymbol{k}_F)| \\ &\approx \hbar^2|\boldsymbol{k}^2 - \boldsymbol{k}_F^2|/2\tilde{m}\end{aligned} \tag{6.114b}$$

which serves to define the renormalized quasi-particle mass $\tilde{m}$ in terms of the original band-structure mass m^*. This quantity has been exhaustively studied for the low density electron gas [6.63] with the results at $U = \infty$ being:

$$\tilde{m} = m^*\left[1 + \frac{8}{15\pi^2}(7\ln 2 - 1)(\boldsymbol{k}_F a)^2\right]$$

in which $a=$ hard sphere radius. We identify $2a$ as the distance of closest approach, i.e., a_0 in the lattice analog. With (6.106) used to eliminate this, and (6.107) to bring in the electron density, we obtain the simple result

$$\tilde{m} = m^*(1 + 0.79\,\rho^{2/3})\,, \tag{6.115}$$

in which $\tilde{m}$ is the quasi-particle effective mass and m^* the unperturbed, band-theoretic effective mass.

Along with a change in inertia, the excited particle above the FS can be scattered to lower energy while an electron within the Fermi sea is excited out of it. This process can conserve energy, and thus give an imaginary contribution to $\tilde{G}_0$. As the amount of phase space depends on how far above the FS we start, it is found that the lifetime τ_k for scattering becomes infinite as $(k - k_F)^{-2}$. More precisely [6.63], with the above substitution for a and $U = \infty$,

$$1/\tau_k = \frac{\hbar}{2m^*}\,1.21\,\rho^{2/3}(k - k_F)^2\,. \tag{6.116}$$

This is not necessarily related to transport properties, as the interactions leading to this scattering do conserve momentum. An identical formula holds for hole quasi-particles.

At a density of 1/8, the increase in mass (6.115) is 20%, at a density of 1/2 (where the low-density theory is unreliable, however) the increase is of 50%. The lifetime effects of quasi-particles near the FS are, according to (6.116), negligible.

6.13 Nagaoka's Model

Figure 6.17 shows the possibility of ferromagnetism at extremely large U for an approximately half-filled band. In view of the usual tendency of a half-filled band toward antiferromagnetism, also shown in this figure but at lower values of the interaction parameter, Nagaoka's mechanism [6.64] is worthy of special note. It predicts what would happen if a sufficient number of electrons were removed from the prototypical antiferromagnet NiO, *viz.*, a phase transition to the ferromagnetic metal. The following calculation will show this is plausible in 3D, whereas it can *never* occur in 1D, where the ferromagnetic phase is rigorously excluded.

Initially, we consider the $U = \infty$ limit. In (6.46) we see that this effectively sets the tendency to antiferromagnetic ordering $\sim W^2/U$ to zero. In this limit, N electrons, one at each site in the crystal, now have perfectly arbitrary spin configurations. The ground state is 2^N- fold degenerate. The introduction of even a single hole changes this dramatically in 2D or 3D (but not for nearest-neighbor hopping in 1D where n.n.n. hopping is required [6.65]). To see this, let us calculate the eigenstates of the hole in a linear chain first. Assume a matrix element-B for moving an electron from a site n to the two neighboring sites $n \pm 1$ in the linear chain. As we know from the tight-binding theory, (6.8), the eigenvalues are then $E(k) = -2B \cos ka$, with $W = 4B$. The infinite potential energy inhibits the motion of all the electrons except those in the vicinity of the hole. Let the initial configuration be

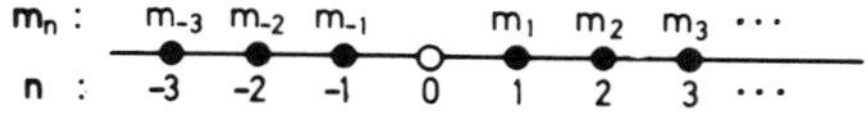

(I)

with the hole indicated by an open circle at $n = 0$ and the electrons by dark circles elsewere. The spins $m_n = \pm 1/2$ of the electrons are a given, arbitrary, set. The initial configuration connects to a linear combination of two

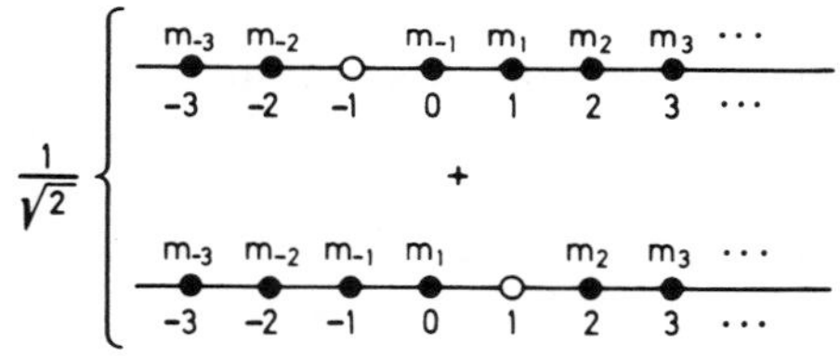

(II)

with matrix element $\sqrt{2}\,B$. This, in turn, connects to I and to a new set of configurations denoted III (where the hole is either at -2 or at $+2$) with matrix element B. This connects back to II, and to a new set, IV, with matrix element B, and so forth. The resulting matrix has the appearance

$$\mathscr{H} = -B \begin{vmatrix} 0 & \sqrt{2} & 0 & & \\ \sqrt{2} & 0 & 1 & & 0 \\ 0 & 1 & 0 & 1 & \\ & 0 & 1 & 0 & \ddots \\ & 0 & & \ddots & \ddots \end{vmatrix} \tag{6.117}$$

bearing some similarity to the Hamiltonian we diagonalized in Chap. 5 for spin waves near the surface of ferromagnets. The eigenvectors

$$V_q = A_q(v_0, v_1, \ldots, v_n \ldots) \tag{6.118a}$$

are readily found

$$v_0 = 2^{-1/2},\ v_n = \cos(qn),\ A_q = (2/N)^{1/2} \tag{6.118b}$$

for $N \to \infty$ and $E(q) = -2B \cos q$. Thus, the local density of states at the origin of the motion:

$$N_0(E) = \sum_q |A_q v_0|^2\, \delta[E - E(q)] \tag{6.119a}$$

$$= \frac{1}{N} \sum_q \delta(E + 2B \cos q) \tag{6.119b}$$

has *precisely* the usual form for free-particle states. As all the results obtained are for any, arbitrary set of the $\{m_n\}$, we see that the extra hole does nothing to promote ferromagnetism. We can proceed to 2 or more holes by use of a determinantal function, and obtain that *regardless of the number of holes*, all the eigenstates in $1D$ are paramagnetic: their energy is independent of the spins in the absence of an applied magnetic field even in this $U \to \infty$ limit.

In two or higher dimensions the configurations which result from moving the hole to some point R_i depend on the path taken, except in the ferromagnetic case. Supposing for the ferromagnetic case a configuration ϕ_a connects to a configuration ϕ_b through two different paths: the matrix element will be -2 (in some units). If in the AF state the two final configurations are different, say $\phi_{b'}$ and $\phi_{b''}$, then the matrix element connecting ϕ_a to $2^{-1/2}(\phi_{b'} + \phi_{b''})$ in the same units will be $-\sqrt{2}$. This difference favors the ferromagnetic state over all other configurations, as we now see with a specific example.

We illustrate with a sq. lattice, starting with the hole at some initial site

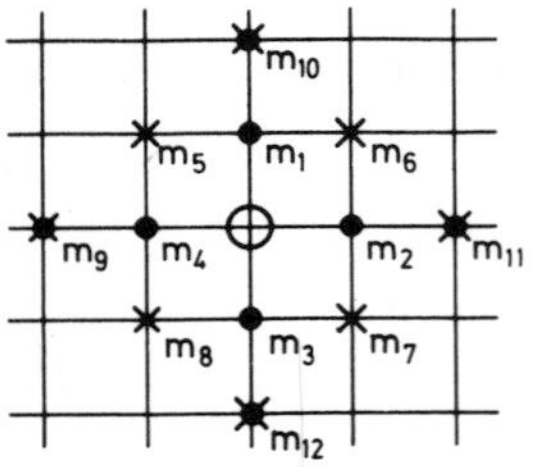

(I)

with spin indices noted, and distinguishing n.n.n. according to whether they are neighbors to a single n.n. (✱) or to two (×).

The state to which (I) connects is the sum over 4 configurations: (II), which, in turn, connect to a linear combination of configurations (III), in which the hole has moved to the n.n.n. spot. It is the motion into the spots marked **X** that creates a new situation. These spots are accessible two ways. For example, the two configurations shown explicitly in (II) connect to the two (of 12 configurations) making up (III). But *they are not necessarily distinct*; if $m_1 = m_4 = m_5$, they are identical. If the m's are random, there is 1/4 probability that

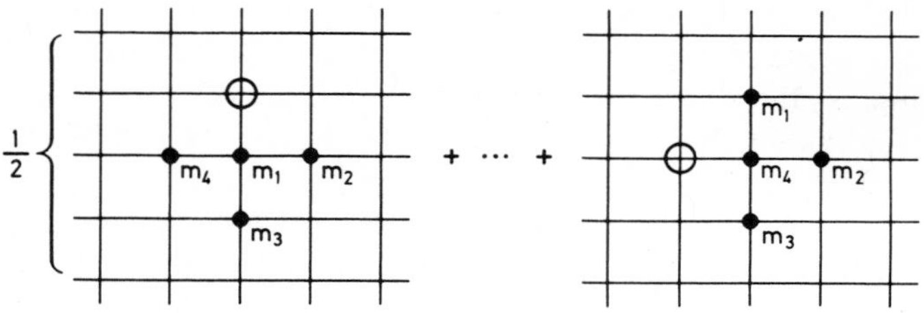

(II)

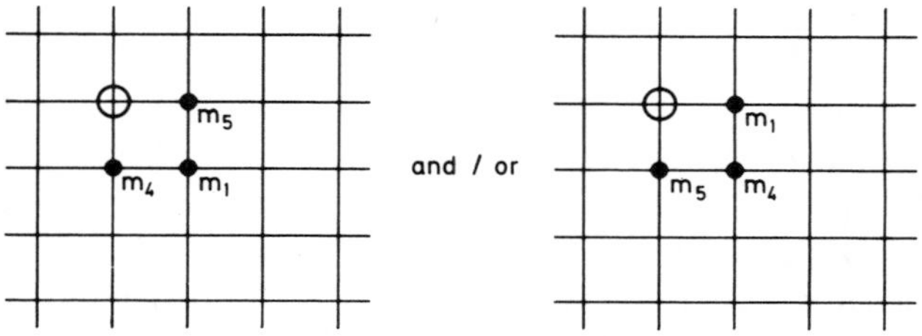

(III)

they are, indeed, identical. If the state is ferromagnetic, they are *necessarily* identical and, finally, if the state is AF all odd subscripted m's are $+$ and even subscripted—, so that they are *never* identical. It is easily seen that the fewer the distinct configurations, the larger the matrix elements. In all cases, $H_{\mathrm{I,II}}$ are $-B\sqrt{4}$, regardless of the m's. But at the second step

$$\begin{aligned} H_{\mathrm{II,III}} &= -B\sqrt{5} \qquad \text{(ferro)} \\ H_{\mathrm{II,III}} &= -B\sqrt{3} \qquad \text{(AF).} \end{aligned} \tag{6.120}$$

Asymptotically, the matrix elements must be

$$H_{M,M+\mathrm{I}} = -2B \qquad \text{(ferro)}$$

and, approximately,

$$H_{M,M+\mathrm{I}} = -2(3/5)^{1/2}B \qquad \text{(AF)} \tag{6.121}$$

to yield a ground state eigenvalue

$$\begin{aligned} &-4B \qquad \text{(ferro)} \\ &-4(3/5)^{1/2}B \qquad \text{(AF)} \end{aligned} \tag{6.122}$$

and something intermediate for the random case. The smoothness of the ferromagnetic state results in the lowest energy. The AF state splinters into many extra nonidentical configurations such as shown in (III) above, and results in smaller matrix elements, hence a ground state which is not nearly so favorable. In 1D where only a single path connects any two states regardless of spins, this tendency is lacking. In 3D, where the number of paths connecting various configurations is greater than in 2D, the tendency to ferromagnetism is substantially increased.

For a small density n_h of holes, the energy favoring ferromagnetic order is thus An_hW, taking the bandwidth W to be proportional to B. On the other hand, at finite U the energy favoring AF near a half-filled band is $A'W^2/U$. With $n_h = \frac{1}{2} - \rho$, we set the opposing energies equal to obtain the phase boundary curve C, Fig. 6.17. By symmetry the equation for a more-than-half-filled band is similar, with the final result being

$$\left|\rho - \frac{1}{2}\right| = \alpha W/U\,, \tag{6.123}$$

α being some lumped constant.

Richmond and *Rikayzen* extended Nagaoka's work in a very interesting way [6.61], formulating a trial function in which all but one of the electron spins are aligned. The odd electron is localized about a particular site, whilst the remaining electrons adjust their motion to the presence of the repulsive potential that it represents. This is exactly soluble by the scattering theory that we have used in many connections throughout this book, and the bound state (if any!) energy is a variational upper bound to the true energy. If a bound state is found, the ferromagnetic state is unstable; thus their calculation also provides an equation for curve C. The results which they obtain agree very well with (6.123).

6.14 Degenerate Bands and Intra-Atomic Exchange Forces

To assess the role of intra-atomic exchange, we here turn to the special case of two degenerate d-like bands, with various Coulomb and exchange integrals. We then assess the situation in the iron series ferromagnetic metals, to see which are the physical parameters to which theory must address itself.

The interaction of a singlet pair within each band is $U_{aa} = U_{bb}$ and by a t matrix, which we approximate by the constant

$$\bar{t}_{aa}(\rho) = \frac{U_{aa}}{1 + U_{aa}(1 - \rho^{1/3})/W} \tag{6.124}$$

For a singlet pair of which one member occupies one band and one the other, there is a similar expression with U_{ab} replacing U_{aa}. It is a simple mathematical identity for Coulomb integrals given in (6.32) to prove $U_{aa} > U_{ab} > 0$. (Intuitively, electrons in two distinct orbitals overlap less, hence have a smaller Coulomb repulsion, than two electron in identical orbitals). Moreover, for triplet pairs (e.g., electrons of parallel spins) in two different bands, this Coulomb integral is further reduced by an exchange correction, (6.41), and becomes $U_{ab} - J_{ab} = U'_{ab}$, which we also write as $U_{ab}(1 - j)$. The parameter j is thus the fractional inter-band exchange parameter, $0 < j < 1$. The ground state singlet energy thus includes 3 distinct interactious:

$$E_0 = N\rho^{5/3}\left\{\frac{18}{5}W + 2\rho^{1/3}[\tilde{\iota}_{aa}(\rho) + \tilde{\iota}_{ab}(\rho) + \tilde{\iota}'_{ab}(\rho)]\right\}. \tag{6.125}$$

It is to be compared to the ferromagnetic state, with half the particles in band a and half in band b but all with spin "up", which has energy

$$E_f = N\rho^{5/3}\left[2^{2/3}\frac{18}{5}W + 4\rho^{1/3}\,\tilde{\iota}'_{ab}(2\rho)\right], \tag{6.126}$$

and with the ferromagnetic state in which all the particles, of spin up, are also in a single band, say a (this represents spin *and* orbital magnetism, and should occur when the perturbations of the solid are too weak to quench the orbital moments of the individual atoms, as in f shells of the rare earths). This state has energy

$$E_{f,0} = N\rho^{5/3}\left(\frac{9}{5}4^{5/3}W\right). \tag{6.127}$$

The case of maximal interband exchange $j = 1$ is special; we compare the energies (6.125–127) and conclude that (6.127), representing spin + orbital magnetism, can never lie lowest. The phase diagram showing the region where the spin-only magnetic moments form is remarkably similar to Fig. 6.16, with only the numerical value of parameters $\rho_c = 0.04$ and $U_{\min}/W = 1$ being different. Doubling the number of bands allows magnetic moment formation at much lower densities and interaction parameter U than previously.

In the physically more plausible cases of partial exchange $j = J_{ab}/U_{ab} < 1$ the situation may be quite different. We compared (6.125–127) by numerical calculation, assuming $U_{ab} = \frac{1}{2}U_{aa} \equiv U$, and found: *spin + orbital magnetism never occurs* for $U/W < 20$ nor for densities far from 1/2, so this case may be eliminated from practical considerations even though it is good to know it exists (in principle) in the atomic limit $W \to 0$. Spin magnetism has itself a restricted range of stability, which depends strongly on j. For $j = 0.5$ or greater, the situation is qualitatively similar to $j = 1$, whereas for smaller values of j (0.2 or 0.1) the regions of stability shrink rapidly and become nonexistent at $j = 0$. Thus, regardless of the strength of the interaction parameter U, the *existence of a magnetic moment* ultimately *depends on the Hun's rule exchange* parameter, as a stabilizing factor. The situation is clearly summarized in Fig. 6.18.

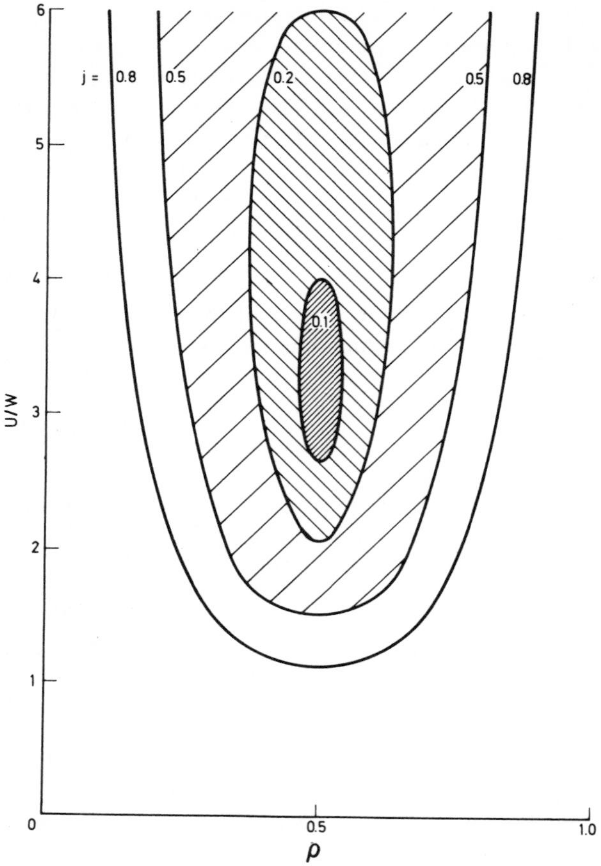

Fig. 6.18. Moment formation for 2 degenerate bands in terms of $U_{aa} = U_{bb} = U$, $U_{ab} = \frac{1}{2}U$, W = bandwidth, ρ = density per atom per band ($1 - \rho$ = density of holes) and exchange parameter $j = J_{ab}/\frac{1}{2}U$. For curve labeled $j = 0.8$ the region lying above the curve is magnetic, and similarly for $j = 0.5$. For $j = 0.1$ or 0.2, the region enclosed by the respective curves is magnetic. There exists a region of spin + orbital magnetism, not shown in this figure, near $\rho = \frac{1}{2}$ and for $U/W > 20$. Spatial ordering of any moments (ferro. vs AF) is not considered in this calculation

The spatial ordering of the moments, once they are created, is a delicate competition between several mechanisms already considered in this chapter, principally, the indirect exchange and Nagaoka mechanisms. But before proceeding, it is prudent to consider to what extent electrons in real materials satisfy the various simplifying assumptions we have made, for example, the effective-mass approximation $E(k) = \hbar^2k^2/2m$ with $m = m^*$ the band structure effective mass before interactions and $= \tilde{m}$ the total mass after the interactions (which are necessarily strong and thus nontrivial) have been incorporated.

The hypothesis of two kinds of d electrons, has been given added credence lately by *Stearns*. Her "95% local" and "5%" itinerant model counters the pure itinerant or pure Heisenberg models, against which substantial evidence has been accumulating. In her view [6.66], most (95%) of the d electrons lie in the relatively flat portions of the band structure, where the high density of states (or small W, large U/W) promotes magnetic moment formation, with the small residual fraction occupying states well described in the effective mass approximation, with effective masses $\sim m_{el}$. To see this, let us examine the Hartree-Fock band structure of iron, as shown in Figs. 6.19 and 20 reproduced from her article, "Why is iron magnetic?" [6.66]. There are shown the flat parts of the d bands, labeled d_{loc} and the itinerant parts, labeled d_{it}, drawn in heavy lines. The latter have a curvature corresponding to approximately the electron free mass m_{el}, and Fermi wave vectors that have been given as [6.69]

	$ak_{F^+}/2\pi$	$ak_{F^-}/2\pi$
Fe	0.50	0.19
Ni	0.65	0.53

(6.128)

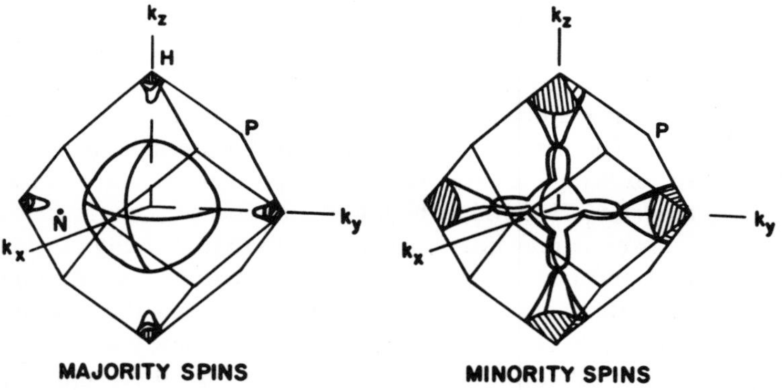

△

Fig. 6.19. Fermi surface of iron, as given in [6.67], here adapted from [6.66]

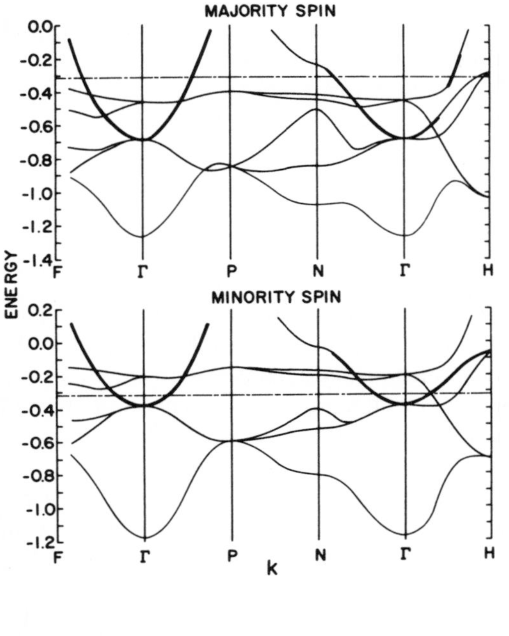

Fig. 6.20. Band structure of iron, for majority and minority electrons as calculated in [6.68] and reproduced in [6.66]

in agreement with tunneling experiments. Various other experiments indicate that the s electrons are only weakly spin polarized, and that the nodes in the Ruderman-Kittel oscillation make this weak indirect exchange mechanism favor antiferromagnetism. It follows that the ferromagnetic alignment arises from the d electrons themselves; but the above wave vectors are all indicative of AF (see Fig. 6.8 f, or Fig. 6.6 for the expected ground state symmetry at the above values of k_F), except the minority one for Fe, k_{F-}. There is thus a mystery to be resolved. Could it be that nearest-neighbor ferromagnetic exchange is sufficiently strong that Heisenberg-like ferromagnetism coexists with the itinerant electron minority? The answer: nearest-neighbor exchange is too weak by one order of magnitude [6.70], so this will not help.

A plausible explanation lies in a combination of physical effects. First, the density of itinerant d electrons is small, and the interactions are large—placing the parameters in the ferromagnetic region of the phase diagram, Fig. 6.17. Second, the indirect exchange mechanism is no longer given by the Ruderman-Kittel formula of (6.56) with R_{ij} an interatomic distance, for the local moments are not point-like but rather spread out, occupying a large fraction of the volume in each unit cell. The paramagnetic Curie temperature measuring the strength of the indirect exchange and the spin wave spectrum are then better given by the jellium model, (6.71, 72), in which both are positive, guaranteeing ferromagnetism. The only remaining question concerns the applicability of what is basically second-order perturbation theory to the interactions between itinerant and localized d-electrons, of equal or greater strength than the relevant energy denominators. The only practical procedure is to calculate the Hartree-Fock energy first, as deduced from the band structure, Fig. 6.20, and add thereto the interaction energies as computed from the $\tilde{t}$'s, as in the 2-band case analyzed above. This total energy is then to be compared to the energy of varying states of polarization, including the antiferromagnetic ordered magnetic phases and the nonmagnetic state. The qualitative arguments given here indicate that a proper calculation will favor the ferromagnetic state of iron. The magnon spectrum, such as we calculate next, should confirm this.

6.15 Magnons in Metals

Once Coulomb and exchange forces have created a magnetic solid, they profoundly affect the elementary excitations. To the quasi-particles and collective modes which are common to all metals, one must add magnons for the magnetic metals. These are the quantized spinwaves which are the property of an ordered magnetic medium. This section is devoted to their properties, especially in ferromagnetic metals.

With the exception of the strong-coupled one-band model, our knowledge of the itinerant ground state is imperfect even for ferromagnets. We cannot simply proceed by computing an excited state, subtracting the ground state energy therefrom, as even small errors of $O(N)$ will obscure effects $O(1)$. A more direct method is required, that of equation of motion. In this procedure, the scattering of a certain type excitation is treated exactly but other terms are neglected ("random phase approximation"). The neglected terms vanish at low density, and the procedure gives satisfactory answers at long wavelengths, therefore one can accept the results with the same confidence (or skepticism!) as for magnons in 3D Heisenberg antiferromagnets—as a semiquantitative, systematic approximation.

For simplicity, we treat the case of a single band in some detail, including the cases—so far ignored—of partial magnetization. We then indicate the extensions of the theory to multi-band cases, and to questions of stability of the assumed ferromagnetic ground state, such as we already examined in connection with the indirect exchange mechanism.

So, let U and ρ be in the range shown in Fig. 6.17 for which the ground state is ferromagnetic. In the Heisenberg model, translational invariance was sufficient to construct the one-magnon states uniquely. Now, it is no longer enough. The degrees of freedom in a metal are sufficiently numerous, that if we only specify that the spin angular momentum must decrease by one unit and the total momentum wave vector must be $\boldsymbol{q}$, there will be a very large number of excitations to satisfy this requirement. We therefore add to these requirements that of a very long or infinite lifetime, and find then that the magnon is a bound state, a linear combination of all the states in which a hole is created in the majority-spin band (taken to be spin "down" henceforth) while a particle, with extra momentum $\hbar\boldsymbol{q}$, is added to the minority spin ("up") sub-band.

The quasi-particle energies are,

$$E_m(\boldsymbol{k}) = E(\boldsymbol{k}) + \frac{1}{N}\sum_{\boldsymbol{k}'} \tilde{\imath}(\boldsymbol{k}\boldsymbol{k}')f_{-m}(\boldsymbol{k}')$$
$$\equiv E(\boldsymbol{k}) + \Delta_m(\boldsymbol{k}) \tag{6.129}$$

where the Fermi function $f_m(\boldsymbol{k})$ is 1 if $E_m(\boldsymbol{k}) \leq E_m(\boldsymbol{k}_{Fm})$ and zero otherwise, as discussed previously. Thus the elementary unit of magnetic excitation must be $a^+_{\boldsymbol{k}+\boldsymbol{q}\uparrow}a_{\boldsymbol{k}\downarrow}|F)$, where $|F)$ is the ground state, and $a_{\boldsymbol{k}m}$ destroys a particle (or creates a hole) in spin sub-band m, and $a^+_{\boldsymbol{k}m}$ creates one. The one-magnon state must be $|q) \equiv \boldsymbol{\Omega}_{\boldsymbol{q}}|F)$, where $\boldsymbol{\Omega}_{\boldsymbol{q}}$ is the operator

$$\boldsymbol{\Omega}_{\boldsymbol{q}} = N^{-1/2}\sum_{\boldsymbol{k}} F_{\boldsymbol{k}} a^+_{\boldsymbol{k}+\boldsymbol{q}\uparrow}a_{\boldsymbol{k}\downarrow}\,. \tag{6.130}$$

The energy of each unit must be,

$$E_{\boldsymbol{k}\boldsymbol{q}} = \frac{\hbar^2}{2m^*}(2\boldsymbol{k}\cdot\boldsymbol{q} + \boldsymbol{q}^2) + \Delta_\uparrow(\boldsymbol{k}+\boldsymbol{q}) - \Delta_\downarrow(\boldsymbol{k}) \tag{6.131}$$

in the effective mass approximation. Each scatters into the others with a strength proportional to $\tilde{t}(\boldsymbol{kk}')$ and to the occupation-number factor. We determine this from the equations of motion; that is, if $\boldsymbol{\Omega}_q$ is a raising operator for $\mathscr{H}$ by an energy $\hbar\omega_q$, then

$$[\mathscr{H}, \boldsymbol{\Omega}_q] = \hbar\omega_q \boldsymbol{\Omega}_q \,. \tag{6.132}$$

The kinetic energy contribution to this commutator yields

$$[\mathscr{H}_0, \boldsymbol{\Omega}_q] = N^{-1/2} \sum_{\boldsymbol{k}} F_{\boldsymbol{k}}(E_{\boldsymbol{k}q}) a^+_{\boldsymbol{k}+\boldsymbol{q}\uparrow} a_{\boldsymbol{k}\downarrow} \tag{6.133}$$

whereas the scattering part yields

$$[\mathscr{H}', \boldsymbol{\Omega}_q] = -N^{-3/2} \sum_{\boldsymbol{k},\boldsymbol{k}'} F_{\boldsymbol{k}'} \tilde{t}(\boldsymbol{kk}')[f_\downarrow(\boldsymbol{k}') - f_\uparrow(\boldsymbol{k}+\boldsymbol{q})] a^+_{\boldsymbol{k}+\boldsymbol{q}\uparrow} a_{\boldsymbol{k}\downarrow} \tag{6.134}$$

omitting terms such as $a^+_{\boldsymbol{k}'+\boldsymbol{q}'m} a_{\boldsymbol{k}'m} a^+_{\boldsymbol{k}+\boldsymbol{q}-\boldsymbol{q}'\uparrow} a_{\boldsymbol{k}\uparrow}$ which are smaller by some power of the density ρ and are neglected in the random-phase approximation. We have identified $a^+_{\boldsymbol{k}m} a_{\boldsymbol{k}m} = \mathfrak{n}_{\boldsymbol{k}m}$ with its average, $f_m(\boldsymbol{k})$, the Fermi function.

Equating terms on both sides of the equation of motion (6.132) we obtain the equations of the amplitudes $F_{\boldsymbol{k}}$

$$F_{\boldsymbol{k}}(E_{\boldsymbol{k}q} - \hbar\omega_q) = N^{-1} \sum_{\boldsymbol{k}'} \tilde{t}(\boldsymbol{kk}')[f_\downarrow(\boldsymbol{k}') - f_\uparrow(\boldsymbol{k}'+\boldsymbol{q})] F_{\boldsymbol{k}'} \,. \tag{6.135}$$

This equation is exactly soluble at $\boldsymbol{q} = \boldsymbol{0}$, by the choice $F_{\boldsymbol{k}} = \text{const.}$, and yields $\omega_0 = 0$. This is an exact result, reflecting the rotational invariance of the magnetic ground state.

For $\boldsymbol{q} \neq \boldsymbol{0}$ this represents a transcendental equation that must generally be solved graphically or numerically. However, in the special case of $\tilde{t}(\boldsymbol{kk}') = \tilde{t} = \text{const}$, there is a very simple solution $F_{\boldsymbol{k}} = A(E_{\boldsymbol{k}q} - \hbar\omega_q)^{-1}$ which leads to the secular equation for the eigenvalue

$$1 = N^{-1} \sum_{\boldsymbol{k}} \frac{\tilde{t}}{E_{\boldsymbol{k}q} - \hbar\omega_q} \,. \tag{6.136}$$

This yields an approximately parabolic magnon $\hbar\omega_q = D\boldsymbol{q}^2$ for $\boldsymbol{q} < \boldsymbol{q}_{\max}$; the maximum q is not at the edge of the BZ but occurs when the denominator of (6.136) becomes complex. This signifies the onset of the scattering regime, in which the bound state does not exist; an attempt to excite a magnon at $\boldsymbol{q} > \boldsymbol{q}_{\max}$ will result only in a broad resonance. Rather than detail this calculation, we examine the case of *two* degenerate bands, which we do calculate in detail. This case has an interesting feature: the elementary units in (6.130) can be $a^+_{\boldsymbol{k}+\boldsymbol{q},a,\uparrow} a_{\boldsymbol{k},a,\downarrow} \pm a^+_{\boldsymbol{k}+\boldsymbol{q},b,\uparrow} a_{\boldsymbol{k},b,\downarrow}$, where a, b are the two bands in question. The (+) combination yields S^+_{tot} at $\boldsymbol{q} = \mathbf{0}$, and thus commutes with the Hamiltonian. It is identified as the zero energy mode which expresses the rotational invariance of our approximation of the magnetic ground state; the (+) mode at $\boldsymbol{q} \neq \mathbf{0}$ is identical to that of a single band worked out above. These are denoted the "acoustic magnons" by analogy with lattice vibrations, where the low-lying spectrum are the acoustic modes. Correspondingly there exist the "optical magnons", the (—) branch, which do not exist for a single band ferromagnet. The optical spectrum start at $2(\Delta_\uparrow - \Delta_\downarrow)$ at $\boldsymbol{q} = \mathbf{0}$ in the two-band case, and increases slightly at higher wave vectors. The equations (with $2m^* = 1$, $\hbar = 1$, and $\boldsymbol{k}_F = 1$) are given below, and the calculated results plotted in Figure 6.21. [The acoustic branch is identical to the solution of the one-band case, (6.136)]. We replace the parameter $\Delta_\uparrow - \Delta_\downarrow$ by the single parameter Δ, and separately consider the cases when $\Delta < 1$ (weak-coupling and partial magnetization), $= 1$ (intermediate), and > 1 (strong-coupling and saturation magnetization).

When $\Delta \geqslant 1$, $n_{k\uparrow} = 0$ and we find the following transcendental equation:

$$\pm \frac{2}{3\Delta} = \frac{1}{2q} L(Q) \qquad \text{where} \tag{6.137}$$

$$Q = \frac{2q}{\Delta + q^2 - \hbar\omega_q} \qquad \text{and} \tag{6.138a}$$

$$L(Q) = \frac{1}{Q^2}\left[\frac{1}{2}(Q^2 - 1)\ln\left|\frac{Q+1}{Q-1}\right| + Q\right]. \tag{6.138b}$$

For $\Delta \leq 1$, $n_{k\uparrow} \neq 0$, and we obtain a slightly more complicated equation:

$$\pm \frac{2}{3}\left[\frac{1 - (1 - \Delta)^{3/2}}{\Delta}\right] = \frac{1}{2q}[L(Q) - (1 - \Delta)L(Q')] \qquad \text{where} \tag{6.139}$$

$$Q' = \frac{2q(1 - \Delta)^{1/2}}{\Delta - q^2 - \hbar\omega_q} \tag{6.140}$$

and (6.138) defines the other parameter and the function $L(Q)$.

The optical magnon mode (—) starts at 2Δ for $\boldsymbol{q} = 0$ and increases somewhat before merging with the continuum. The acoustic (+) branch starts at $\hbar\omega_0 = 0$ for $\boldsymbol{q} = 0$ and increases approximately $\sim Dq^2 + O(q^4)$, with the parabolic approximation $\hbar\omega_q \sim Dq^2$ improving in relative accuracy as Δ is increased. Expansion of the equations leads to a formula for D

$$D = \frac{1 + (1 - \Delta)^{3/2} - 4/5[1 - (1 - \Delta)^{5/2}]/\Delta}{1 - (1 - \Delta)^{3/2}} \qquad \text{for} \quad \Delta \leq 1 \tag{6.141a}$$

and

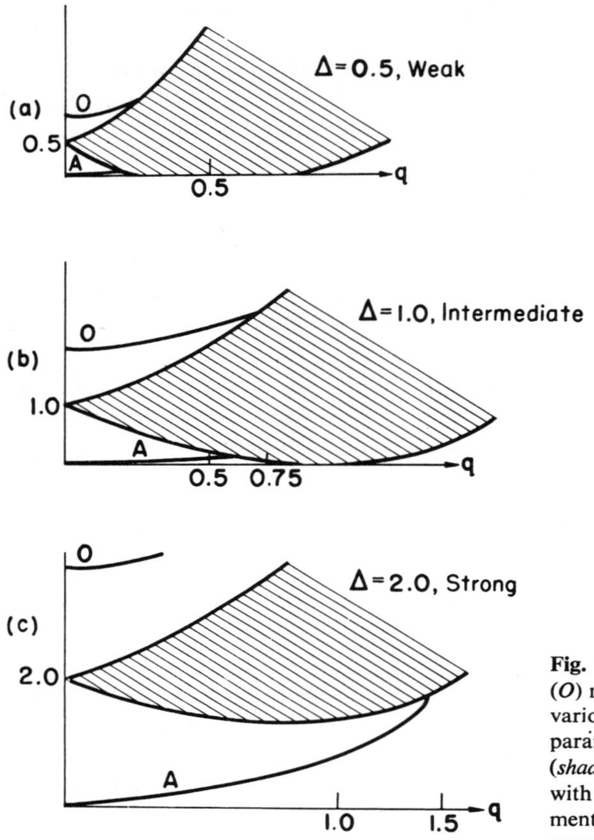

Fig. 6.21. Acoustic (*A*) and optical (*O*) magnons in the band theory, for various values of Δ, the "Stoner gap parameter. Continuum indicated (*shading*) is for elementary excitations with spin flip; continuum for elementary excitations without spin flip remains the same as shown in Fig. 6.5

$$D = 1 - \frac{4}{5\Delta} \quad \text{for} \quad \Delta \geqslant 1 . \tag{6.141b}$$

The acoustic branch enters the continuum and "dies" at $q_{\max} = 0.75\Delta$ for $\Delta \geq 1$.

Antiferromagnetism occurs when spin-down electrons start to fill the Brillouin zone, as may be seen in the following demonstration for strong coupling. Assume *that* $\Delta \geq 1$, and *every state in the spin-down zone is filled, while every state in the spin-up zone is empty.* The eigenvalue equation can be expanded in power of $E(\boldsymbol{k}+\boldsymbol{q}) - E(\boldsymbol{k}) \equiv w(\boldsymbol{k}, \boldsymbol{q})$, a procedure certainly valid for small $\boldsymbol{q}$. We make use of the assumption $E(-\boldsymbol{k}) = E(\boldsymbol{k})$ to prove,

$$\sum_{\boldsymbol{k}} w^{2p+1} \equiv 0 \qquad p = 0, 1, 2, \ldots, \qquad \text{all } \boldsymbol{q}, \tag{6.142}$$

which together with the identity $\sum w^{2p} \geq 0$ readily establishes the desired result

$$\hbar\omega_0 = 0 \geq \hbar\omega_{\boldsymbol{q}} \qquad \text{all} \quad \boldsymbol{q} \neq 0 \,. \tag{6.143}$$

Not only does a collective magnon mode exist at every $\boldsymbol{q}$ in the Brillouin zone, but the $\boldsymbol{q} = \mathbf{0}$ mode is a *maximum*, and the ferromagnetic state must be unstable against the emission of any number and any type of magnons. The new ground state must then be an antiferromagnetic, or a spiral spin configuration of the type previously discussed.

The antiferromagnetic behavior sets in even before the spin-down Brillouin zone is completely filled by electrons, in the neighborhood of a half-filled zone, although the precise point at which it occurs must be calculated numerically. It is akin to the antiferromagnetism of insulators. Whereas the exclusion principle has the effect in the ferromagnetic state of preventing the kinetic motion of electrons from atom to atom, such motion is not prohibited in the antiferromagnetic configurations. The difference with insulators, is that the "hopping" (band, or kinetic, energy) represents real transitions in the metallic state, but only virtual transitions in the insulator.

Connection between the band theory and the indirect exchange theory is very natural and easy to establish. For when one band is supposed narrow, the Hund's rule splitting of Kramers' degeneracy usually will send the entire spin-up band above the Fermi level, whereas the entire spin-down band will remain below this energy. This explains why narrow bands most likely lead to integral numbers of Bohr magnetons, just as in a localized electron theory. If this narrow band interacts with a broader band, the broader band will be only slightly polarized. Whenever perturbation theory is applicable, *something like* the Ruderman-Kittel formula must result therefrom. If perturbation theory is not applicable, then the formulas derived in this section and the previous one provide a fair initial approximation to a strong-coupling theory, and should be used instead. Instead of two degenerate bands, one should consider the two dissimilar bands—one very narrow and the other delocalized, and the resultant magnon spectrum. The magnons are constituted out of the pair excitations,

$$a^{+}_{\boldsymbol{k}+\boldsymbol{q},r',\uparrow}a_{\boldsymbol{k},r,\downarrow} \tag{6.144}$$

with r, r' labeling the respective bands; there are thus 4 such operators for each $\boldsymbol{k}$, $\boldsymbol{q}$; but the linear combinations are no longer the obvious $\pm$ ones, due to lack of degeneracy. Nevertheless, the coefficients can be readily obtained and the eigenvalue problem solved if the relevant scattering amplitudes $\tilde{t}_{r,r'}$ are constants, independent of $\boldsymbol{k}$, $\boldsymbol{k}'$. Such a calculation, with the correct band structure such as in Fig. 6.20, is required for a semi-quantitative comparison of theory with such experiments as have been performed on the iron series metals. This is one of the presently unfinished tasks of the type being addressed by *Hubbard* [6.55] and others. We now turn to questions of the stability of magnetic moments in metals.

6.16 Marginal Magnetism of Impurities

A magnetic impurity atom does not necessarily retain its spin or any finite fraction thereof when it is immersed in a nonmagnetic metal, although the tendency to do so is greatest when the magnetic orbitals are isolated from the host metal by virtue of small size (the case for *f*-shell magnetism) or disparity in energies (lack of resonance with the conduction band). Consequently, there have been studies to determine the properties of marginally magnetic solute atoms, to establish the key physical parameters involved in the preservation of a moment. This work, initiated by *Friedel, Anderson, Wolff* and taken up by many others[13] has been evolving over two decades. The question to be answered concerns the magnitude of the overlap between the magnetic shell and the nonmagnetic conduction band, which so reduces the effective interaction $t(U, \rho)$ that the spin disappears.

This question is somewhat confused by the Kondo effect, for if a moment were to appear, it would so couple to the conduction band electrons that the resultant spin might remain zero. So it is not clear under what circumstances one expects a discontinuity in the ground state parameters with a change in the interaction parameters, and under what circumstances one does not. In this section, we shall indicate some of the methods for calculating the criteria for the appearance of a localized moment. The Kondo effect, principally interesting because of the finite T effects, will be treated in the appropriate chapter of the second volume.

The method we present below is illustrative of the simplifications that can be made in the study of an impurity in a 3D solid. First, it is only required to study s waves, for the higher angular momenta have little if any overlap with a point impurity at the origin. Second, the conduction band can be treated in the leading approximation, in which the correct density of states is preserved near the band edges. This is achieved by assuming that the matrix element connecting one quasi-spherical shell about the origin to the next distant shell is a constant, $-W/4$. For example, if the central atom is a regular host atom, the Hamiltonian matrix for the s states is merely

$$\mathscr{H}_0 = \left|\begin{matrix} 0 & -\frac{1}{4}W & 0 & \cdots & \\ -\frac{1}{4}W & 0 & -\frac{1}{4}W & \cdots & 0 \\ 0 & -\frac{1}{4}W & 0 & -\frac{1}{4}W & \cdots \\ \vdots\; 0 & \vdots & \vdots & \vdots & \end{matrix}\right| \tag{6.145}$$

[13] The topic is reviewed from many points of view in *Magnetism*, see [6.41]. An earlier, still valuable review was given by *Heeger* in *Solid State Physics*, see [6.38].

taking the center-of-gravity of the band to be at zero. It is, in fact, possible to prove that $\mathscr{H}_0$ can always be put into tri-diagonal form such as shown above for *arbitrary* band structure [6.71], but the matrix element connecting the central $n = 0$ site to the first $n = 1$ shell is not necessarily equal to that which connects the $n = 1$ to the $n = 2$, etc. However, for *any* band structure in 3D, the matrix elements become $-W/4$ asymptotically. It is a great simplification to consider them constant down to $n = 0$, so let us examine this case in detail. By inspection, the eigenvectors of (6.145) have components $u_n = (2/N)^{1/2} \sin k(n + 1)$, i.e.,

$$\boldsymbol{u} = \left(\frac{2}{N}\right)^{1/2} (\sin k, \sin 2k, \dots, \sin Nk) \tag{6.146}$$

with $n = 0$ labeling the origin, and the sample assumed to be spherical in shape, in shape, of radius Na. Thus, the k's are determined by the vanishing of $\sin(N + 1)k$, and are

$$k = p\pi/(N + 1), \qquad p = 1, 2, \dots \tag{6.147}$$

and the energies are,

$$E_k = -\frac{1}{2} W \cos k \tag{6.148}$$

with the bandwidth W, as desired. The local density of states, as measured at the origin, is

$$N(E) = \frac{2}{N} \sum_k \sin^2 k \, \delta\left(E + \frac{1}{2} W \cos k\right) \tag{6.149a}$$

$$= (2/\pi) \int_0^\pi dk \sin^2 k \, \delta\left(E + \frac{1}{2} W \cos k\right) \tag{6.149b}$$

$$= (8/\pi W^2)\left(\frac{1}{4} W^2 - E^2\right)^{1/2}. \tag{6.149c}$$

It is semicircular in shape, and has the proper square root dependence on the energy near the band edges, characteristic of all energy bands in 3D; the approximation of constant matrix element sacrifices only the van Hove singularities.

Problem 6.5. Derive the density of states (6.149c) from (6.149a), filling in the indicated steps and integration.

It requires only a slight modification to incorporate the changes due to an impurity at the origin. The $n = 0$ to 1 matrix element can be written $-\lambda W/4$, with λ generally $\ll 1$ for a compact magnetic shell; this feature (small λ) is the essence of Anderson's model magnetic impurity [6.72], whereas the *Wolff* model

[6.73, 74] takes $\lambda = 1$; both also include a two-body potential at the $n = 0$ site. In the Hartree-Fock approximation, we can even take this into account by an effective potential $\varepsilon_m \equiv V + U\langle \mathfrak{n}_{0,-m}\rangle$ acting only on the central site, where $V =$ one body potential (localizing the orbital with respect to the c.o.g. of the conduction band), U the parameter characterizing the two-body forces, and $\langle \mathfrak{n}_{0,m}\rangle$, with $m = \uparrow$ or $\downarrow$ (or $\pm 1/2$), the occupation of the impurity orbital, averaged in the ground state. This is to be determined self-consistently as we shall see. The Hamiltonian now has the matrix representation

$$\mathscr{H} = \begin{vmatrix} \varepsilon & -\lambda\frac{1}{4}W & \cdots\cdots & \\ -\lambda\frac{1}{4}W & 0 & -\frac{1}{4}W & 0 \\ 0 & -\frac{1}{4}W & 0 & \ddots \\ & \vdots & \vdots & \end{vmatrix} . \tag{6.150}$$

If, as a further idealization we assume a half-filled band, the energy ε at the impurity site has the effect of lifting an existing symmetry between holes and electrons and "charging up" the impurity orbital relative to the host atoms. A compensating charge must thus exist in the conduction band of the metal, in the vicinity of the impurity. Similarly, if the impurity acquires a net spin magnetic moment by the mechanism we shall shortly investigate, then a compensating spin polarization of the conduction electrons in the neighborhood of the impurity should cancel it, resulting in a net singlet state. The calculation of these compensations requires many-body techniques, whereas (6.150) does not.

The eigenstates of (6.150) can be specified by a phase shift for $n \geq 1$ but that leaves the amplitude at $n = 0$ undetermined. Specifically, the equations for $n \geq 2$ are solved by taking the eigenvalue to be $-\frac{1}{2}W\cos k$ as before, together with

$$u_n = (2/N)^{1/2} \sin(kn + \theta_k), \qquad n \geq 1 \tag{6.151}$$

and $u_0 = (2/N)^{1/2} A_k$, with A_k and θ_k to be determined by the equations at $n = 0$ and 1. At $N + 1$, the requirement $\sin(kN + \theta_k) = 0$ yields the slightly shifted k's

$$k = \frac{p\pi - \theta_k}{N} = \frac{p\pi - (\theta_k - k)}{N+1} = k_0 - \frac{\delta k}{N+1} \tag{6.152}$$

referring to the evenly spaced k's previously found in (6.147) as k_0, and defining the conventional *phase shift* $\delta_k = \theta_k - k$.

The eigenvalue equation at $n = 1$ yields

$$-\lambda\frac{1}{4}WA_k - \frac{1}{4}W\sin(2k + \theta_k) = -\frac{1}{2}W\cos k\sin(k + \theta_k)$$

which is solved by inspection by setting

$$-\frac{1}{4}\lambda WA_k = -\frac{1}{4}W\sin\theta_k\,. \tag{6.153}$$

The equation at $n = 0$ is

$$\varepsilon A_k - \lambda\frac{1}{4}W\sin(k + \theta_k) = -\frac{1}{2}W\cos k\,A_k\,. \tag{6.154}$$

These two equations in the two unknowns A and θ are to be solved, with the resulting θ then used in (6.152) to yield the new k's.

It is more convenient to use $e^{ik} \equiv z$ than k as the independent variable, and $e^{i\theta}$ than θ as the dependent variable, for (6.153, 154) are simple binomials in the exponentiated functions. Thus, complex variables impose themselves in this problem.

$$e^{2i\theta(z)} = \frac{4\varepsilon z + W(z^2+1) - \lambda^2 W}{4\varepsilon z + W(z^2+1) - \lambda^2 W z^2} \quad \text{or} \tag{6.155}$$

$$\theta_k = \tan^{-1}\left[\frac{\lambda^2 W\sin k}{4\varepsilon + W(2-\lambda^2)\cos k}\right] \quad \text{and}$$

$$A_k = \frac{\lambda W\sin k}{[(4\varepsilon + W(2-\lambda^2)\cos k)^2 + (\lambda^2 W\sin k)^2]^{1/2}}\,. \tag{6.156}$$

There can be *no* bound state as long as $[4\varepsilon + W(2-\lambda^2)\cos k]$ does not vanish, i.e., as long as the inequality

$$2|\varepsilon| + \frac{1}{2}W\lambda^2 < W \tag{6.157}$$

is satisfied.

The quantity of greatest interest is the occupation of each spin component at the impurity, given by

$$\langle n_{0m}\rangle = \frac{2}{N}\sum_{k<k_F}(A_k)^2$$
$$= (2/\pi)\int_0^{k_F} dk\,\frac{(\lambda W\sin k)^2}{(4\varepsilon_m + W(2-\lambda^2)\cos k)^2 + (\lambda^2 W\sin k)^2}$$

$$= (2\lambda^2\pi)^{-1} \int_{-k_F}^{+k_F} dk(1 - \cos 2\theta_k) \tag{6.158}$$

in which we use the fact that θ is odd in k, and $\varepsilon_m \equiv V + U\langle \mathfrak{n}_{0,-m}\rangle$. For the special case of a 1/2-filled band this integral is exactly calculated by complex variables. With $\supset$ a contour of half the unit circle, we have

$$\langle \mathfrak{n}_{0m}\rangle = (2\lambda^2\pi i)^{-1} \int \frac{dz}{z}[1 - e^{2i\theta(z)}]\,. \tag{6.159}$$

As there are no poles within the unit circle with the exception of $z = 0$, this contour can be deformed into the line integral up the imaginary axis $\uparrow$ with the origin excluded, and evaluated in terms of logarithms. One finds

$$\langle \mathfrak{n}_{0m}\rangle = \frac{1}{2} - \frac{1}{2\pi i(1-\lambda^2)} \int_0^1 \frac{dy}{y}\left[\frac{1+y^2}{y^2 - c_m iy - b} - \text{c.c.}\right] \tag{6.160}$$

where

$$c_m = \frac{4}{W(1-\lambda^2)}(V + U\langle n_{0,-m}\rangle), \quad b = 1/(1-\lambda^2)\,.$$

The *Wolff* model [6.73] corresponds to $\lambda = 1$, *Anderson*'s model [6.72] to the limit $\lambda \to 0$. Let us discuss Wolff's model first. The physics is particularly simple. One atom differs from the others only in the energy of its atomic orbital (V) relative to the others, and in the magnitude of the intra-atomic Coulomb potential U (assumed zero on the rest of the lattice). The bond connecting it to its neighbors is unexceptional, hence $\lambda = 1$. In the absence of the potentials, the occupation of the "impurity" is $\langle \mathfrak{n}_{0m}\rangle = 1/2$ for both values of m. So now let us define the "electron occupation number defect" δn_m:

$$\delta n_m \equiv \frac{1}{2} - \langle \mathfrak{n}_{0m}\rangle\,, \tag{6.161}$$

put $\lambda^2 = 1$ in (6.160) and obtain a somewhat simpler relation

$$\delta n_m = +\frac{1}{2\pi i}\int_0^1 \frac{dy}{y}\left[\frac{1+y^2}{1 + \frac{4}{W}\left(V + \frac{1}{2}U - U\delta n_{-m}\right)iy} - \text{c.c.}\right]. \tag{6.162}$$

Performing the integrals, one obtains two equations in the unknowns $\delta n_\uparrow$ and $\delta n_\downarrow$

$$\delta n_m = \frac{1}{\pi}[a_m^{-1} + (1 - a_m^{-2})\tan^{-1} a_m] \tag{6.163}$$

with

$$a_m = \frac{4}{W}\left(V + \frac{1}{2}U\right) - \frac{4U}{W}\delta n_{-m}$$
$$\equiv \frac{4U}{W}(\mu - \delta n_{-m}) \,, \qquad (6.164)$$

defining a new parameter μ.

Although derived under the assumption that no bound state exists, these formulas are now serendipitously valid for the entire range of parameters because of analytical continuation. Thus the appearance or disappearance of bound states turns out not to be a significant consideration in the appearance or disappearance of a local moment. We have introduced the parameter $\mu = (V + \frac{1}{2}U)/U$ in (6.164), not to be confused with the Fermi energy which is *always* zero in this section and in the next. As for the magnetic moment: nonmagnetic solutions have $\delta n_\uparrow = \delta n_\downarrow$, magnetic solutions have $m = \delta n_\uparrow - \delta n_\uparrow \neq 0$. At $\mu = 0$ symmetry dictates $\delta n_\uparrow = -\delta n_\downarrow = \pm m/2$ for the magnetic solution; insert this into (6.163) to obtain

$$(\mu = 0) \quad x = \frac{2}{\pi}\left(\frac{2U}{W}\right)[x^{-1} + (1 - x^{-2})\tan^{-1}x] \qquad (6.165)$$

writing $x = 2Um/W$. This equation always has the trivial solution $x = 0$, but in addition it may have a nontrivial solution if the slope of the right-hand-side exceeds 1 at the origin. This yields

$$(\mu = 0) \quad \frac{4}{3\pi} \cdot \frac{4U}{W} > 1 \qquad \text{or} \qquad 4U/W > 3\pi/4 = 2.3562 \qquad (6.166a)$$

as the necessary condition for a local moment to form. At $\mu \neq 0$ one requires a larger ratio $4U/W$ for the moment to persist. Expanding (6.163) to third order in δn_m yields a parabolic relation valid for small μ

$$(|\mu| \ll 1) \quad \frac{4U}{W} > \frac{3\pi}{4}\left[1 + \frac{6}{5}\left(\frac{3\pi}{8}\mu\right)^2 + \cdots\right]. \qquad (6.166b)$$

Numerically, one may solve for the simultaneous solutions of (6.163) or one may converge to them by *relaxation*: insert an initial value for $\delta n_\downarrow$, calculate $\delta n_\uparrow$ by (6.163), reinsert the result into (6.163) to obtain a new $\delta n_\downarrow$, etc. This procedure is found to always converge to the magnetic solution whenever it exists. When there is no magnetic solution, it quickly converges to the nonmagnetic solution $\delta n_\uparrow = \delta n_\downarrow$. This type of behavior characterizes the magnetic solution as a "stable fixed point" of the coupled nonlinear equations, and the trivial solution as an "unstable fixed point". An independent confirmation can be found in a calculation of the energy associated with each solution. Where a magnetic solution exists, the energy is a minimum at the proper value of m (and $-m$), and a local maximum at $m = 0$.

The results of numerical calculation of (6.163) can then be put in graphical form, as shown in Figure 6.22. In the figure we also indicate the regions where bound states form: above the band for $\mu > 0$ and below it for $\mu < 0$. Such bound states give the magnetic impurity the aspect of an isolated one-electron atom. But there is ample parameter space for a magnetic solution without any bound state detaching itself from the continuum, and this is of course the new feature.

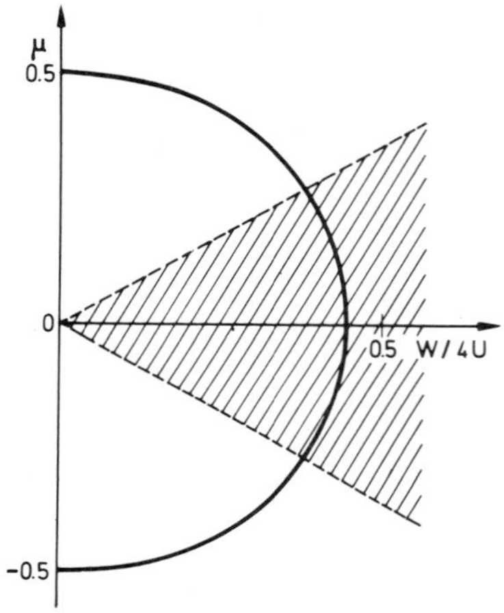

Fig. 6.22. Parameter space in the Wolff model. Magnetic solutions occupy the approximately semicircular region. Shading indicates regions free of one-body bound states. Above this, *one* level will detach itself and lie above $\frac{1}{2}W$, the maximum permissible energy in the continuum; below this, it will lie below $-\frac{1}{2}W$, the lowest energy of the continuum

Anderson's model is characterized by two parameters, one of which is the same μ and the other is denoted Δ

$$\Delta \equiv \frac{1}{4} W\lambda^2/U \,. \tag{6.167}$$

We solve it first, then provide explanations. In the limit $\lambda^2 \to 0$, the amplitudes (6.156) are seen to yield a Lorentzian probability centered at the impurity level ε; only that part lying below the Fermi level will contribute to the occupation of the impurity orbital. The result:

$$\delta n_m = \frac{1}{\pi} \tan^{-1} \left(\frac{\mu - \delta n_{-m}}{\Delta} \right). \tag{6.168}$$

We calculate some of the physical properties from the numerical solution of these coupled equations, much as we did for (6.163). Some of the quantities of interest are: the net electron occupation-number defect $\delta n = \delta n_\uparrow + \delta n_\downarrow$ in the magnetic and nonmagnetic states, shown in Fig. 6.23; the magnetization as a function of μ for several values of Δ shown in Figure 6.24; and noting that m is a maximum at $\mu = 0$, the dependence of m on Δ at $\mu = 0$ in Fig. 6.25. The symmetry of the solutions with respect to μ is maintained throughout, of course, as it was in the Wolff model previously. $\mu = 0$ is called "symmetric Anderson model".

The electron probability distribution at the impurity site, the Lorentzian distribution characterizing Anderson's model, is shown in Fig. 6.26. The asymmetric magnetic configuration is degenerate: one can interchange spins up and

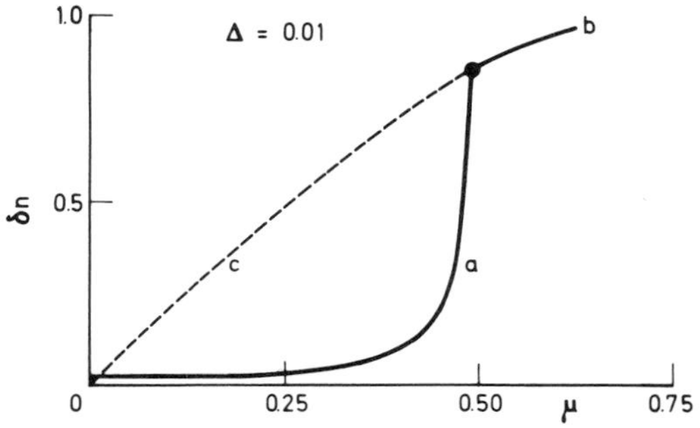

Fig. 6.23. Electron deficit (or charge accumulation) $\delta n = \delta n_\uparrow + \delta n_\downarrow$ on the impurity's orbitals as a function of the net effective potential μ defined in the text. The magnetic solution (*curve a*) is stable until approximately $\mu = \frac{1}{2}$; at larger μ the nonmagnetic solutions yield curve b. The nonmagnetic solution at smaller potentials, curve c, is unstable. The magnetic solution shows a great stability against charge accumulation compared to the (unstable) nonmagnetic solution. This rather typical curve was calculated for $\Delta = 0.01$. For $\mu \to -\mu$, $\delta n \to -\delta n$

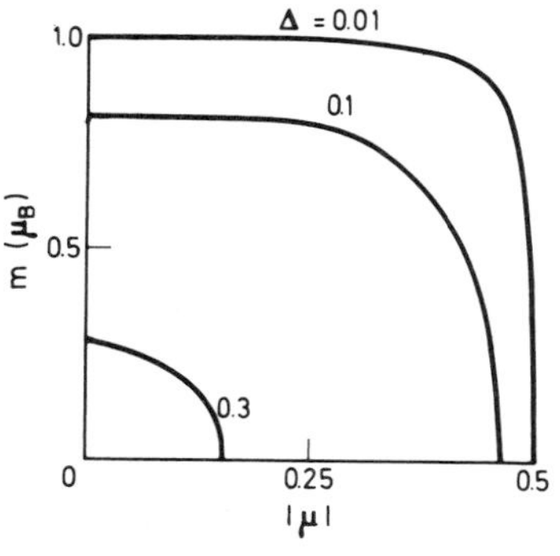

Fig. 6.24. Magnetization (in Bohr magnetons as function of effective potential parameter μ defined in the text. For $\mu \gtrsim \frac{1}{2}$ there is no magnetic solution, at smaller values the curves $m(\mu)$ depend on the kinetic energy parameter Δ, the general tendency being shown by the three curves for increasing Δ

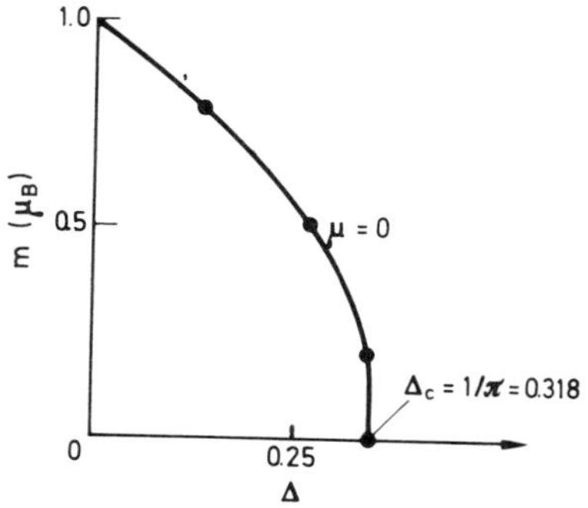

Fig. 6.25. The curve $m(\Delta)$ at the most favorable potential, $\mu = 0$, i.e. $V + \frac{1}{2}v = 0$, $v > 0$, the so-called "symmetric Anderson model".

down to obtain a distinct, yet degenerate state; a magnetic field applied to the impurity site only, would lift this degeneracy by an amount linear in the applied field. Thus, we know that a magnetic moment is associated with the state.

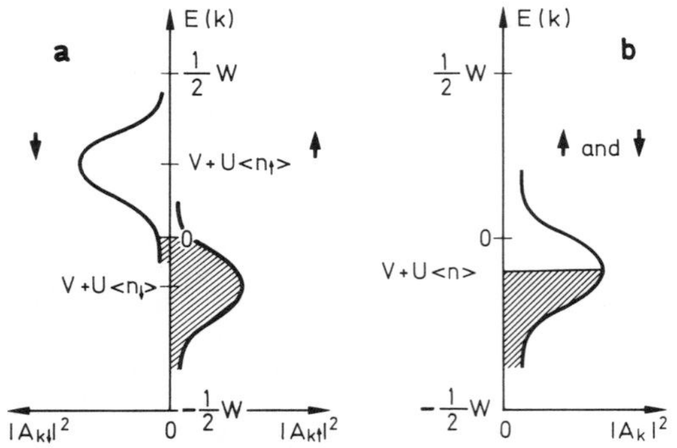

Fig. 6.26 a,b. (Amplitude)^2 of waves at impurity in cases (**a**) moment exists and (**b**) no magnetic moment for the same Anderson model impurity. The approximately Lorentzian shaped curves apply for the case $\lambda \ll 1$. Magnetic case (**a**) is degenerate: a solution with $\uparrow$ and $\downarrow$ interchanged is equally valid. *Note:* $V < 0$, $V + U > 0$ are assumed

The original physical basis for Anderson's model was the following: a magnetic d-orbital (shell index $n = 0$ in our nomenclature) is weakly connected ($\frac{1}{4}\lambda W$) to an s-orbital (shell index $n = 1$) which then connects to the rest of the lattice in the usual way. The 2-body forces and the "weak link" characterize the magnetic shell. After the exact Hartree-Fock solution given here, we seek some insight into 2-body correlations in the following Section.

6.17 Correlations and Equivalence to *s-d* Model

By solving the Hartree-Fock equations, we determined the range of parameters for which an impurity atom, imbedded in a nonmagnetic metal, acquires a localized moment. Now we consider the correlations required to re-establish the rotational invariance of the model, and to understand the concomitant polarization of the conduction electrons near the localized moment. For the symmetry-breaking Hartree-Fock solutions, which offer us a choice of two ground states (Fig. 6.26a with majority spins "up" or "down"), are obviously just the first step toward the correct singlet ground state.

De Gennes [6.75] and *Schrieffer* and *Wolff* [6.76] have established the correspondence between the magnetic solution of Anderson's model and the *s-d* exchange model, have shown that the effective exchange parameter is AF in sign and determined its magnitude from the microscopic parameters such as λ, W,

U, etc. Their principal motivation: the *s-d* model being explicitly rotationally invariant, no further demonstrations are required. Moreover, the Kondo effect and other thermodynamic and dynamic phenomena are fairly well understood for the *s-d* model but have not yet been thoroughly studied for the models of the preceding section for which one must go to much higher orders in the perturbation theory. In fact, a direct calculation of the Kondo effect starting with Anderson's model requires an accuracy better than $O(\lambda^6)$ and is reasonably complicated [6.77].

For maximum simplicity, we treat only the symmetrical model $\mu = 0$, $(V + \frac{1}{2}U) = 0$, but the reader can proceed to the more general case with no essential modifications. We assume U is large so that $m \sim 1$, and follow the Schrieffer-Wolff type analysis. We then show how to extract the effective exchange parameter, not just in the Anderson model, but for arbitrary λ including the Wolff limit $\lambda = 1$.

To start, suppose there is precisely one electron of spin "up" on the impurity site (the zeroth shell, denoted $r = 0$) with the rest of the electrons in the normal Fermi sea on the shells $r = 1, 2, \ldots$ Then the matrix element for the scattering of an electron $k\downarrow$, of spin "down" from near the Fermi level, into the localized $r = 0$ shell, is $\lambda W(2/N)^{1/2}$, at the cost of an additional excitation energy $\frac{1}{2}U$. In second order, this process restores the electron to the conduction band at k'. Thus, a scattering from $k\downarrow$ to $k'\downarrow$ has occurred, with amplitude $-(\lambda W)^2(2/N)/(U/2)$. Conduction electrons of spin "up" do not benefit from this scattering mechanism, which is forbidden to them by the exclusion principle. Thus, we have found a spin-dependent scattering mechanism that can be written as the sum of a potential and a spin term

$$V_{\text{eff}}(2/N) \sum_{kk'} (c^*_{k\uparrow} c_{k'\uparrow} + c^*_{k\downarrow} c_{k'\downarrow}) - J_z S_z \frac{1}{2}(2/N) \sum_{kk'} (c^*_{k\uparrow} c_{k'\uparrow} - c^*_{k\downarrow} c_{k'\downarrow}) \tag{6.169}$$

with $2/N$ instead of the usual $1/N$ to take care of the different amplitudes in our shell formalism, and the k, k' restricted to the neighborhood of the Fermi $k_F = \frac{1}{2}\pi$. S_z is a counter: $+1/2$ for an electron with spin "up" at $r = 0$, and $-1/2$ for spin "down". For no $k\uparrow$ scattering to occur, we require

$$V_{\text{eff}} - \frac{1}{4} J_z = 0$$

and for the $k\downarrow$ scattering to yield the correct result, we need

$$V_{\text{eff}} + \frac{1}{4} J_z = -(\lambda W)^2 / \frac{1}{2} U \, .$$

This leads to the following value for the exchange constant and effective potential

$$J_z = -4\lambda^2 W^2/U = -16W\Delta \qquad \text{and} \qquad V_{\text{eff}} = -4W\Delta \, . \tag{6.170}$$

A similar reasoning yields the spin-flip scattering terms. In the intermediate state, 2 electrons of opposite spin occupy the $r = 0$ shell. If instead of the spin "down" electron returning to the Fermi sea, the spin "up" electron did so, we would have a scattering from $k\downarrow$ to $k'\uparrow$. This is described by

$$-J_\perp S^- c^*_{k'\uparrow} c_{k\downarrow}(2/N)$$

with S^- indicating that the counter has gone from $+1/2$ to $-1/2$, and the negative sign from the Pauli principle (the ordering of two electrons has been interchanged). The magnitude of $J_\perp$ is again given by second-order perturbation theory: $-16W\Delta$, and is identical to J_z. The inverse process provides us with the Hermitean conjugate of the above term, and completes our effective *s-d* exchange Hamiltonian

$$\mathscr{H} = \text{K.E.} + J\boldsymbol{S}\cdot\boldsymbol{s}_c + \frac{1}{4}J\mathfrak{n} \tag{6.171}$$

where $\boldsymbol{s}_c$ and $\mathfrak{n}$ are the local spin-density and particle-density operators, restricted to the neighborhood of k_F (more or less) and carrying the additional factor 2 if characterized by the $1D$ set of k's.

For a more quantitative study at finite λ, we can start by the Hartree-Fock solution of the preceding section and then add in higher-order terms if desired. The proper of an AF exchange parameter is that the conduction sea close to the local moment is polarized antiparallel to it. Then the Friedel (i.e., Ruderman-Kittel) oscillations periodically reverse the polarization as one proceeds further. We shall verify these features explicitly.

At the rth shell, $r = 1, 2, \ldots$ the magnetization $m(r)$ is

$$\begin{aligned} m(r) &= \frac{2}{N}\sum_{k<k_F}[\sin^2(rk + \theta_{k\uparrow}) - \sin^2(rk + \theta_{k\downarrow})] \\ &= \frac{1}{N}\operatorname{Re}\{\sum e^{2irk}(e^{2i\theta_{k\downarrow}} - e^{2i\theta_{k\uparrow}})\}\,. \end{aligned} \tag{6.172}$$

With $z = \exp(ik)$ and $\theta(z)$ given in (6.155) we follow the procedure of (6.159) et seq. to evaluate the above as the following integral:

$$\begin{aligned} m(r) = {} & \frac{\lambda^2(-1)^r}{2\pi i(1-\lambda^2)}\int_0^1 \frac{dy}{y} y^{2r}(1+y^2) \\ & \times \{[(y^2 - c_\downarrow iy - b)^{-1} - (y^2 - c_\uparrow iy - b)^{-1}] - \text{c.c.}\} \end{aligned} \tag{6.173}$$

with c_m, b defined in (6.160). The main difference with the $r = 0$ calculation is the newly appeared factor λ^2 and the $(-y^2)^r$. The results, plotted in Fig. 6.27, show that the first shell is polarized antiparallel to the $r = 0$ impurity shell, therefore imply the effective AF coupling. The change of sign at each successive shell reflects the half-filled band, $k_F = \pi/2$. To obtain J_{eff} by this procedure,

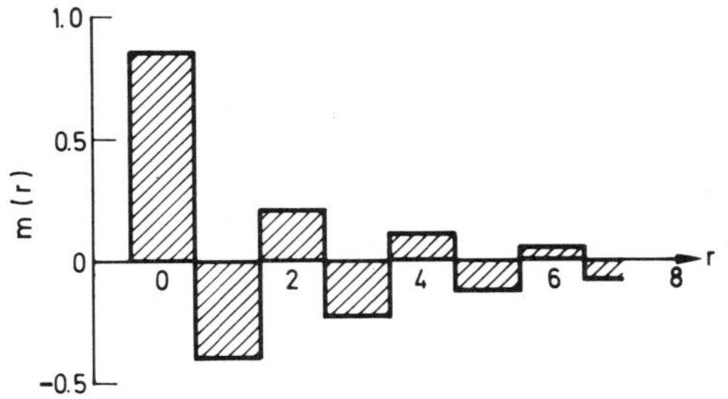

Fig. 6.27. Schematic Hartree-Fock result: magnetization $m(r)$ of the rth shell relative to magnetic shell at $r = 0$, according to (6.173)

evaluate the spin oscillations that would be caused by an equivalent *s-d* interaction [restricted to $J_z S^z s_c^z(0)$ for fair comparison with the Hartree-Fock]. The latter is also a point-scatterer which is positive for spins up and negative for spins down, and can be analyzed by the same set of equations above. Thus, we re-evaluate $m(r)$ which is once more given in (6.173), with however a different set of parameters: $\lambda = 1$ and $c_\uparrow = 0$, $c_\downarrow = \frac{1}{2}J_z$, corresponding to the appropriate exchange and potential scattering. Denote the result, $m_{sd}(r)$. The best fit to the equation

$$m_{sd}(r - 1) = m(r) \tag{6.174}$$

will yield the value of J_z (*alias* J_{eff}) in terms of the more "microscopic" parameters V, U, λ, W. Such is the object of active, on-going research by some of the most sophisticated techniques developed in the many-body problem [6.77]. The goal is to explain the figure and Table on p. 36, starting from first principles. Is it possible that we physicists are now struggling more and more to obtain less and less? Only History will judge, while our present task—the study of thermal properties continues in Volume 2.

Bibliography

Taken as a whole, the hundreds of references annotating the textual material in this book aready constitute a broad bibliography. We have consistently referred to review articles and compendia wherever possible, to lighten the search into the literature by our reader. However, this had to be balanced against a desire to cite the original research for this too is important, to credit the innovators.

Among the continuing series of publications in magnetism, we cite here a few culled from out of many:

G.T. Rado and H. Suhl have edited a series of volumes under the title *Magnetism: a Treatise on Modern Theory and Materials* (Academic New York) in which review articles by various authors outline the main developments in the field. A look at the latest volume of this on-going series, and at the latest published proceedings of the bi-annual *International Magnetism Conference* will quickly bring a technically minded reader up-to-date in the field. More classical references include vols. 18/1 and 18/2 of the *Handbuch der Physik,* ed. by S. Flügge (Springer, Berlin, Heidelberg, New York 1968) which are entirely dedicated to the topic of magnetism, and a textbook:

A. Herpin: *Théorie du Magnétisme* (Presses Univ. de France, Paris 1968).

The microscopic details of the interactions between the magnetic ion and the lattice in which it is imbedded are thoroughly explored in

H. J. Zeiger, G. W. Pratt: *Magnetic Interactions in Solids* (Clarendon, Oxford 1973) containing appendices on symmetry properties and group theory.

Many books and articles have appeared recently on the topic of phase transitions, particularly those of magnetic systems. A selected bibliography for this rapidly evolving field will be included in the second volume of this book. For the present, we include only readings which are designed to supplement or illuminate the list of references already given in the first chapter of the present volume, devoted to the history of the theory of magnetism. It is a topic not covered often or well in the usual scientific literature, and many of the references we have cited will be, perforce, unfamiliar to the reader. We therefore follow with brief descriptions and annotations on some of the more interesting books we have read, with suggestions for further reading.

J. B. Porta: *Natural Magick,* Naples, 1589, 1st English Edition 1658, with the second edition of 1669 being "even more scarce than the first, whose coiled and damaged examples testify to the fact that it was used and worn out by the practical man and in the laboratory." From reprint of 1st ed., Basic Books, Inc., New York, 1957 preface. It is a colorful treatise on "the causes of wonderful things" such as "counterfeiting gold", the "wonders of the loadstone", "beautifying women", "cookery", etc.

W. Gilbert: *On the Magnet* (in Latin *De Magnete*), London, 1600 (transl. for tercentenary of Gilbert club, London, of which Lord Kelvin was president). This 1900 edition edition was a private one limited to 250 copies, but a faithful, indeed luxurious reproduction was made available by Basic Books, Inc., New York, 1958. The reader should shun an inferior translation by P. F Mottelay, 1893, a bibliographer whose main contribution was the thorough but uncritical *Bibliographical History of Electricity and Magnetism*, London, 1922, covering the period up to and including Faraday.

E. Whittaker: *A History of the Theories of Aether and Electricity* (Harper and Row, New York 1960) in 2 volumes, provides a most complete scientific and bibliographical history of electricity and magnetism from earliest times up to and including the advent of quantum mechanics. Free use of equations is delightful for the scientific reader but renders this work inaccessible to the layman, who may turn to

F. Cajori: *A History of Physics*, (Dover, New York 1962) or the interesting and diversified *Histoire de la Science*, Maurice Daumas, Ed., Encyclop. de la Pléiade, Paris, 1957.

J. C. Maxwell: *A Treatise on Electricity and Magnetism*, the famous and complete account of the theory (before the discovery of the electron). Published in 1874, it is available as a reprint by Dover, New York, 1954 as are Maxwell's collected works. These should be read, for if we are to follow Maxwell's own expressed advice: it "is of great advantage to the student of any subject to read in the original ... for science is always most completely assimilated when it is found in its nascent state. Every student of science should, in fact, be an antiquary in his subject." Some polishing and pruning transformed the voluminous *Treatise* into J. H. Jeans' *The Mathematical Theory of Electricity and Magnetism,* Cambridge Univ. Press, Cambridge, 1907. One follows subsequent history with the various editions of "Jeans": the second, revised to include "motion of the electrons", 1911, the fourth, revised to include the theory of relativity, 1919, and finally the fifth, of 1925, incorporating the quantum theory of Bohr.

J. A. Ewing: *Magnetic Induction in Iron and other Metals* (Electrician Publ. Co, London 1900) covers the more practical, technical aspects of magnetism. It was Ewing who, for example, coined the word "hysteresis". The modern successor to this work is

R. M. Bozorth's *Ferromagnetism* (Van Nostrand, Princeton 1951) one thousand pages of collected facts and figures about magnetic materials and including a *chronological bibliography* covering the period 1850–1950.

The influence of modern quantum theory is first seen in the proceedings of the 6th Solvay Congress, Brussels, October 1930 on the topic of *Magnetism*,

Gauthier-Villars, Paris, 1932. The contributions of W. Pauli: "L'Electron Magnétique", and of P. Weiss and W. Heisenberg (all in French) are a milestone. The next milepost is

J. H. Van Vleck: A Survey of the Theory of Ferromagnetism, Rev. Mod. Phys. **17**, 27 (1945). A post-war explosion in numbers of journals and publications made this simple review the last one capable of being written by one person within a reasonably small number of pages. The student of recent history of magnetism must now turn to conference proceedings or compendia.

Finally:

J. C. Slater: *Quantum Theory of Molecules and Solids*, Vol. 1. McGraw-Hill, New York 1963). Critical analysis and thorough textbook for exchange within the molecular theory. An earlier, related work is

P. O. Löwdin: Quantum Theory of Cohesive Properties of Solids, Adv. Phys. **5**, 1 (1956) discussed the matrix analysis of the nonorthogonality problem. Zeiger and Pratt, *op. cit.* afford a detailed examination of the exchange processes as does

J. Goodenough: *Magnetism and the Chemical Bond* (Interscience Publ. Co., New York 1963).

References

Chapter 1

1.1 Lucretius Carus: *De Rerum Natura*, 1st century B.C. References are to *vv.* 906 ff., in the translation by Th. Creech, London, 1714
1.2 Pliny, quoted in W. Gilbert: *De Magnete*, trans., Gilbert Club, London, 1900, rev. ed., Basic Books, New York, 1958, p. 8
1.3 John Baptista Porta: *Natural Magick* (Naples, 1589), reprint of 1st English ed. (Basic Books, New York 1957) p. 212
1.4 P. A. Schilp (ed.): *Albert Einstein: Philosopher-Scientist*, Vol. I (Harper & Row, New York 1959) p. 9
1.5 W. Gilbert: *De Magnete*, trans., Gilbert Club, London 1900, rev. ed. (Basic Books, New York 1958) p. 3
1.6 J. Dryden: From "epistle to Doctor Walter Charleton, physician in ordinary to King Charles I"
1.7 Nathaniel Carpenter, Dean of Ireland, quoted in P. F. Mottelay: *A Bibliographical History of Electricity and Magnetism* (London 1922) p. 107
1.8 Spoken by Sagredus in Galileo's: *Dialogo sopra i due massimi sistemi del mondo Tolemaico e Copernicano*, 1632
1.9 *Opera di Galileo Galilei*, Ed. Nazionale, Firenze, 1890–1909, Vol. V, p. 232
1.10 Marin Mersenne: "Cogitata physico-mathematica" (1644), the part entitled "Tractatus de Magnetis proprietatibus"
1.11 J. Keill: *Introductio ad Veram Physicam*, 1705 (transl., 1776)
1.12 Although this work has never been translated from the Latin, a good account of it is given by Père René Just Haüy in his "Exposition raisonnée de la théorie de l'électricité et du magnétisme", 1787.
1.13 Almost all of Coulomb's memoirs are collected in a single work published by the *Société francaise de physique* in 1884, *Collection des Mémoires Relatifs à la Physique*, Vol. 1
1.14 *Mémoire lu à l'Institute le 26 prairial, an 7, par le citoyen Coulomb*, Mémoires de l'Institut, Vol. III, p. 176
1.15 S.D. Poisson: *Mémoire sur la Théorie du Magnétisme*, Mémoires de l'Academie, Vol. V, p. 247
1.16 F. Cajori: *A History of Physics* (Dover, New York 1962) p. 102
1.17 J. B. Priestley: *History of Electricity* (London, 1775) p. 86
1.18 *Collection des Mémoires Relatifs à la Physique*, Soc. Franc. de Phys., 1884, Vol. II, pp. 141, 144
1.19 A. Einstein: *The Method of Theoretical Physics* (Oxford, 1933)
1.20 J. C. Maxwell: *A Treatise on Electricity and Magnetism*, 1873 (reprinted by Dover, New York 1954)
1.21 H. Adams: *The Education of Henry Adams* (Random House, New York 1931)
1.22 G. Johnstone Stoney: Trans. Roy. Dub. Soc. **4**, 583 (1891)
1.23 For a more complete account of the birth of the electron, see D. L. Anderson: *The Discovery of the Electron* (Van Nostrand, Princeton, N. J. 1964)

1.24 "Mathematical and Physical Papers of W. Thomson, Lord Kelvin", Vol. V (Cambridge, 1911)
1.25 A. Sommerfeld: *Electrodynamics* (Academic, New York 1952) p. 236
1.26 P. Curie: Ann. Chim. Phys. **5** (7), 289 (1895), and *Oeuvres* (Paris 1908)
1.27 P. Langevin: Ann. Chim. Phys. **5** (8), 70 (1905), and J. Phys. **4** (4), 678 (1905)
1.28 P. Weiss: J. de Phys. **6** (4), 661 (1907)
1.29 J. H. Van Vleck: *The Theory of Electric and Magnetic Susceptibilities* (Oxford 1932) p. 104
1.30 H. Goldstein: *Classical Mechanics* (Addison-Wesley, Reading, Mass. 1951)
1.31 W. Gerlach, O. Stern: Z. Phys. **9**, 349 (1922)
1.32 A. H. Compton: J. Franklin Inst. **192**, 144 (1921)
1.33 G. Uhlenbeck, S. Goudsmit: Naturwiss. **13**, 953 (1925)
1.34 See Compt. Rend., September, 1923, and also De Broglie's thesis. For this work, he was awarded the 1929 Nobel prize in physics
1.35 W. Heisenberg, M. Born, P. Jordan: Z. Phys. **35**, 557 (1926)
1.36 M. Born, N. Wiener: J. Math. Phys. (M.I.T.) **5**, 84 (1926)
1.37 E. Schrödinger: Ann. Phys. **79** (4), 734 (1926)
C. Eckart: Phys. Rev. **28**, 711 (1926)
In these references, and in much of the narrative in the text, we follow a truly fascinating account given by
Sir E. Whittaker: *A History of the Theories of Aether and Electricity*, Vol. II (Harper & Row, New York 1960)
1.38 P. Jordan, E. Wigner: Z. Phys. **47**, 631 (1928)
1.39 P.A.M. Dirac: Proc. Roy. Soc. **117**A, 610 (1928)
1.40 P.A.M. Dirac: Proc. Roy. Soc. **126**A, 360 (1930)
1.41 J.C. Slater: Phys. Rev. **34**, 1293 (1929)
1.42 Quoted by J. H. Van Vleck in a talk, "American Physics Becomes of Age", Phys. Today **17**, 21 (1964)
1.43 We thank Ch. Perelman for this insightful observation.
1.44 E. Ising: Z. Phys. **31**, 253 (1925)
1.45 E. C. Stoner: *Magnetism and Matter* (Methuen, London 1934) p. 100
1.46 F. Bloch: Zeit. Phys. **61**, 206 (1930)
1.47 J. C. Slater: Phys. Rev. **35**, 509 (1930)
1.48 W. Pauli: In *Le Magnétisme*, 6th Solvay Conf. (Gauthier-Villars, Paris 1932) p. 212
1.49 L. Néel: Ann. Phys. (Paris) **17**, 64 (1932); J. Phys. Radium **3**, 160 (1932)
1.50 G. H. Wannier: Phys. Rev. **52**, 191 (1937)
1.51 J. C. Slater: Phys. Rev. **52**, 198 (1937)
1.52 J. C. Slater: Phys. Rev. **49**, 537, 931 (1936)
R. Bozorth: Bell Syst. Tech. J. **19**, 1 (1940)
1.53a E. C. Stoner: Proc. Roy. Soc. **169**A 339 (1939)
1.53b A. Hubert: *Theorie der Domänewände in geordneten Medien*, Lecture Notes in Physics, Vol 26 (Springer, Berlin, Heidelbesg, New York 1974) is a recent review. Quantum aspects are first treated in
A. Antoulas, R. Schilling, W. Baltensperger: Solid State Commun. **18**, 1435 (1976); and
R. Schilling: Phys. Rev. B**15**, 2700 (1977)
1.54 C. Kooy, U. Enz: Philips Res. Rept. **15**, 7 (1960)
1.55 A. H. Bobeck: Bell Syst. Tech. J. **46**, 1901 (1967)
1.56 A. H. Eschenfelder: *Magnetic Bubble Technology*, Springer Series in Solid State Sciences, Vol. 14 (Springer, Berlin, Heidelberg, New York 1980)
1.57 M. Prutton: *Thin Ferromagnetic Films* (Butterworth, Washington 1964)
R. Soohoo: *Magnetic Thin Films* (Harper & Row, New York 1965)
1.58 M. Blois: J. Appl. Phys. **26**, 975 (1955)
1.59 R. Sherwood et al.: J. Appl. Phys. **30**, 217 (1959)

1.60 A. Thiele: Bell Syst. Tech. J. **48**, 3287 (1969)
A. Thiele et al.: Bell Syst. Tech. J. **50**, 711, 725 (1971)
1.61 H. Callen, R. M. Josephs: J. Appl. Phys. **42**, 1977 (1971)
1.62 M. Pomerantz, F. Dacol, A. Segmuller: Phys. Rev. Lett. **40**, 246 (1978), Physics Today **34**, 20 (1981)
1.63 L. Onsager: Phys. Rev. **65**, 117 (1944)
1.64 L. Onsager: Nuovo Cimento Suppl. **6**, 261 (1949)
1.65 Quoted from E. Montroll, R. Potts, J. Ward: "Correlations and Spontaneous Magnetization of the Two-Dimensional Ising Model" in Onsager celebration issue of J. Math. Phys. **4**, 308 (1963)
1.66 N. Mermin, H. Wagner: Phys. Rev. Lett. **17**, 1133 (1966)
1.67 H. E. Stanley, T. A. Kaplan: Phys. Rev. Lett. **17**, 913 (1966)
1.68 J. Kosterlitz, D. Thouless: J. Phys. **C6**, 1181 (1973)
1.69 J. Friedel: Can. J. Phys. **34**, 1190 (1956); J. Phys. Rad. **19**, 573 (1958)
1.70 P. W. Anderson: Phys. Rev. **124**, 41 (1961)
1.71 J. Kondo: Prog. Theor. Phys. **28**, 846 (1962); **32**, 37 (1964)
1.72 K. G. Wilson: The renormalization group: Critical phenomena and the Kondo problem. Rev. Mod. Phys. **47**, 773 (1975)
1.73 D. Mattis: Phys. Rev. Lett. **19**, 1478 (1967)
1.74 K. Hepp: Solid State Commun. **8**, 2087 (1970)
1.75 N. Andrei: Phys. Rev. Lett. **45**, 379 (1980)
1.76 P. B. Wiegmann et al.: Phys. Lett. **81A**, 175, 179 (1981)
1.77 J. Kogut: Revs. Mod. Phys. **51**, 659 (1979)
1.78 A. Heeger: Solid State Phys. **23**, 283 (1969)
1.79 A. Clogston: Phys. Rev. **125**, 541 (1962)

Chapter 2

2.1 P. A. M. Dirac: Proc. Roy. Soc. **112A**, 661 (1926)
2.2 W. Heisenberg: Z. Phys. **38**, 441(1926)
2.3 J. Blatt, V. Weisskopf: *Theoretical Nuclear Physics* (Wiley, New York 1952) p. 136
2.4 P. A. M. Dirac: Proc. Roy. Soc. **123A**, 714 (1929)
2.5 J. H. Van Vleck: *The Theory of Electric and Magnetic Susceptibilities* (Oxford 1932)
2.6 E. Lieb, D. Mattis: Unpublished work (1962)
2.7 W. Heitler, F. London: Z. Phys. **44**, 455 (1927)
2.8 R. S. Mulliken: Phys. Rev. **43**, 279 (1933)
2.9 J. C. Slater: *Quantum Theory of Molecules and Solids* (McGraw-Hill, New York 1963)
2.10 C. Herring: Rev. Mod. Phys. **34**, 631 (1962)
2.11 J. C. Slater: Phys. Rev. **35**, 509 (1930)
2.12 D. R. Inglis: Phys. Rev. **46**, 135 (1934)
2.13 T. Arai: Phys. Rev. **134**, A824 (1964)
2.14 E. Harris, J. Owen: Phys. Rev. Lett. **11**, 9 (1963)
2.15 D. S. Rodbell, I. S. Jacobs, J. Owen, E. A. Harris Phys. Rev. Lett. **11**, 10 (1963)
2.16 W. J. Carr: Phys. Rev. **92**, 28 (1953)
2.17 F. Takano: J. Phys. Soc. Jpn. **14**, 348 (1959)
2.18 P. O. Löwdin, et al.; J. Math. Phys. **1**, 461 (1960)
2.19 T. L. Gilbert: J. Math. Phys. **3**, 107 (1962)
2.20 J. Calais, K. Appel: J. Math. Phys. **5**, 1001 (1964)

Chapter 3

3.1 A. R. Edmonds: *Angular Momentum in Quantum Mechanics* (Princeton Univ. Press, Princeton, N. J. 1957)
3.2 E. U. Condon, G. Shortley: *The Theory of Atomic Spectra* (Cambridge Univ. Press, New York 1935)
3.3 E. Goto, H. Kolm, K. Ford: Phys. Rev. **132**, 387 (1963)
3.4 P. A. M. Dirac: Proc. Roy. Soc. (London) **A123**, 60 (1931)
3.5 J. Schwinger: "On Angular Momentum", U.S. Atomic Energy Commission Rpt. NYO-3071 (1952) reprinted in L. Biedenharn and H. Van Dam (eds.): *Quantum Theory of Angular Momentum* (Academic, New York 1965)
3.6 T. Holstein, H. Primakoff: Phys. Rev. **58**, 1048 (1940)
3.7 J. Villain: J. de Phys. **35**, 27 (1974)
3.8 P. Jordan, E. Wigner: Z. Phys. **47**, 631 (1928)
3.9 D. Mattis, S. Nam: J. Math. Phys. **13**, 1185 (1972)

Chapter 4

4.1 W. Pauli: Z. Phys. **31**, 765 (1925)
4.2 G. E. Uhlenbeck, S. Goudsmit: Naturwissenschaften **13**, 953 (1925)
4.3 W. Pauli: Z. Phys. **43**, 601 (1927)
4.4 J. C. Slater: Phys. Rev. **34**, 1293 (1929)
4.5 E. Lieb, D. Mattis: Phys. Rev. **125**, 164 (1962)
4.6 E. Lieb: Phys. Rev. **130,** 2518 (1963)
4.7 F. Hund: *Linienspektren und periodisches System der Elemente* (Berlin 1927)
4.8 V. Fock: Z. Phys. **61**, 126 (1930)
4.9 J. C. Slater: Phys. Rev. **35**, 210 (1930)
4.10 E. U. Condon, G. Shortley: *The Theory of Atomic Spectra* (Cambridge Univ. Press, New York 1935) pp. 174–179
4.11 *American Institute of Physics Handbook*, 2nd ed. (McGraw-Hill, New York 1963) pp. 7–21
4.12 J. S. Griffith: *The Theory of Transition-Metal Ions* (Cambridge Univ. Press, New York 1961) H. F. Schaefer (ed.): *Methods of Electronic Structure Theory* (Plenum, New York 1977)
4.13 E. A. Abbott: Flatland (Grant Dahlstrom, Pasadena, California, 1884)
4.14 E. Lieb, D. Mattis: Unpublished work (1962)
4.15 R. D. Mattuck: *A Guide to Feynman Diagrams in the Many-Body Problem*, 2nd ed. (McGraw-Hill, New York 1976)

Chapter 5

5.1 J. Van Kranendonk, J. H. Van Vleck: Rev. Mod. Phys. **30**, 1 (1958)
5.2 F. Keffer, T. Oguchi: Phys. Rev. **117**, 718 (1960)
5.3 M. H. Cohen, F. Keffer: Phys. Rev. **99**, 1128, 1135 (1955)
5.4 R. Griffiths: Phys. Rev. **176**, 655 (1968)
5.5 M. Wortis: Phys. Rev. **132**, 85 (1963)
5.6 N. Fukuda, M. Wortis: J. Phys. Chem. Sol. **24**, 1675 (1963)
5.7 M. Wortis: Phys. Rev. **138A**, 1126 (1965)
5.8 F. J. Dyson: Phys. Rev. **102**, 1217, 1230 (1956)
5.9 R. Boyd, J. Callaway: Phys. Rev. **138A**, 1621 (1965)
K. Hepp: Phys. Rev. *B5*, 95 (1972)

5.10 E. W. Montroll: "Lattice Statistics", in *Applied Combinatorial Mathematics*, ed. by E. F. Beckenback(Wiley, New York 1964)
5.11 E. Lieb, D. Mattis: J. Math. Phys. **3**, 749 (1962)
5.12 E. Lieb, T. Schultz, D. Mattis: Ann. Phys. (N. Y.) **16**, 407 (1961) Appendix B
5.13 D. Mattis: Phys. Rev. **130**, 76 (1963)
5.14 T. Nagamiya, K. Yosida, R. Kubo: Adv. Phys. (Philos. Mag. Suppl.) **4** (13), 1 (1955)
M. Steiner, J. Villain, C. Windsor: Adv. Phys. (Philos. Mag. Suppl.) **25**, 87 (1976)
5.15 S. Katsura: Phys. Rev. **127**, 1508 (1962)
5.16 T. Niemeijer: Physica **36**, 377 (1967)
5.17 M. Bevis, A. Sievers, J. Harrison, D. Taylor, D. Thouless: Phys. Rev. Lett. **41**, 987 (1978)
5.18 D. R. Taylor: Phys. Rev. Lett. **42**, 1302 (1979)
5.19 H. Bethe: Z. Phys. **71**, 205 (1931)
5.20 L. Hulthén: Ark. Met. Astron. Fysik **26**A, Na. 11 (1938)
5.21 R. Orbach: Phys. Rev. **112**, 309 (1958)
5.22 L. R. Walker: Phys. Rev. **116**, 1289 (1959)
5.23 J. Des Cloizeaux, J. J. Pearson: Phys. Rev. **128**, 2131 (1962)
M. Fowler: Phys. Rev. **B17** 2989 (1978); J. Phys. **C11**, L977 (1978)
M. Fowler, M. Puga: Phys. Rev. **B18**, 421 (1978)
A. Ovchinnikov: Sov. Phys. JETP **29**, 727 (1969)
5.24 R. B. Griffiths: Phys. Rev. **133**, A768 (1964)
5.25 C. N. Yang, C. P. Yang: Phys. Rev. **150**, 321 (1966)
5.26 R. J. Baxter: Ann. Phys. (N. Y.) **70**, 323 (1972)
5.27 J. D. Johnson, B. McCoy: Phys. Rev. **A6**, 1613 (1972)
5.28 J. D. Johnson: Phys. Rev. **A9**, 1743 (1974)
5.29 T. Yamada: Prog. Theor. Phys. Jpn. **41**, 880 (1969)
5.30 J. C. Bonner, B. Sutherland, P. Richards: AIP Conf. Proc. **24**, 335 (1975)
5.31 P. Hohenberg, W. Brinkman: Phys. Rev. **B10**, 128 (1974)
5.32 M. Steiner, J. Villain, C. Windsor: Adv. Phys. 25, 87 (1976)
G. Müller, H. Beck, J. Bonner: Phys. Rev. Lett. **43**, 75 (1979)
M. Puga: Phys. Rev. Lett. **42**, 405 (1979)
5.33 R. Birgeneau, G. Shirane: Phys. Today, Dec. (1978) p. 32
H. Mikeska, W. Pesch: J. Phys. C**12**, L37 (1979).
S. Satija, G. Shirane, Y. Yoshizawa, K. Hirakawa: Phys. Rev. Lett. **44**, 1548 (1980)
5.34 J. Johnson, S. Krinsky, B. McCoy: Phys. Rev. **A8**, 2526 (1973)
A. Luther, I. Peschel: Phys. Rev. B**9**, 2911 (1974); **B12**, 3908 (1975)
H. Fogedby: J. Phys. **C11**, 4767 (1978)
5.35 P. W. Anderson: Phys. Rev. **86**, 694 (1952)
5.36 H. L. Davis: Phys. Rev. **120**, 789 (1960)
5.37 J. Oitmaa, D. D. Betts: Can. J. Phys. **56**, 897 (1978)
R. Jullien *et al.* Phys. Rev. Lett. **44**, 1551 (1980)
5.38 T. Oguchi: J. Phys. Chem. Sol. **24**, 1649 (1963)
5.39 R. Bartkowski: Phys. Rev. B**5**, 4536 (1972)
5.40 D. Mattis: Phys. Rev. Lett. 4**2**, 1503 (1979)
5.41 J. Oitmaa, D. Betts: Phys. Lett. **68**A, 450 (1978)
5.42 W. P. Wolf: Rep. Prog. Phys. **24**, 212 (1961)
5.43 T. Nakamura, M. Bloch: Phys. Rev. **132**, 2528 (1963)
5.44 R. Jellito: Z. Naturforsch. **19a**, 1567, 1580 (1964);
R. F. Wallis, A. A. Maradudin, I. Ipatova, A. K. Klochikhin: Solid State Commun. **5**, 89 (1967):
D. L. Mills, A. A. Maradudin: J. Phys. Chem. Solids **28**, 1855 (1967);
L. Dobrzyinski, D. L. Mills. Phys: Rev. **186**, 538 (1969);
E. Ilisca, E. Gallais: J. de Phys. **33**, 811 (1972). See also

R. De Wames, T. Wolfram: Phys. Rev. **185**, 720 (1969) and for AF: ibid, 762, and for case of additional n.n.n. couplings
I. Harada, O. Nagai. J. Phys. Soc. **42**, 738 (1977); J. Appl. Phys. **49**, 2144 (1978); Phys. Rev. B**19**, 3622 (1979);
J. Banavar, F. Keffer: Phys. Rev. B**17**, 2974 (1978)
5.45 J. Kosterlitz, D. Thouless: J. Phys. C**6**, 1181 (1973)
5.46 J. Kosterlitz: J. Phys. C**7**, 1046 (1974)
5.47 J. Scott Russell: Proc. Roy. Soc. Edinburgh (1844) p. 319
5.48 J. Tjon, J. Wright: Phys. Rev. B**15**, 3470 (1977)
5.49 K. Nakamura, T. Sasada: Phys. Lett. A**48**, 321 (1974)
5.50 M. Lakshmanan, T. Ruijgrok, C. Thompson: Physica **84**A, 577 (1976)
5.51 L. A. Takhtajan: Phys. Lett. **64**A, 235 (1977)
5.52 H. Fogedby: "Theoretical Aspects of Mainly Low-Dimensional Magnetic Systems", notes from the Inst. Laue-Langevin, Grenoble, France (March 1979); J. Phys. A **13**, 1467 (1980)
5.53 H. Mikeska: J. Phys. C**11**, L29 (1978), C**13**, 2913 (1980)
K. Leung, D. Hone, D. Mills, P. Riseborough, S. Trullinger: Phys. Rev. B**21**, 4017 (1980)
J. José, P. Sahni: Phys. Rev. Lett. **43**, 78 (1978), erratum **43**, 1843 (1978)

Chapter 6

6.1 J. C. Slater: Rev. Mod. Phys. **25**, 199 (1953)
6.2 C. Herring: "Exchange Interactions Among Intinerant Electrons", in *Magnetism*, ed. by G. Rado, H. Suhl (Academic, New York 1966)
T. Moriya: J. Magn. Magn. Mat. **14**, 1 (1979)
6.3 J. Slater, G. Koster: Phys. Rev. **94**, 1498 (1954)
6.4 E. Wohlfarth, J. Cornwell: Phys. Rev. Lett. **7**, 342 (1961)
6.5 H. Ehrenreich, H. Philipp, D. Olechna: Phys. Rev. **131**, 2469 (1963)
J. C. Phillips: Phys. Rev. **133**, A1020 (1964)
L. F. Mattheiss: Phys. Rev. **134**, A970 (1964)
C. Wang, J. Callaway: Phys. Rev. B**15**, 298 (1977)
D. R. Penn: Phys. Rev. Lett. **42**, 921 (1979)
L. Kleinman *et al.*: Phys. Rev. B**22**, 1105 (1980)
6.6 E. Fawcett: Adv. Phys. (Philos. Mag. Suppl.) **13**, 139 (1964)
6.7 R. E. Peierls: *Quantum Theory of Solids* (Oxford Univ. Press, Oxford 1955) p. 148
6.8 M. Glasser: J. Math. Phys. **5**, 1150 (1964); erratum J. Math. Phys. **7**, 1340 (1966)
6.9 A. W. Saenz, R. O'Rourke: Rev. Mod. Phys. **27**, 381 (1955)
6.10 J. H. Van Vleck: Nuovo Cimento **6**, (Ser. X, Suppl. 3) 857 (1957)
D. Kojima, A. Isihara: Phys. Rev. B**20**, 489 (1979)
6.11 F. Bloch: Zeit. Phys. **57**, 545 (1929)
6.12 E. P. Wigner: Trans. Faraday Soc. **205**, 678 (1938)
6.13 J. C. Slater: Phys. Rev. **49**, 537, 931 (1936); **52**, 198 (1937)
6.14 E. Lieb, D. Mattis: Phys. Rev. **125**, 164 (1962)
6.15 B. T. Matthias, R. Bozorth, J. H. Van Vleck: Phys. Rev. Lett. **7**, 160 (1961)
6.16 W. Kohn: Phys. Rev. **133**, A171 (1964)
6.17 H. A. Kramers: Physica **1**, 182 (1934)
6.18 P. W. Anderson: Solid State Phys. **14**, 99 (1963); in *Magnetism*, Vol. 1, ed. by G. Rado, H. Suhl (Academic, New York 1964) Chap. 2
6.19 M. A. Ruderman, C. Kittel: Phys. Rev. **96**, 99 (1954)
6.20 N. Bloembergen, T. J. Rowland: Phys. Rev. **97**, 1679 (1955)
6.21 K. Yosida: Phys. Rev. **106**, 893 (1957)

6.22 W. Marshall, T. Cranshaw, C. Johnson, M. Ridout: Rev. Mod. Phys. **36**, 399 (1964)
6.23 F. Smith: Phys. Rev. Lett. **36**, 1221 (1976)
6.24 G.S. Rushbrooke: J. Math. Phys. **5**, 1106 (1964)
6.25 A. Heeger, A. Klein, P. Tu: Phys. Rev. Lett. **17**, 803 (1966)
6.26 R. J. Elliott: Phys. Rev. **124**, 346 (1961)
H. Miwa, K. Yosida: Prog. Theor. Phys. (Kyoto) **26**, 693 (1961)
T. A. Kaplan: Phys. Rev. **124**, 329 (1961)
6.27 D. Lyons, T. Kaplan: Phys. Rev. **120**, 1580 (1960); **126**, 540 (1962)
6.28 E. F. Bertaut: in *Magnetism*, Vol. 3, ed. by G. Rado, H. Suhl (Academic, New York 1963) Chap. 4
6.29 D. Mattis: Phys. Rev. **130**, 76 (1963)
6.30 H. Bader, R. Schilling: Phys. Rev. **B19**, 3556 (1979); **B20**, 1977 (1979)
6.31 J. Reitz, M. B. Stearns: J. Appl. Phys. **50** (3), 2066 (1979)
6.32 S. Methfessel, private communication.
6.33 T. A. Kaplan, D. H. Lyons: Phys. Rev. **129**, 2072 (1963)
6.34 J. Kondo: Prog. Theor. Phys. (Kyoto) **32**, 37 (1964)
6.35 W. Meissner, G. Voigt: Ann. Physik **7**, 761, 892 (1930)
6.36 J. P. Franck, F. D. Manchester, D. L. Martin: Proc. Roy. Soc. (London) **A263**, 494 (1961)
6.37 M. D. Daybell, W. Steyert: Phys. Rev. Lett. **18**, 398 (1967)
6.38 A. J. Heeger: "Localized Moments and Nonmoments in Metals: the Kondo Effect", in *Solid State Physics*, Vol. 23, ed. by F. Seitz, D. Turnbull, H. Ehrenreich (Academic, New York 1969) p. 284 and Fig. 28 on p. 380
6.39 D. Cragg, P. Lloyd: J. Phys. **C12**, L215 (1979)
D. Mattis: Phys. Rev. Lett. **19**, 1478 (1967)
6.40 K. Yosida: Phys. Rev. **147**, 223 (1966)
6.41 K. Yosida, A. Yoshimori: in *Magnetism*, Vol. 5, ed. by G. Rado, H. Suhl (Academic, New York 1973) p. 253
6.42 M. Daybell: "Thermal and Transport Properties", in Ref. 6.41
6.43 K. Binder, K. Schröder: Phys. Rev. **B14**, 2142 (1976)
6.44 L. R. Walker, R. E. Walstedt: Phys. Rev. Lett. **38**, 514 (1977), Phys Rev. **B22**, 3816 (1980)
6.45 D. Mattis: Phys. Lett. **56A**, 421 (1976); err. **60A**, 492 (1977)
D. Sherrington: Phys. Rev. Lett. 41, 1321 (1978)
W. Y. Ching, D. L. Huber: Phys. Rev. **B20**, 4721 (1979)
6.46 J. M. Luttinger: Phys. Rev. Lett. **37**, 778 (1976)
6.47 S. F. Edwards, P. W. Anderson: J. Phys. **F5**, 965 (1975), (nearest-neighbor model)
6.48 D. Sherrington, S. Kirkpatrick: Phys. Rev. Lett. **35**, 1792 (1975); Phys. Rev. **B17**, 4384 (1978), (long-range interaction model)
6.49 P. Reed: J. Phys. **C12**, L799 (1979) and references therein L. Morgenstern, K. Binder: Phys. Rev. Lett. **43**, 1615 (1979), (for 2D)
6.50 G. Toulouse: Commun. Phys. **2**, 115 (1977)
S. Kirkpatrick: Phys. Rev. **B16**, 4630 (1977)
E. Fradkin et al.: Phys. Rev. **B 18**, 4789 (1978)
6.51 L. Marland, D. Betts: Phys. Rev. Lett. **43**, 1618 (1979)
6.52 R. Palmer, C. Pond: J. Phys. **F9**, 1451 (1979)
6.53 J. M. Kosterlitz, D. J. Thouless, R. C. Jones: Phys. Rev. Lett. **36**, 1217 (1976)
6.54 M. L. Mehta: *Random Matrices* (Academic, New York 1967)
D. C. Mattis, R. Raghavan: Phys. Lett. **75A**, 313 (1980)
6.55 S. H. Liu: Phys. Rev. **B17**, 3629 (1978)
V. Korenman, J. Murray, R. Prange: Phys, Rev. **B16**, 4032, 4048, 4058 (1977)
J. Hubbard: Phys. Rev. **B19**, 2626; **B20**, 4584 (1979)
6.56 B. Matthias, R. Bozorth: Phys. Rev. **109** 604 (1958)
6.57 S. Pickart, H. Alperin, G. Shirane, R. Nathans Phys. Rev. Lett. **12**, 444 (1964)

6.58 J. Hubbard: Proc. Roy. Soc. **A266**, 238 (1963); **A277**, 237 (1964); **A281**, 401 (1964); also see [6.2]
6.59 J.R. Schrieffer: *Theory of Superconductivity* (Benjamin, New York 1964)
6.60 V. M. Galitski: Sov. Phys. JETP **7**, 104 (1958)
J. Kanamori: Prog. Theor. Phys. (Kyoto) **30**, 275 (1963)
6.61 P. Richmond, G. Rickayzen: J. Phys. **C2**, 528 (1969)
6.62 E. Lieb, F. Wu: Phys. Rev. Lett. **20**, 1445 (1968)
6.63 R. Mattuck: *A Guide to Feynman Diagrams in the Many-Body Problem*, 2nd ed. (McGraw-Hill, New York 1976) p. 183 (p. 156 in the 1967 ed.)
6.64 Y. Nagaoka: Phys. Rev. **147**, 392 (1966)
J. Sokoloff: Phys. Rev. **B3**, 3826 (1971)
6.65 D. Mattis, R. Pena: Phys. Rev. **10**, 1006 (1974)
6.66 M. B. Stearns: Phys. Today (April 1978) pp. 34–39
6.67 A. V. Gold, L. Hodges, P. Panousis, R. Stone: Int. J. Magn. **2**, 357 (1971)
6.68 K. Duff, T. Das: Phys. Rev. **B3**, 192, 2294 (1971)
6.69 M. B. Stearns: J. Magnetism Magn. Mater. **5**, 167 (1977)
6.70 R. Stuart, W. Marshall: Phys. Rev. **120**, 353 (1960)
6.71 R. Haydock, V. Heine, M. Kelly: J. Phys. **C5**, 2845 (1972)
6.72 P. W. Anderson: Phys. Rev. **124**, 41 (1961)
6.73 P. A. Wolff: Phys. Rev. **124**, 1030 (1961)
6.74 J. Friedel: Can. J. Phys. **34**, 1190 (1956)
6.75 P. G. de Gennes: J. Phys. **23**, 510 (1962)
6.76 J. R. Schrieffer, P. A. Wolff: Phys. Rev. **194**, 491 (1966)
6.77 D. Hamann: Phys. Rev. **B2**, 1373 (1970)
P. B. Wiegmann: Phys. Lett. **80A**, 163 (1980)
H. Fukuyama, A. Sakurai: Progr. Theor. Phys. **62**, 595 (1979)
H. Krishna-murthy, J. Wilkins,
K. G. Wilson: Phys. Rev. **B21**, 1003–1083 (1980)

Subject Index

Dynamics of Solids and Liquids by Neutron Scattering

Editors: S. W. Lovesey, T. Springer

1977. 156 figures, 15 tables. XI, 379 pages
(Topics in Current Physics, Volume 3)
ISBN 3-540-08156-9

Contents:
S. W. Lovesey: Introduction. – *H. G. Smith, N. Wakabayashi:* Phonons. – *B. Dorner, R. Comès:* Phonons and Structural Phase Transformations. – *J. W. White:* Dynamics of Molecular Crystals, Polymers, and Adsorbed Species. – *T. Springer:* Molecular Rotations, and Diffusion in Solids, in Particular Hydrogen in Metals. – *R. D. Mountain:* Collective Modes in Classical Monoatomic Liquids. – *S. W. Lovesey, J. M. Loveluck:* Magnetic Scattering.

Hydrogen in Metals I

Basic Properties

Editors: G. Alefeld, J. Völkl

1978. 178 figures, 31 tables. XVI, 426 pages
(Topics in Applied Physics, Volume 28)
ISBN 3-540-08705-2

Contents:
G. Alefeld, J. Völkl: Introduction. – *H. Wagner:* Elastic Interaction and Phase Transition in Coherent Metal-Hydrogen Alloys. – *H. Peisl:* Lattice Strains Due to Hydrogen in Metals. – *T. Springer:* Investigation of Vibrations in Metal Hydrides by Neutron Spectroscopy. – *A. C. Switendick:* The Change in Electronic Properties on Hydrogen Alloying and Hydride Formation. – *F. E. Wagner, G. Wortmann:* Mössbauer Studies of Metal-Hydrogen Systems. – *W. E. Wallace:* Magnetic Properties of Metal Hydrides and Hydrogenated Intermetallic Compounds. – *K. W. Kehr:* Theory of the Diffusion of Hydrogen in Metals. – *R. M. Cotts:* Nuclear Magnetic Resonance on Metal-Hydrogen Systems. – *K. Sköld:* Quasielastic Neutron Scattering Studies of Metal Hydrides. – *H. Kronmüller:* Magnetic Aftereffects of Hydrogen Isotopes in Ferromagnetic Metals and Alloys. – *J. Völkl, G. Alefeld:* Diffusion of Hydrogen in Metals. – *A. Seeger:* Positive Muons as Light Isotopes of Hydrogen.

Hydrogen in Metals II

Application-Oriented Properties

Editors: G. Alefeld, J. Völkl

1978. 162 figures, 21 tables. XXII, 387 pages
(Topics in Applied Physics, Volume 29)
ISBN 3-540-08883-0

Contents:
G. Alefeld: Introduction. – *T. Schober, H. Wenzl:* The System NbH(D), TaH(D), VH(D): Structures, Phase Diagrams, Morphologies, Methods of Preparation. – *E. Wicke, H. Brodowsky, H. Züchner:* Hydrogen in Palladium and Palladium Alloys. – *B. Baranowsky:* Metal-Hydrogen Systems at High Hydrogen Pressures. – *R. Wiswall:* Hydrogen Storage in Metals. – *B. Stritzker, H. Wühl:* Superconductivity in Metal-Hydrogen Systems. – *H. Wipf:* Electro- and Thermotransport of Hydrogen in Metals. – *Ch. A. Wert:* Trapping of Hydrogen in Metals.

Springer-Verlag
Berlin
Heidelberg
New York